TOXIC EXPOSURES

PHIL BROWN

Toxic Exposures

CONTESTED ILLNESSES
AND THE ENVIRONMENTAL
HEALTH MOVEMENT

COLUMBIA UNIVERSITY PRESS   NEW YORK

COLUMBIA UNIVERSITY PRESS
Publishers Since 1893
New York   Chichester, West Sussex
Copyright © 2007 Columbia University Press
All rights reserved
Library of Congress Cataloging-in-Publication Data
Brown, Phil.
Toxic exposures : contested illnesses and the environmental health movement /
    Phil Brown.
        p.   cm.
Includes bibliographical references and index.
ISBN-13: 978-0-231-12948-0 (cloth : alk. paper)
1. Environmentally induced diseases. 2. Asthma—Etiology. 3. Breast—Cancer—
    Etiology. 4. Persian Gulf Syndrome—Etiology. I. Title.
[DNLM: 1. Environmental Exposure—adverse effects. 2. Asthma—etiology.
    3. Breast Neoplasms—etiology. 4. Environmental Health—trends. 5. Persian
    Gulf Syndrome—etiology. 6. Public Policy. WA 30.5 B878t 2007]
RB152.5.B76   2007
615.9'02—dc22   2006034124

Columbia University Press books are printed on permanent and durable
    acid-free paper.
This book was printed on paper with recycled content.
Printed in the United States of America
c 10 9 8 7 6 5 4 3 2
Designed by Lisa Hamm

References to Internet Web sites (URLs) were accurate at the time of writing. Neither the author nor Columbia University Press is responsible for URLs that may have expired or changed since the manuscript was prepared.

To my wife, Ronnie Littenberg; my daughter, Liza Littenberg-Brown; and my son, Michael Littenberg-Brown

CONTENTS

Foreword by Lois Gibbs   ix

Preface: Toxic Exposures and the Challenge of Environmental Health   xiii

Acknowledgments   xxv

List of Abbreviations   xxxiii

1	Citizen-Science Alliances and Health Social Movements: Contested Illnesses and Challenges to the Dominant Epidemiological Paradigm	1
2	Breast Cancer: A Powerful Movement and a Struggle for Science	43
3	Asthma, Environmental Factors, and Environmental Justice	100
4	Gulf War–Related Illnesses and the Hunt for Causation: The "Stress of War" Versus the "Dirty Battlefield"	139
5	Similarities and Differences Among Asthma, Breast Cancer, and Gulf War Illnesses	180
6	The New Precautionary Approach: A Public Paradigm in Progress	202
7	Implications of the Contested Illnesses Perspective	228
8	Conclusion: The Growing Environmental Health Movement	272

Notes   281

Bibliography   311

Index   339

FOREWORD
LOIS GIBBS

Toxic Exposures: Contested Illnesses and the Environmental Health Movement is a modern version of the John Snow story. This book describes the struggle of families victimized by chemical exposures and their partners in science to attain public recognition of the connection between disease and environmental exposures. From this effort was born a new movement that crosses scientific, political, disease-related, health, and cultural boundaries.

Like John Snow, the scientist in 1890s London who identified contamination in the public drinking water well as the source of cholera that was raging through the city, today's environmental health scientists face many difficult challenges. Snow not only had to convince other scientists and community leaders that the public well was causing the problem but also had to overcome many complex obstacles to scientific investigation. The obstacles facing modern environmental health investigators are even more complex. First, there is the lack of scientific understanding of the body's interaction with chemicals; second, there is the lack of studies that provide clear evidence linking cause and effect in humans, for most of the chemicals in use; and third, there is the enormous financial interest of multibillion-dollar corporations that want to avoid identifying any link between their chemicals and products and adverse health effects.

Despite these obstacles, there have been major advances in our understanding of the connection between adverse health impacts and environmental chemicals. The environmental health and environmental justice movements have contributed enormously to this understanding. The collaboration of community leaders from different sectors of social change organizations with life and social scientists brought attention to and raised public awareness of these issues by combining human-interest stories and credible scientific information.

These movements grew from individual communities seeking information and justice to broad coalitions of activists, scientists, and civic groups who sparked a public conversation about the links between environmental chemicals and health. Their work has moved society to ask fundamental questions about how decisions are being made about public health. This book describes how the new energy focused on creating systemic change—a paradigm shift—from yesterday's presumption that we can manage health risks from environmental chemicals to a new recognition that we cannot.

Leaders in the movement no longer accept their future welfare being decided by the root question of risk management: How much exposure/risk can humans and/or the earth absorb without damage? Instead, a common-sense approach has been introduced that asks: How much chemical exposure can we avoid? This is the vision and goal of leaders within the community health and justice movements. They understand that achieving it is only possible through a nationwide public health mandate by the American people.

I came to this understanding through my personal twenty-eight-year journey, filled with joy, love, disappointment, laughter, anger, tears, and finally a commitment to change the way things are to the way things should be. Love Canal in Niagara Falls, New York, was my backyard, where mothers comparing notes about their families' health discovered a cluster of birth defects and other diseases among our children. Living around a leaking toxic waste site with 20,000 tons of chemicals seemed the logical cause of these problems. To test our theory, we collaborated with a number of scientists to interview residents and map the diseases in relation to the dump. This exercise not only showed a clear increase in diseases but also a unique pattern of disease clustered in low-lying areas where the leaking chemicals pooled. Despite the obvious evidence, the health authorities denied any connection. We were told that there were so many illnesses in our community due to an unfortunate random clustering of genetically defective people.

As this book describes, the families at Love Canal were not alone in their quest for answers. Anne Anderson, of Woburn, Massachusetts, followed a similar path. When she realized that several families in her neighborhood had children with leukemia, she worked with local scientists to create a map of the disease and found a link to a drinking water well. There too the authorities dismissed the issue as an angry mother looking to blame her child's illness on something and denied any connection.

The stories of Love Canal and Woburn helped bring the issue of environmental chemicals adversely affecting human health to the forefront of public attention. Different cases emerged with similar patterns of controversy, scientific debate, and serious financial implications for governments and corporations. The public disclosure of these stories changed the understanding of chemical health risks until it was no longer credible for health investigators to dismiss the evidence.

The scientific controversy, though traumatic to the local communities, was often the reason these stories made news. The explanation so often given by the health authorities was so unbelievable that it provided more support and credibility for the local activists. In Tucson, Arizona, for example, the health authorities investigating testicular cancer in teenage boys told a community leader that it was due to spicy foods. In Jacksonville, Arkansas, parents who lost many of their children before they reached six months of age were told it wasn't the dioxin in their water causing these deaths, but rather illiterate mothers feeding the babies soap.

At the root of the controversy is the enormous financial impact of linking adverse public health outcomes with environmental chemical exposures. In John Snow's time, there was no business or chemical industry lobby, or concern about personal injury cases against the polluter. Nor were there government concerns about cleanup costs, job loss, and economic growth. But like Snow, the world will never be so innocent and naïve again.

Today costs play a critical role in such health studies, though this is rarely publicly acknowledged. A connection made between a disease and chemical exposures is going to cost millions, maybe billions. To avoid this, our government has adopted the operating principle that toxic chemicals are presumed innocent of harming human health until proven guilty. This places the burden of proving an illness was caused by chemical exposure on the victims. Although this is unfair, grassroots groups have become sophisticated health investigators. When teamed with scientists and health professionals, they have had a powerful impact on the public's understanding of environmental health risks, corporate practices, and government policies.

Toxic Exposures clearly demonstrates how laypeople rise to the challenge with scientific evidence linking chemical exposure and disease, even when multiple official studies fail to make such connections. Because the evidence of harm is usually not enough to trigger official and corporate action to remedy it, the efforts to protect public health must become a social and political movement. This realization was the driving force behind the formation of the environmental health and environmental justice movements. Leaders understood the need to join together to achieve change. Groups began to link across issues, diseases, and cultural, social, and geographical boundaries. This book describes the development of this large social change network.

Phil Brown shows how the network emerged slowly and naturally. At first, there were individual leaders from contaminated communities and sick families. Then came the realization that those most often affected were low-income communities and/or communities of color, and those individuals joined with local groups. Later the movement expanded to include researchers, scientists, issue groups, and allies most people would least expect to join and play an important role, including disease-related groups like Breast Cancer Survivors, the Endometriosis and Learning Disabilities associations, the Vietnam and Gulf War veterans, the Parent Teachers Association, the National Association for the Advancement of Colored People, and the American Public Health Association.

This new movement has expanded beyond the notion that just making the link between cancer and environmental chemicals will somehow lead to change. We now know better. The tobacco industry is a good example of how slowly change happens when corporate interests are at stake. As this book clearly demonstrates, it took decades for activists to learn this lesson and to seek ways to create the political pressure to change how society makes decisions about chemicals and public health.

The new environmental health movement is one of the most exciting social justice movements in the country. It is a David-versus-Goliath effort to protect the innocent and prevent further harm by replacing a corporate-controlled structure that accepts chemical exposures as a way of doing business with a system that seeks to define how much chemical exposure we can avoid. This precautionary movement is creating new businesses, products, and opportunities.

Toxic Exposures: Contested Illnesses and the Environmental Health Movement shows us how community-driven health studies, conducted with scientists and researchers, are discovering innovative ways to move society away from chemical dependency and toward protecting our families, regardless of income or color, from becoming victims.

PREFACE

TOXIC EXPOSURES AND THE CHALLENGE OF ENVIRONMENTAL HEALTH

IN THE midst of rising incidence of many environmentally induced diseases, rampant increases in hazardous exposures to multiple toxic substances, and government and corporate attacks on regulation and safety, a massive challenge is being made to established ways of looking at health and the environment. Many people are critical of the long-dominant biomedical model that emphasizes the centrality of genetic makeup and individual lifestyle practices. The critics come from many venues: communities with disease clusters that are likely caused by toxic waste sites or air pollution sources; women facing an ever-increasing breast cancer incidence rate; people of color concerned by the asthma epidemic and the many unequal burdens of environmental degradation they shoulder; and scientists upset at their colleagues' failure to address environmental health as well as at corporate attempts to shape the research enterprise in its favor, often with outright dishonesty.

Before the past three decades of widespread activism and concern about toxic exposures, people had little opportunity to get information and action they needed. Few sympathetic professionals were available; the scientific knowledge base was weak; government agencies were largely unprepared; laypeople were not listened to as bearers of useful knowledge; and ordinary people lacked their own resources and organizations for discovery and action. The situation has changed a great deal. Although boundless problems remain, average people have shown that they can achieve many things: buyout of contaminated areas, economic settlements from polluting companies, control and abolition of dangerous chemicals, government and corporate toxics use reduction, health monitoring for people in toxics-affected areas, regulation of oil refinery flaring, substitution of cleaner emission buses, participation in decision making about siting of hazardous facilities, membership on peer review panels for environmental health research, and a host of other actions.

I show here how citizen action can make a difference in calling attention to toxic exposures and in putting forth ways to remediate and prevent them, especially when it takes the form of flexible social movements. I demonstrate that scientists and citizens can work together in a variety of ways and that health-based social movements are a new and important political phenomenon that empowers citizens, advances science, and guides policymakers.

In combination, laypeople and professionals have created a milieu in which they identify many diseases and conditions as *contested illnesses,* involving scientific disputes and extensive public debates over environmental causes. In the eyes of these critical laypeople and professionals, traditional science, business, government, media, and other social institutions join together to hold a variety of *dominant epidemiological paradigms* for contested illnesses. These paradigms are views on the nature of disease causation and impact that are widely held by many parties in science, medicine, government, public life, industry, and medical philanthropy. They are based largely on the notion of individual causes and often of personal responsibility. Moreover, they ignore the toxic effects of industrial production, poor access to care, many forms of social deprivation, and other institutional features of our stratified society. In opposition, challengers pose *public paradigms,* patterns of public action and attitudes not just for a specific scientific argument, but also for a broad perspective underlying current scientific and societal processes. In this way, they offer opposing points of view such as environmental causation and reclaim scientific involvement as a popular concern that should not be hidden by a professional world of *scientized* views (seemingly objective notions of science that frame political and moral questions in scientific terms, that limit public participation in decision making and thus ensure that the latter becomes the purview of experts, and that delegitimize the importance of those questions that may not be conducive to scientific analysis), where expertise wins solely by its overwhelming power.

Just as these critical challengers understand the limits of an individual causation model, they also know that action by individual people will not change the situation. Thus, they have formed activist organizations and social movements to pursue an alternative pathway that focuses on environmental factors in disease and sees the struggle for that point of view as a part of overall democratic participation in the society. By participating with *critical epidemiologists* (epidemiologists who consider race, class, economics, and political power as significant factors in disease) and other sympathetic scientists in *citizen-science alliances* (collaborations between scientists and laypeople, typically when those laypeople are part of a toxic waste group or other environmental group), these citizens have had powerful impacts. Their work has led to new scientific data, major shifts in scientific thinking, valuable challenges to science policy, and a vibrant environmental health movement that has consolidated the best characteristics of the civil rights, women's, environmental, environmental justice, and other movements.

In this book, I focus on how these challenges take shape around asthma, breast cancer, and Gulf War illnesses. These three examples provide the opportunity to examine illness contestation across a large set of concerns. These diseases and conditions have been important in terms of medicine, public health, science, politics, social movements, and public awareness, and they provide a good vehicle for analyzing the fundamental problems our society faces regarding the creation of environmental health and a just and healthy society. Environmental justice—an approach that highlights race and sometimes class inequalities in the manifold experiences of our environments, including toxic exposures, land use, transportation, urban development, parkland, food security, housing, sanitation, and education—is a key component of this effort, integral to growing the environmental health movement. For ease and variety, I alternate between using the terms *environmental justice movement*, *environmental health and justice movement*, and *environmental health movement*.

In separate chapters, I examine the social, scientific, and policy-related discoveries of diseases and their putative causes and focus on the relationship between science and social movements. By examining asthma, breast cancer, and Gulf War illnesses, I am able to explore the relative importance of several components for each disease or condition: the strength of the science base, public awareness and perceptions of risk, and the sources of support for the environmental causation hypothesis. Gulf War–related illnesses are weak in each component, and the claimants therefore have the least success in their efforts to get recognition, research, and care. Environmental causation of asthma is better established than for breast cancer, but the severity of the disease is less, making for less public concern. The political clout of the asthma activists is less developed than that of the breast cancer activists, and although asthma activism is powerful, it is not as well developed as breast cancer activism. The power of the social movement ultimately trumps other components, and we see breast cancer activism as the best endowed, most widely received, and most intensely organized of the movements. This comparison offers a central lesson about the power of health social movements.

The terms and concepts I have briefly mentioned here—and I promise to explain them in depth soon—are the elements of a *contested illnesses perspective* that goes beyond traditional notions of lay involvement to argue that the role of organized social movements and social movement organizations is crucial in the process of recognizing and acting on diseases and conditions of known or potential environmental causation. This perspective places much emphasis on both political-economic and ideological factors as determinants of the contestation around these diseases and conditions. By using the lens of the dominant epidemiological paradigm, and by focusing on the structure and alterations of public understanding, scientific knowledge, and public policy, my perspective offers a new way to look at health and the environment—and, by extension, at the relationship between people, knowledge, power, and authority.

These concepts and terms are clarified in chapter 1, but here I want to say how I arrived at this interest because that story provides a glimpse into all the questions and issues I deal with in this book.

HOW I GOT HERE: MY ENTRY INTO ENVIRONMENTAL HEALTH

I never intended to study environmental health, but came to it serendipitously. My earliest research was on mental health policy and on mental patients' rights. In the midst of a two-year research project in that field, I got a jump start in environmental health by studying the Woburn childhood leukemia crisis, which led to my writing *No Safe Place: Toxic Waste, Leukemia, and Community Action*.[1] During a two-year research leave in 1984–1986 at the Massachusetts Mental Health Center, part of the Harvard University Department of Psychiatry, I was involved with both the Laboratory of Social Psychiatry and the Program in Psychiatry and the Law. In the weekly meeting of the latter group, I was intrigued when psychiatrist Edwin Mikkelsen reported on his interviews with the Woburn families who were suing W. R. Grace Chemicals and Beatrice Foods for contaminating municipal water wells, a practice that had led to a large number of leukemia cases, mostly in children. Ed Mikkelsen had been retained by attorney Jan Schlictmann to demonstrate that the families suffered psychological damage. He recounted a story that went beyond medical interviews and exams, extending to a series of public-health investigations prodded by local residents who had discovered this disease cluster.

The Woburn residents, without prior activist histories or public-health knowledge, had educated and organized themselves in an incredibly effective way. Their efforts made national attention, putting the Woburn case alongside Love Canal as a key example of toxic waste organizing and community-initiated research. Ed Mikkelsen asked for help in thinking about sociological approaches, and we shortly came up with the idea for a book. My thoughts immediately went to a book I had read some years earlier, Adeline Gordon Levine's *Love Canal: Power, Politics, and People*. I was amazed at the Love Canal residents' efforts to determine environmental health effects and to trace them to specific contaminants. My first impulse on reading *Love Canal* was to call this approach *popular epidemiology*, though at the time I had no other situations on which to hang this term. As soon as Ed Mikkelsen and I began to talk about the case, I knew that my term was indeed a concept that might explain a new approach to environmental activism. In the process of writing the Woburn book, I discovered a small but growing number of passionate and intelligent academics who were studying other communities facing toxic waste contamination and found myself in the company of committed scholars.

Woburn residents had for decades complained about dishwasher discoloration, foul odor, and bad taste in the water. Private and public laboratory assays had indicated the presence of organic compounds. The first lay detection efforts were begun by Anne Anderson, whose son Jimmy had been diagnosed with acute lymphocytic leukemia in 1972. Anderson knocked on doors and put together information during 1973–1974 about other cases. She hypothesized that the alarming leukemia incidence was caused by a waterborne agent. In 1975, she asked state officials to test the water but was told that testing could not be done at an individual's initiative. The 1979 discovery of 184 unmarked barrels in a vacant lot led to the state Environmental Protection Agency (EPA) taking water samples from municipal wells, showing wells G and H to have high concentrations of known animal carcinogens, especially trichloroethylene and tetrachloroethylene. Well G had forty times the EPA's maximum tolerable trichloroethylene concentration. As a result, the state closed both wells.

A few weeks later an engineer who worked for the state EPA drove past the nearby Industri-plex construction site and thought he saw violations of the Wetlands Act. A resultant federal EPA study found dangerous levels of lead, arsenic, and chromium. Instead of sharing this information with the affected community, the EPA told neither the town officials nor the public, who learned it only months later from the local newspaper. Anne Anderson's pastor, Reverend Bruce Young, to whom she had come for help, was initially distrustful of Anderson's theory, but came to similar conclusions once the newspaper broke the story. Along with a few leukemia victims, he placed an ad in the Woburn paper seeking people who knew of childhood leukemia cases. Working with John Truman, Jimmy Anderson's doctor, Young and Anderson prepared a questionnaire and plotted the cases on a map. Six of the twelve cases were closely grouped in East Woburn. Over the years, they identified more cases, locating twenty-eight cases over a longer period, 1965 to 1980; sixteen of those people died. In January 1980, Young, Anderson, and twenty others formed For a Cleaner Environment to galvanize community support, deal with government, work with professionals, and engage in health studies.

Jimmy Anderson died in January 1981, and five days later the Centers for Disease Control and Department of Public Health (DPH) study was released, stating that there were twelve cases of childhood leukemia in East Woburn, when only slight more than five were expected. Yet the DPH argued that the case-control method (twelve cases, twenty-four controls) failed to find characteristics that differentiated victims from nonvictims and that with sparse environmental data prior to 1979 no linkage could be made to the water supply.

The conjuncture of Jimmy Anderson's death and the DPH's failure to implicate the wells led the residents to criticize official scientific studies. They received help when Harvard School of Public Health biostatisticians Marvin Zelen and Steven Lagakos worked with For a Cleaner Environment members to design a health study focusing on child leukemia, birth defects, and reproductive disorders. They trained 235 volunteers to collect data on adverse pregnancy outcomes and childhood disorders from 5,010 interviews, covering 57 percent of Woburn residences with telephones.

During this period, the state Department of Environmental Protection's hydrogeological investigations found that the bedrock in the affected area was shaped like a bowl, with wells G and H in the deepest part. The contamination source was not the Industri-Plex site as had been believed, but rather facilities of W. R. Grace and Beatrice Foods. This discovery led eight families of leukemia victims to file a $400 million suit in May 1982 against those corporations. A smaller company, Unifirst, was also sued but quickly settled before trial.

The trial was separate from the health study, but soon became mired in a contentious struggle over facts and science. In collaboration with consultant physicians and scientists, the families accumulated further evidence of health effects. In February 1984, the For a Cleaner Environment and Harvard data were made public. Childhood leukemia was significantly associated with exposure to water from wells G and H. Children with leukemia received an average of 21.1 percent of their yearly water supply from the wells, compared to 10.6 percent for children without leukemia. Controlling for risk factors during pregnancy, the investigators found that access to contaminated water was associated with perinatal deaths and some birth defects (deaths since

1970; eye and ear anomalies; and central nervous system, chromosomal, and oral cleft anomalies). For childhood disorders, water exposure was associated with kidney, urinary, and respiratory diseases. But because the judge broke the trial into various components, the families never got their community-catalyzed research admitted as evidence. In July 1986, a federal district court jury did find Grace had negligently dumped chemicals; Beatrice Foods was absolved. An $8 million out-of-court settlement with Grace was reached in 1986. The families filed an appeal against Beatrice, based on suppression of evidence, but the appeals court rejected it in 1990, and the Supreme Court declined to hear the case.

Throughout these cases, Woburn activists had continually to defend their data. In 1995, the DPH issued a draft report for public comment that claimed no environmental basis for reproductive disorders. Upon examining the research design, For a Cleaner Environment activists and their scientific colleagues found that the DPH had analyzed only a brief time period, which was too late to capture many of the earlier effects. However, the DPH found and reluctantly acknowledged a dose-response relationship between childhood leukemia and maternal consumption of water from the contaminated wells G and H. Because the leukemia cluster was the primary problem, this DPH admission was quite a vindication for the families.

Through this long process, Woburn had achieved national recognition as a toxic waste case that sparked many other communities to action. The case had prompted the country's most complex community environmental health survey, and it was a public drama: Jonathan Harr's book *A Civil Action* was a bestseller, and the resultant film starring John Travolta as attorney Jan Schlictmann was a box office hit.[2] The popular-epidemiology approach that began at Love Canal and continued at Woburn became a major influence on environmental health activism.

In trying to understand what was happening in Woburn, I was fortunate to find other social scientists looking at similar issues. Their interest in environmental health took them to the very origins of the field of environmental sociology—Kai Erikson's 1972 monograph *Everything in Its Path: The Destruction of Community in the Buffalo Creek Flood*. A lake of coal mining sludge, contained by a poorly constructed and inadequately maintained dam, swept down a Kentucky hollow and destroyed whole villages, killing 125 people and wounding many others.[3] It also left immense psychological scars on the residents of the coal mining hamlets. Erikson was called by the plaintiff's lawyers to write a report on the damage done to the residents of the Appalachian community that was so thoroughly destroyed by corporate malfeasance.

Buffalo Creek was not a toxic crisis, but nevertheless served as the first book-length community study of human-caused environmental disaster.

Erikson used the residents' eloquent descriptions to fashion an emotionally powerful, sociologically astute account, tying together the shock of individual trauma and the collective loss of communality. His study was particularly significant in showing the centrality of community effects, in highlighting both mental and physical health outcomes, and in situating the human-made disaster in the cultural, social, and historical context of the community. It was a piece of sociological research in the service of the affected people.

The rich legacy continued with Adeline Levine's *Love Canal: Science, Politics, and People,* which recounted the story of a buried waste site in a small suburb of Niagara Falls, the environmental disaster it produced, and the resultant crisis that made this incident into the sentinel case in the development of the toxic waste movement and, by extension, of the whole modern environmental movement.[4] With Erikson's and Levine's works as background, a core of sociological researchers expanded this understanding of toxic waste crises (although a couple of the scholars were not sociologists by degree, their writings were very much in a sociological vein). At the same time as I was studying Woburn, Michael Edelstein examined a water-contamination episode in Legler, New Jersey, and wrote about it in *Contaminated Communities: The Social and Psychological Impacts of Residential Toxic Exposure;* and Henry Vyner studied how radiation and other invisible environmental contaminants caused psychological damage in addition to physical disease.[5] Lee Clarke's *Acceptable Risk? Making Decisions in a Toxic Environment* detailed the Binghamton, New York, state office building fire; Steve Kroll-Smith and Stephen R. Couch's *The Real Disaster Is above Ground: A Mine Fire and Social Conflict* studied an underground mine fire in Centralia, Pennsylvania; Steven Picou examined oil spills in *Social Disruption and Psychological Stress in an Alaskan Fishing Community: The Impact of the Exxon Valdez Oil Spill.*[6] Michael Reich's *Toxic Politics: Responding to Chemical Disasters* compared the Seveso, Italy, dioxin explosion, the Michigan polybrominated biphenyl cattle-feed contamination, and the polychlorinated biphenyl contamination of cooking oil in Japan. Martha Balshem's *Cancer in the Community: Class and Medical Authority* looked at the people's perception of hazard in a Philadelphia working-class neighborhood.[7] Taken together, these social scientific studies brought a different kind of attention than was coming from environmental scientists. These scholars emphasized the collective damage, the attendant social movement mobilization, and the impact of structural political-economic forces behind the crises and their attempted resolution.

These studies recounted stories not told in the routine scientific literature, offering a rich texture of personal experiences and community effects. They emphasized the democratic rights of individuals and communities to learn about the hazards and disasters befalling them and to achieve remediation, compensation, and justice. My experience in studying Woburn is just one of a broad range of phenomena around contested illnesses and toxic waste incidents, but I aim in this book to synthesize my work, the work of those scholars I have mentioned here, and the work of others who have followed them. I use the concepts I have developed to show the lineage of an environmental health and justice movement, with roots in Rachel Carson's *Silent Spring,* the studies I mentioned, and many other strains of environmental health research now going on. At the heart of those studies and central to my approach are the community-centered nature of environmental health problems and the activist model of community-based participatory research that leads to a new democratic science. Just as Erikson's Buffalo Creek book was sociological research in the service of the affected people, so too were these other works, and this aim represents the core of my worldview.

As an environmental activist and an activist-scholar situated in the highly organized Boston area, I was nourished by an epicenter of political activities and cultural ferment. The National Toxics Campaign, Toxics Action Center (then called the Massachusetts Campaign to Clean Up Hazardous Waste), and Massachusetts Public Interest Group made the area one of the nation's leading centers for toxic waste activism and environmental health concerns. I came to know many local movement groups and leaders, many of whom I met at the annual Toxics Action Center conferences that brought together hundreds of people from toxic waste and other environmental groups in Massachusetts and increasingly from around New England. Those connections and the trusted reputation I had in the local environmental health community allowed me very special access to the research sites on which this book centers. The connections I had with several groups led me to choose research sites where I could observe the way activists and scientists collaborated. When I approached these groups, which were already overloaded with requests for internships and research projects, I found a warm welcome. Alternatives for Community and Environment in Boston provided a fine example of how an environmental justice group organized around environmental causes of asthma. West Harlem Environmental Action is another environmental justice group for which asthma is an important focus. It was not originally a research site, but I added it later for interviews and a short series of observations. Silent Spring Institute in Newton, Massachusetts, offered a perfect location to study environmental causes of breast cancer. The Boston Environmental Hazard Center, a collaborative project of the Department of Veterans' Affairs Medical Center and the Boston University School of Public Health was a valuable location to study Gulf War illnesses. Because I was concerned with how our society can reduce toxic exposures that cause so many diseases, I also studied the Toxic Use Reduction Institute, a state agency in Lowell, Massachusetts. Individuals involved in both the Toxic Use Reduction Institute and other departments and centers at the University of Massachusetts–Lowell were getting the Precautionary Principle Project started. At the same time, breast cancer and other environmental health activists were eagerly pursuing the precautionary principle. As a result, it was logical to include this group and then to include the Precautionary Principle Project and its successor, the Alliance for a Healthy Tomorrow, as research sites (I say more about these groups in chapter 1).

I have sought to practice what Sheldon Krimsky terms *advocacy science*.[8] At each of my research sites, I had access through knowing key people in the organizations who trusted my research capacities and sensibilities by virtue of my past work in related environmental health areas. During the research process, further trust developed so that these organizations and other organizations in their orbits asked me to be involved in their work in various ways. Alternatives for Community and Environment suggested I give feedback on my observations to its staff. We also collaborated on a research proposal and on a series of environmental ethics forums. The Boston Environmental Hazard Center asked me to be on the science board for a research project it was proposing. The Precautionary Principle Project, which was informally connected to the Toxic Use Reduction Institute through overlapping members, asked me to be a workshop facilitator at its 2002 international conference on the precautionary principle; when the project later transformed itself into a broader group, the Alliance for a Healthy Tomorrow, this group asked me and my graduate students, Brian Mayer and Laura Senier, to collaborate on several things: developing and performing a pesticide awareness survey, working on a project to examine environmental factors in autism, and participating in meetings to develop communications projects with scientists.

More recently, the Toward Tomorrow Project, run by the Lowell Center for Sustainable Production (which has been a key player in the development of the precautionary principle), invited me to be on the advisory board for a national project that interviews elders of the environmental movement and related movements in order to develop a proactive environmental health agenda. Because of my asthma research and a project on labor-environment coalitions, the Boston Urban Asthma Coalition and Massachusetts Coalition for Occupational Safety and Health asked me, Laura Senier, and Brian Mayer to evaluate a program to provide safer cleaning alternatives for custodians in Boston public schools. Silent Spring Institute asked me and one of my graduate students, Sabrina McCormick, to collaborate on a research project that included a presentation at the American Public Health Association annual conference and preparation of a companion journal article.[9] That project sought to demonstrate a long historical legacy of community involvement in health research, which would help justify the continuation and strengthening of community participation in current research on environmental factors in breast cancer. Most recently, my colleague at Brown, Rachel Morello-Frosch, and I have been pleased to partner with Silent Spring Institute and Communities for a Better Environment (a California environmental justice group) in an environmental justice project funded by the National Institute of Environmental Health Sciences and the National Science Foundation that links breast cancer awareness advocacy with environmental justice activism.

I have been fortunate to work with the Toxics Action Center for many years in a variety of capacities, which has been a very significant source of the information and ideas presented here. I have led workshops at some of the group's annual conferences. One of my graduate students, Rebecca Gasior Altman, worked closely with the center to interview local community activists with whom it had worked in order to understand their future activist trajectories. Rebecca's work was so helpful to the center that its members asked her to make conference presentations and to be part of special summer organizing sessions for activists. I have been involved with the Toxics Action Center and with other activists and scholars in a study of how community groups deal with health research, funded by the National Institute of Environmental Health Sciences. For several years, I have worked alongside the Toxics Action Center to build an environmental health and justice network in Rhode Island and to collaborate in depth with some local toxics groups such as the Environmental Neighborhood Awareness Committee of Tiverton, Hartford Park Residents Association, Alton Community Action, Bradford Citizens to Stop Pollution, and Concerned Residents of Reservoir Triangle. I have also helped the Toxics Action Center gain support for this Rhode Island work from the Cox Foundation and have helped Rhode Island Legal Services get a federal EPA grant to work in tandem with these efforts.

These organizations' confidence in me as a researcher was based on my providing them with collaborative relationships worthy of their trust. In that process, I have learned a great deal about humane and ethical methods of research and social change. These groups welcomed me into their meetings, educational activities, and public organizing, and I shared with them my personal beliefs and experiences. I learned how to see problems through their eyes and to appreciate the daily life of their work. And I brought my students and colleagues into these settings, where they were excited to learn and serve, while being themselves accepted. Through this give and take of academics and community groups coparticipating in research, education, and organizing, I learned how we can always expand our research communities in more inclusive ways. These aspects of community-based participatory research have helped me in my work and offer lessons to scientists, policymakers, and activists for democratizing science through an innovative and rapidly spreading environmental health movement.

As I am putting finishing touches on this book in September 2006, I think about one of the very recent successes of this movement. In my capacity as director of the Community Outreach Core of Brown University's Superfund Basic Research Program, I began working in June 2005 with our community partner, the Environmental Neighborhood Awareness Committee of Tiverton, to get state support for home loans to the one hundred homes located on the extremely contaminated land of a working-class community. Coal gasification waste is so severe that the town imposed a digging and construction moratorium, and homeowners cannot get small home equity loans to fix a leaking roof or to replace a heating furnace because their property has no value. We sought either a state mandate for banks and other lenders to provide loans or a state agency program. We researched this concern nationally, finding no comparable program, and we hoped our proposal would be a model bill for other states. We also engaged students who were in the "Environmental Justice" senior seminar that I cotaught with Rachel Morello-Frosch to become involved in this project. Working with sympathetic state legislators (Rep. Joseph Amaral and Sen. Walter Felag) and legislative staff, we crafted a bill, and in July 2006 we succeeded in passage of the Environmentally Compromised Home Opportunity program. It will make $500,000 available, through Rhode Island Housing, for low-interest home-improvement loans of up to $25,000 for homes with toxic contamination. Governor Donald Carcieri carried out a ceremonial signing of the Environmentally Compromised Home Opportunity Revolving Loan Fund Act at Neighborhood Awareness leader Gail Corvello's house in Tiverton on Friday, July 21. This kind of project fulfills one part of a larger set of needs for the Tiverton residents and a small part of a national set of needs, but it is also happening all over the country and throughout the world. It is what the environmental health and justice movement can do and what it can show science, government, business, and the public.

ACKNOWLEDGMENTS

I AM grateful to many people and institutions for their support of the research and practice that has led to this book. In 1998, I wrote a proposal for research on contested illnesses to be funded by the Robert Wood Johnson Foundation (Contested Illnesses—Disputes over Environmentally Induced Disease, Grant no. 036273) and the National Science Foundation (Citizen-Science Alliances in Contested Environmental Diseases, SES-9975518). That was the genesis of this phase of a career-long project. Sol Levine was the first program director of the Robert Wood Johnson Foundation's Investigator Awards in Health Policy Research, an exceptional grant program that elicited groundbreaking work in health policy. The program transcended traditional boundaries of research in what it funded, and it opened the door to many creative scholars. Sol encouraged me to apply, but by the time I did so, he had sadly passed away. Alvin R. Tarlov was the new national program director, and Barbara Kivimae Krimgold was the deputy director. Later, David Mechanic took over as program director, with Lynn Rogut as deputy director. Once peer reviewers made decisions on who received grants, these people in the leadership positions gave me a powerful boost through their interest in my work. For instance, they all invited me to make presentations at their home institutions and always had time to talk about my work under way.

At the National Science Foundation, Rachelle Hollander was the long-time program officer of the Ethics and Values Studies Program, subsequently transformed into the Science and Society Program. Rachelle had been an early scholar in the area of lay knowledge in science, and her leadership of the program played a key role in so much scholarship in the areas I study and in related areas. She was always an enthusiastic supporter of my work, and though she has moved elsewhere in the National Science Foundation, we continue to be involved through another grant to study the ethical and social implications of nanotechnology.

So I began with great support from these two foundations, and the luxury of starting out with two grants made me think that I could build something larger than just an individual project. In the coming two years, I was to bring in two doctoral students supported by these grants, and I had just taught a seminar with graduate students and advanced undergraduates on the subject matter that led to the project. Why not, then, have a research group? So I floated the idea, and the Contested Illnesses Research Group began. Sabrina McCormick, the first doctoral student in the project, was joined by undergraduates Meadow Linder, Theo Luebke, and Joshua Mandelbaum, and by a faculty member at nearby Providence College, Stephen Zavestoski.

My original goal for the group was to write four articles and then a book. Because of my wonderful collaborators, the four articles turned into twenty, and I received numerous other grants in related areas, extending the contested illnesses project into new phases that have stretched forward to the present and will continue in some years from now. Only with my wonderful colleagues would this blossoming be possible, and I thank them deeply.

From its beginnings to the present, the Contested Illnesses Research Group has included a talented array of faculty, graduate students, and undergraduates. It has been a true home for intellectual and passionate pursuit of knowledge and social change—a place to hone arguments, develop ideas, pursue concepts and theories, write and read through multiple paper drafts, write grants, provide encouragement and support to many aspects of people's lives, guide people through theses and dissertations, help people get their first faculty jobs, and show others in and outside Brown how to work together collectively. Crystal Adams, Carrie Alexandrowicz, Rebecca Gasior Altman, Mara Averick, Sarah Fort, Margaret Frye, Angela Hackel, Elizabeth Hoover, Meadow Linder, Theo Luebke, Joshua Mandelbaum, Brian Mayer, Sabrina McCormick, Rachel Morello-Frosch, Heleneke Mulder, Laura Senier, Pamela Webster, and Stephen Zavestoski have graced the meeting room over these years. In addition to working with me on many projects, group members patiently read and discussed my book chapters, making sure that I took the best of our collective experience in shaping my volume. They gave so much attention to this process, and I am extraordinarily grateful for such a venue to present my ideas and have them reshaped.

Stephen Zavestoski provided additional leadership to the early years of the research team, sharing his enthusiasm, theoretical richness, and methodological strength. He was always available to do interviews, participate in conferences, and guide students in the group and in the seminars that gave students field experiences with the organizations I was studying. Steve was the person I could always count on to give me a reality check about new ideas and directions and was himself always coming up with new topics and foci for us. We worked extensively in developing concepts, such as the dominant epidemiological paradigm, that would become central for my approach. We worked together to develop the new approach to health social movements as well, shepherding it through conferences, workshops, and panels.

In 2000, Rachel Morello-Frosch began the process of considering a faculty position at Brown, and she finally arrived in 2002. I have experienced an amazing collaboration with her since then. We have written articles and grants together, embarked on major new projects, and cotaught an environmental justice seminar. Rachel has shaped the continued development of the health social movement approach and imbued it with a strong environmental justice focus. Her connections to many environmental justice groups, the environmental breast cancer movement, and other environmental groups has enriched our joint work and has shown me so many new ways of looking at the world. Rachel has helped direct the Contested Illnesses Research Group and has made it possible for us to work extensively with local environmental justice and toxic waste groups. She is the most encouraging and forthcoming colleague I could hope for, and her imprint is strongly on this book.

Sabrina McCormick, the first research assistant in my group, brought a powerful interest in and devotion to the environmental breast cancer movement. She made important contacts with many groups that enabled me to get far more knowledge about this movement than I hoped for. She also brought a multifaceted sensitivity to this project through her documentary film work. Brian Mayer, the next research assistant to come on board, brought a substantial interest in toxics reduction and facilitated strong contacts with groups in that area. His interest in the labor movement helped shape the important focus of my group on labor-environment coalitions. As our work on household exposure and body burden developed, Rebecca Gasior Altman stepped up with a deep appreciation for the importance of biomonitoring and forged some great connections in that area. Laura Senier brought new energy to our collaboration with many community groups, ranging from labor to health to toxic wastes, and has been exceptionally helpful in bringing forth the kind of knowledge that comes from such partnerships. These four past and present graduate students represented my group in speaking at conferences and other public venues, and in doing so they helped gain more allies and supporters for our collaborations. They coauthored many of the articles stemming from my project, some as first authors, and each offered his or her talents to improve the writing and to help shape concepts developed in this book.

Research of this sort requires dedicated support from people and groups whom I was studying or with whom I was collaborating. Many people facilitated access to research sites, welcoming me into both private and public worlds, trusting me to be part of their organizational life, and giving me the legitimacy to access other people and groups. Silent Spring Institute has been the most consistent group of environmental scientist-activists anywhere and provides a model both for citizen-science alliances and for visionary research. In research partnerships and other interactions with Silent Spring, I have expanded my horizons and learned much that shows up in these pages. Julia Brody and Ruthann Rudel graciously provided me access to research Silent Spring Institute. They eventually became amazing colleagues and have taught me so much about breast cancer, endocrine disrupters, precautionary approaches, and citizen-science alliances. At Alternatives for Community and Environment, Klare Allen, Penn Loh, and Jodi Sugerman-Brozan opened their doors to me and my students. At the Boston Environmental Hazard Center, Richard Clapp, David Ozonoff, Susan Proctor, and Roberta White welcomed me. At the Toxic Use Reduction Institute, Mike Ellenbecker and Ken Geiser were gracious hosts. At the Precautionary Principle Project, Joel Tickner did the same. Joel has also been a constant discussant of so many of the ideas I worked on in this book, as well as a strong inspiration. And at West Harlem Environmental Action, Swati Prakash and Peggy Shepard paved the way for me. Many of these people and groups later became research collaborators, closing the circle.

Many colleagues read drafts of the articles I wrote from the Contested Illnesses Research Group before I submitted them to journals. These people are the ones who point out how to build from your strengths, make you realize your shortcomings before you "go public," and suggest new avenues to take. I thank Chloe Bird, Julia Brody, Richard Clapp, Adele Clarke, Peter Conrad, Elizabeth Cooksey, Stephen R. Couch, Charles Engel, Steve Epstein, Susan Ferguson, Ken Geiser, Lois Gibbs, David Hess, Polly Hoppin, Patricia Hynes, Maren Klawiter, Sheldon Krimsky, Steve Kroll-Smith, Penn Loh, Jim Mahoney, Gerald Markowitz, Kelly Moore, Laura Potts, Lindsay Prior, Susan Proctor, Dianne Quigley, David Rosner, Ruthann Rudel, Suzanne Snedeker, Jodi Sugerman-Brozan, Joel Tickner, Simon Wessely, and Sally Zierler.

I also guest-edited two special journal issues that helped to advance many of the ideas in this book. Alan Heston, editor of *Annals of the American Academy of Political and Social Science,* invited me to edit an issue titled *Health and the Environment* that appeared in 2002, and Jonathan Gabe supervised the process by which *Sociology of Health and Illness* selected Stephen Zavestoski and me to edit the 2004 annual special issue, *Social Movements in Health.* Daniel Cress and Daniel Myers were the co-organizers of the special meeting of the Collective Behavior and Social Movements Section held prior to the American Sociological Association conference in 2002. They and others on the planning committee gave us a valuable venue for leading a two-session workshop that helped develop my focus on health social movements.

Peter Conrad, a wonderful friend and insightful colleague, has been a long-term discussant of so many ideas in this book, especially concerning health social movements. His thoughtful support of my research and writing has been a mainstay for me. He has always supported scholars in expanding the purview of medical sociology in order to incorporate other fields and to enrich all sociology as a result. This goal has been very important to my work at the interface of medical sociology and environmental sociology.

Once I got a near-final book draft together, I relied on three friends and colleagues to go over it with a sharp analytic eye: Richard Clapp, David Hess, and Kelly Moore. They were enthusiastic and thorough in their intense reading of the manuscript, full of good ideas, and supportive of the project. Dick Clapp, a longtime dear friend and colleague, has been for decades the most consistent supporter of my work in health and the environment, and his aid on the public-health and epidemiological issues is much larger than just this careful attention to the draft. Dick has also been a model for merging academics and advocacy, which has so much been a part of my life and which is a significant thread that runs throughout this book. David Hess brought particularly sharp attention to the area of health social movements through various talks and conferences he organized in recent years at Rensselaer Polytechnic Institute. From his vantage point in a major center of science and technology studies, David also was an astute commentator on the science studies elements of the book. Kelly Moore brought to this review task the lens of her sociology training and work in a combination of health social movements and science studies. Her level of engagement with my written words was spectacular, and it was a privilege to have someone give the book such a close reading.

I am very grateful for research grants that have enabled the work that has led to what I write here. In addition to the first two grants I mentioned at the opening of these acknowledgments, the National Science Foundation provided funding for "Blue and Green Shades of Health: The Social Construction of Health Risks in the Labor and Environmental Movements" (SES-035069) and for "The Research Right-to-Know: Ethics and Values in Communicating Research Data to Individuals and Communities" (SES 0450837). The National Institute of Environmental Health Sciences provided support for "Reuse in RI: A State-Based Approach to Complex Exposures" (Superfund Basic Research Program, 1 P42 ES013660–01) and "Linking Breast Cancer Advocacy and Environment Justice" (1 R25 ES013258–01). In terms of the latter, Shobha

Srinivasan, former program officer of the institute's Environmental Justice: Partnerships for Communication program, has been an amazing facilitator of the kind of collaborative work in which I engage, and her fine work in running the program gave me the opportunity to expand my horizons by engaging with many other grantees at the annual conferences and in between.

A backstage of activists and organizations has always provided a scaffolding for the kind of work I pursue. The Toxics Action Center has provided New England's six states with one of the country's finest environmental health organizations, an incubator of grassroots activism, a forum where I could meet many people and organizations, and a group with whom I have been very involved in a number of community projects. Matt Wilson, for many years the center's director, and Alyssa Schuren, the current director, have been gems. Lois Gibbs and the Center for Health, Environment, and Justice have provided a similar experience and motivation on the national level, and I have been honored to collaborate with them on some projects. The Alliance for a Healthy Tomorrow has impelled a statewide movement that pioneered the precautionary principle and gave me a continuing connection to affecting public policy, especially at the governmental level. In the Superfund Basic Research Program at Brown, for which I direct the Community Outreach Core, I have been enriched through working in major partnerships with the Environmental Neighborhood Awareness Committee of Tiverton and the Woonasquatucket River Watershed Council. Through Dianne Quigley's Collaborative Initiative for Research Ethics in Environmental Health and in tandem with the Toxics Action Center, I helped form the Providence Environmental Justice Education Forum. Through that process, I gained much by working with Rhode Island Legal Services, Hartford Park Residents Association, and various toxic waste groups throughout Rhode Island. Communities for a Better Environment in Oakland, California, and its northern California director Carla Perez have been a pleasure to collaborate with in the breast cancer and environmental justice project.

Grants to other people with whom I collaborate have also been important: "Community Environmental Health Research: Finding Meaning" was funded by the National Institute of Environmental Health Sciences (R25 ES012084–01), and "Short Courses for Environmental Health Research Ethics" was funded first by the National Institute of Allergy and Infectious Diseases (T15 49650) and later by the National Heart, Lung, and Blood Institute (T15 HL69792–01).

At Brown University, I am grateful for funding assistance from the Graduate School, the Undergraduate Teaching and Research Assistant Program, and the Department of Sociology.

I have been exceptionally fortunate over two years to have extensive research assistance on the book from Mara Averick. She tracked down many articles, checked endless facts, helped synthesize material, offered skillful advice on style and flow, and carefully edited all the chapters through multiple drafts. I appreciate so much her dedication to this book effort.

John Michel, my editor at Columbia University Press, was a great supporter of this project, and offered much valuable guidance. After his death, Patrick Fitzgerald shepherded the book through completion. Leslie Kriesel and Marina Petrova helped throughout the production process. Everyone at the Press has been a delight to work with.

Underlying everything I have written about here is an environmental health and environmental justice movement composed of countless people all over our country. Their actions as residents and workers dealing with toxic hazards in neighborhoods and workplaces, their efforts as scientists and scholars in advancing environmental health science, their generosity as progressive foundation staff, and their cooperation as government professionals in carrying out the true role of government—to help the population—are what makes good things happen. Thanks to all of them.

My family deserves much praise. My wife, Ronnie Littenberg, has been discussing with me for decades many of the ideas that have taken shape in this book. She has read many article drafts to help improve them, helped me think about innovative ideas for grant proposals, and been a constant support. My children, Michael Littenberg-Brown and Liza Littenberg-Brown, have been the joys of my life and have listened to and been excited by much of this work from the time they were in elementary school. I thank them above all.

ABBREVIATIONS

ACE	Alternatives for Community and Environment
ACS	American Cancer Society
ALA	American Lung Association
ALS	amyotrophic lateral sclerosis
BEHC	Boston Environmental Hazard Center
CDC	Centers for Disease Control and Prevention
DDE	dichloro diphenyl dichloroethylene
DES	diethylstilbestrol
DDT	dichloro diphenyl trichloroethane
DEP	dominant epidemiological paradigm
DOD	Department of Defense
DOE	Department of Energy
DPH	Department of Public Health
EBCM	environmental breast cancer movement
EPA	Environmental Protection Agency
EWG	Environmental Working Group
FDA	Food and Drug Administration

GIS	geographical information system
GWRIs	Gulf War–related illnesses
HHS	Department of Health and Human Services
HUD	Department of Housing and Urban Development
MUPS	medically unexplained physical symptoms
NAAQS	National Ambient Air Quality Standards
NBCC	National Breast Cancer Coalition
NCI	National Cancer Institute
NIEHS	National Institute of Environmental Health Sciences
NIH	National Institutes of Health
NSR	New Source Review
PAHs	polycyclic aromatic hydrocarbons
PBDEs	polybrominated diphenyl ethers
PCBs	polychlorinated biphenyls
PVC	polyvinyl chloride
REEP	Roxbury Environmental Empowerment Project
TURI	Toxic Use Reduction Institute
VA	Department of Veterans Affairs
VSOs	veteran service organizations
WE ACT	West Harlem Environmental Action

TOXIC EXPOSURES

|1|

CITIZEN-SCIENCE ALLIANCES AND HEALTH SOCIAL MOVEMENTS

CONTESTED ILLNESSES AND CHALLENGES TO THE DOMINANT EPIDEMIOLOGICAL PARADIGM

MANY DISEASES and conditions involve considerable dispute. Sometimes that dispute centers on whether medicine, science, government, and business believe the disease exists. At other times, there is disagreement over the way to treat the disease and whether people have equal access to treatment. In other cases, the conflict concerns how to study the cause or exacerbation of the disease or both. When we turn to the intersection of health and the environment, we find numerous contested illnesses that involve scientific disputes and extensive public debates over environmental causes. These environmental health problems are among society's most disputed health issues.

DETERMINING ENVIRONMENTAL HEALTH PROBLEMS

What exactly is meant by *environmental health problems*? The fullest definition would include the totality of hazards and health effects found in our living and working conditions: bacteria and viruses in human waste; animal vectors for infectious diseases; surface-water and groundwater pollution; air pollution from fires, vehicle exhaust, and incineration; chemical and petroleum product spills and explosions; and disasters such as floods, hurricanes, landslides, and fires (which may be natural, human caused, or human exacerbated). But that definition is broad enough to encompass virtually all disease-causing factors. It is more sensible to focus on the health effects caused by toxic substances in people's immediate or proximate surroundings (soil, air, water, food, household goods), a definition that mirrors most research and policy on environmental health. These effects are chemical- and radiation-related symptoms and diseases that impact groups of people in workplaces and communities. I term them *environmentally induced diseases* because they are in some part caused by environmental factors, even though environmental factors may interact with genetic predispositions or with some personal behaviors.

Focusing our definition of environmental health problems on toxic substances makes sense for several reasons. Toxic exposure has engendered much conflict, policymaking, legislation, public awareness, media attention, and social movement activity. It leads to disputes between laypeople and professionals, between citizens and governments, and among professionals. Because of the pressure from activists, especially when there is a localized health crisis, research on toxic health effects has spurred research that might not otherwise be done. Furthermore, toxic exposures demonstrate interesting and ongoing examples of how people determine what are social problems and how they engage in political contestation concerning environmentally induced diseases.

To give a sense of how contentious these disputes are, it is useful to note that the federal Environmental Protection Agency (EPA) has taken more than twenty years to release a risk assessment of dioxins, even though that class of chemicals has been one of the most studied in terms of toxic effects. Consider that dioxins appear in so many forms and through so many processes—from the bleaching of virtually all the paper pulp used in paper manufacturing to the burning of all trash in incinerators, the production of many industrial and agricultural chemicals, explosions in chemical plants such as Seveso (Italy), and the spraying of Agent Orange to deforest Vietnam during the war, to offer only a few examples. The political, economic, military, and ideological implications of a definitive statement of health risks from dioxins and the tightening of regulations that would follow are tremendous—and this is only one of the many classes of toxic substances with which people are concerned.

Environmental health is so strongly contested because the hazards identified by laypeople and some scientists are crucial parts of the modern economy, and the challengers seek to level the playing field by having corporations be more responsible. In addition, environmental health activism poses a philosophical challenge by arguing for a people-centered society rather than one dominated by profit seeking without regard for consequences. This activism also criticizes the centrality of individual responsibility, preferring to locate most responsibility in the underlying social structure and social institutions (economic arrangements, power relations, government agencies, corporations, and industry associations). Further, such activism criticizes all levels of government for not adequately protecting the public, even when clearly mandated to do so.

The debates over environmentally induced diseases bear witness to important problems in our society—problems concerning health status, social inequalities, corporate involvement in both causes and cover-ups of disease, governmental responsibility for monitoring health hazards and for controlling corporate practices, and public belief that much science and technology are producing problems rather than solving them. In thinking about these issues, we need to ask: Why do certain diseases have high visibility? How are causes of disease ascertained? Why is there so much controversy over disease causation? Why are some suspected causes given more attention than others?

These questions can be only partially answered by focusing on the internal mechanisms of science and medicine. To answer them fully, we must examine the many parties involved in disputes over disease. This examination leads us to look at the role of social movements and lay advocacy, as well as at the interaction between lay, professional, and government worlds. We know that scientific perspectives and lay perspectives (especially activist ones) may not agree in many cases, and we must then ask: What are the implications of disagreement? How do scientific, medical, and policy communities deal with fighting disease in light of this disagreement? Who defines which diseases are most important and which toxics get addressed?

PUBLIC OPINION ON ENVIRONMENTAL CAUSATION OF DISEASE

The Pew Charitable Trusts, one of the nation's major health foundations, sponsored a project called Health-Track, subsequently renamed the Trust for America's Health. Health-Track's mission was to educate the public, the health care system, and the government about the widespread impact of environmental factors in disease and the glaring lack of a health-tracking system to monitor diseases and to push for a nationwide biomonitoring program.

As part of its overall program, Health-Track understood the need to demonstrate widespread public concern over environmental factors through a public survey, so it conducted one on April 20–30, 2000, which interviewed 1,565 men and women nationwide (with a +/−3 percent margin of error). The results are striking, as shown in table 1.1. For sinus and allergy problems and for childhood asthma, a majority of respondents felt that environmental factors play a major role, and more than 90 percent of people agreed that the environment played some role. For birth defects and childhood cancer, two recently rising categories of disease, well more than 33 percent believed in a major environmental role, and close to 90 percent believe in a major or a minor role. For breast cancer, 25 percent believed in a major role, and 39 percent in a minor role, thus giving us well more than three-quarters of those surveyed believing in an environmental role in several major health issues.

TABLE 1.1 Public Beliefs in Environmental Causation of Disease

THINKING ABOUT SOME SPECIFIC ILLNESS, DO YOU THINK ENVIRONMENTAL FACTORS PLAY A MAJOR ROLE, MINOR ROLE, OR NO ROLE AT ALL IN CAUSING IT?

	MAJOR ROLE	MINOR ROLE	NO ROLE AT ALL	MAJOR AND MINOR ROLE TOTAL
Sinus and allergy problems	62%	28%	6%	90%
Asthma in children	53%	31%	8%	84%
Birth defects	39%	39%	12%	78%
Cancer in children, such as leukemia	37%	34%	14%	71%
Colds and flus	32%	44%	18%	77%
Breast cancer	25%	39%	18%	64%
Brain tumors	25%	34%	17%	59%
Learning disabilities	23%	39%	24%	62%
Infertility	22%	40%	17%	62%
Prostate cancer	19%	36%	21%	55%
Parkinson's disease	12%	29%	21%	41%

Note: Numbers do not total 100% because of nonresponse.

Source: Adapted from Princeton Research Associates, for Health-Track, *National Survey of Public Perceptions of Environmental Health Risks* (Princeton, N.J.: Princeton Research Associates, June 2000).

These figures are in sharp contrast to most governmental perspectives (as represented in government publications, reports, and funding allocations) and to most medical research literature, which does not place much emphasis on environmental causation. This disparity can be seen as part of a frequently observed phenomenon in which laypeople make higher estimates of risks and hazards than do professionals.[1] There is some basis for believing that laypeople make overestimates because they do not have the same access to actual disease or mortality data as do professionals. Laypeople also overestimate dreaded outcomes, such as radiation hazards, compared to more mundane hazards such as auto accidents. These estimates are also influenced by the public's perceptions of their control over exposures (e.g., the driving of cars versus electromagnetic fields from power lines). At the same time, however, laypeople's everyday experiences may validate such overestimates. First, they are often the first to notice disease increases and clusters, and they often have to fight to get recognition and action. Second, they face corporate and governmental resistance to the recognition of environment health effects, making them think that there are indeed problems that have not yet been adequately acknowledged. Third, laypeople often point to environmental and occupational factors that are later found to be very accurate and that become part of the core of medical knowledge and governmental regulation. Fourth, laypeople focus on the proximate concerns they see in their everyday lives and worries.

In fact, it is likely that many of the clusters noticed by laypeople are valid observations. Rather than the problem being people's overestimates based on proximity and familiarity, perhaps it is that science simply has not yet seen the clusters identified by laypeople and has not figured out appropriate methods to quantify hazards, exposures, and clusters. Indeed, there is growing evidence that laypeople are correct. Health-Track pointed out that our society spends enormous amounts on health without the accompanying gains in health status that we see in other advanced industrial societies.[2] One reason for this outcome is that as a society we do not do a good job of monitoring health status and the factors implicated in disease. The excellent health registries of Scandinavian countries enable them to track many generations of people for disease, providing high-quality methods for disaggregating genetic factors from nongenetic factors. One example of how health tracking can help is seen in the Swedish response to polybrominated diphenyl ethers (PBDEs), fire retardants that have been found in high levels in breast milk. In Scandinavian countries, data on everyone in each country are collected from cradle to grave and are centrally located and accessible to researchers. Once Swedish authorities noticed the persistent and bioaccumulative properties of PBDEs, Swedish scientists and regulators looked at longitudinal data that had been gathered from three decades of sampling breast milk to track exposure to dioxin, polychlorinated biphenyls (PCBs), and other pollutants that accumulate in body fat. Alarmed by the rapid increase in PBDE body burden, Sweden banned a number of PBDEs.

Health tracking also alerts us to the increase over time in other exposures and in diseases and conditions that are known to be or are potentially caused by environmental toxins. Supporters of monitoring in the United States seek a higher quality of research capacity, though for them higher quality mainly means better integration of data on pollution emissions and exposures, along with some health outcome surveillance information. The infrastructure to collect such data does not exist here, mainly because we do not have any centralized system of health care. Advocates of better health tracking point to the need for universal health care not only to get the centralized, systematized infrastructure that such an excellent health-tracking system would really require, but to bring our society into the modern world of health care.

The evidence for public concern extends beyond opinion poll data. Devra Davis's 2002 book *When Smoke Ran Like Water: Tales of Environmental*

Deception and the Battle Against Pollution was nominated for the National Book Award, also signifying the popularity of such topics. Environmental contamination resonates widely with the public. We see this interest in the popularity of movies such as *Erin Brokovich* (2000, starring Julia Roberts as a legal researcher turned activist uncovering Pacific Gas and Electric's cover-up of a major chromium contamination episode) and *A Civil Action* (1998, starring John Travolta and focusing on the legal aspects of the Woburn childhood leukemia case that I discussed in the preface), and in widely viewed TV specials such as Bill Moyers's 2001 *Trade Secrets,* which showed how the chemical industry hid data on how their products injure people.

But the most important evidence for public concern over toxics is the large number of instances in which citizens form community groups to deal with existing toxic contamination or to guard against the potential for contamination from new sitings of incinerators, dumpsites, or other sources. The popularity of books and movies about toxic contamination would not be possible without the actual mobilization of people, and this environmental health movement has become one of the most important social movements of our time. Environmental justice activists have built on this movement, expanding their concern to other hazards and inequalities. They have strengthened the grassroots struggle for environmentalism and environmental health by highlighting how social inequality and discrimination adversely impact health. They have also broadened strategies to improve environmental health by moving away from a focus on technical fixes in the policy arena to an emphasis on community organizing that explicitly links social justice and public health. In the process, they have helped reshape the environmental health movement to address racial and class inequalities and to take on a more totalizing political perspective. To understand how broad this concern is, consider that the Center for Health, Environment, and Justice (the national resource organization started by Love Canal organizer Lois Gibbs) has worked with more than ten thousand groups since its founding. This book views advances in environmental health as stemming primarily from social movements and social movement organizations such as the Center for Health, Environment, and Justice, the Toxics Action Center, and hundreds of environmental justice and environmental health groups around the country.

THE NATURE OF ILLNESS CONTESTATION

We might expect that a general public belief in the presence and etiology of a disease would be accepted once there was sufficient medical and scientific research. But such acceptance is possible only when the disease or its putative cause is not controversial. For example, even the powerful evidence linking tobacco and lung cancer was initially insufficient to produce general social consensus. The tobacco industry and its allies marshaled much financial and political support to deny the evidence. By using their extensive financial resources to perform research on the mechanisms of cancer instead of on the causes, by focusing on individual characteristics that might have correlations to cancer, and by carrying out focused advertising campaigns, tobacco companies created a public controversy in order to confuse the issue and create ambiguity. That effort postponed for decades the full medical recognition of the problem and stalled government action. Sufferers and their allies had to fight politically rather than rely exclusively on scientific evidence, despite its strength.[3]

Likewise, the dangers of lead were known early in the twentieth century, yet lead producers and the paint companies that used lead as a major product component deliberately hid data on lead's health effects while advertising lead as a wonderful substance. The evidence of its ill effects was so overwhelming that many countries banned lead paint in the 1910s and 1920s, yet the United States waited until 1970. Scientists such as Herbert Needleman, who researched the dangers of lead on children, were branded as troublemakers and hounded by scientific colleagues.[4] Science alone could not produce appropriate regulation to reduce lead poisoning. It took a concerted social movement beginning in the 1960s, pioneered by black and Latino rights groups (most notably the Black Panther Party and the Young Lords Organization).

We see similar problems with most environmental and occupational diseases. A broadly acceptable social definition, a new consensus by experts, and often a social movement are necessary to achieve a belief in the existence of the disease and its social and environmental causation. This situation is vastly different from the past, when professionals originated ideas and controlled the process of problem identification. Virtually all cases of contaminated communities are detected by lay discovery, largely because affected populations tend to be the first to notice sentinel health events or disease clusters that may indicate systematic or new environmental health problems. In addition, scientists and government agencies do not usually carry out routine surveillance that would detect such problems. Even routine surveillance is insufficient; for example, a state cancer registry may be mandated to publish annual reports of cancer excesses by town and city and may routinely tell the local health department, but this approach does not guarantee that residents will be made aware of higher than average rates of cancer where they live. Even when asked by communities to look into a problem, state agencies do not do enough.

For example, a survey of all fifty state public health department responses to lay cancer cluster reports found that there were an estimated 1,300–1,650 such reports in 1988, a large number for short-staffed agencies. Many health departments discouraged citizens, however, sometimes requesting extensive data from them before they would go further with their investigations. And they often merely gave routine responses emphasizing the lifestyle causes of cancer, the fact that one of three Americans will develop some form of cancer, and that clusters occur at random.[5]

Disputes over environmental causation are not the only places to find illness contestation, either. For sociologists, contestation concerning health is a very common topic of study. Based on the work of Elliot Freidson, one of the major scholars of both medical sociology and the sociology of professions, we understand the frequent clash of views between lay and professional perspectives.[6] We know that biological disease is experienced in the diverse ways that people experience illness, differing according to race, class, gender, religion, ethnicity, and locality. The literature is full of lay-professional conflicts over etiology, appropriate treatment, and combinations of self-care and alternative care alongside mainstream medicine. Medical sociologists are therefore steeped in the tradition of contestation. This contestation usually involves diseases or health situations that present a clear challenge to major social institutions. The environmentally induced diseases I discuss here are clear examples because the social definition brings with it a critique of corporate practice and government policy. Other examples include iatrogenic diseases (diseases caused by the practice of medicine) for which pharmaceutical firms have withheld data on side effects and unnecessary surgery such as hysterectomies.

PUBLIC REACTION TO ENVIRONMENTAL HEALTH EFFECTS

By not being receptive at earlier stages to known or potential environmental causation, science and government may lose precious time in discovering, treating, and preventing environmentally induced diseases. As a result, they are likely to lose public trust. Precautionary principle advocates speak in terms of "late lessons from early warnings" in reference to situations where knowledge of chemicals' toxicity were ignored until those effects had become very widespread. There are enough such examples that environmental activists increasingly feel that government, science, and business should have learned by now to be more cautious when examining aspects of the environment that have health effects. These aspects include well-known hazards with clear health effects, such as asbestos, which causes the lung disease asbestosis and the lung cancer mesothelioma; lead, which causes learning disabilities; benzene, which causes leukemia; PCBs used as electrical insulators, which cause chloracne; and diethylstilbestrol, a drug shown to be ineffective in its intended goal of preventing miscarriages, yet long used until evidence mounted of its role in causing vaginal cancer in users' daughters.[7] In all these cases, credible evidence of dangers existed long before any action was taken, yet policymakers and scientists are still not applying these lessons in the present.

We have further examples from the history of the occupational safety and health movement, where workers and their unions have long been the first line of discovery of harmful effects of unsafe machinery and production processes, as well as of a variety of toxins, including coal dust, cotton mill dust, silica, lead, radiation, and agricultural pesticides. Occupational health issues have often brought to light the economic trade-offs whereby profits have frequently been placed ahead of the safety of the producers and consumers of industrial and agricultural goods. In many cases, corporations denied the existence of known health effects and instead blamed the victim. For example, company doctors at the Johns Manville Company told workers with asbestosis or mesothelioma (a rare lung cancer that is a signature disease of asbestos) that they merely had emphysema from smoking.[8] Here we see how the corporate world can play on the same individual responsibility theme that permeates science and government. The more those institutions emphasize personal control over hazards, the more the corporate claim of individual responsibility is legitimized.

Much of what toxic waste activists have learned about such corporate trade-offs has come from looking at the occupational health world. Yet these activists have dealt with a different set of issues as well because they have called into question exposures that often occur at lower levels than occupational exposures and are without an organizational apparatus, such as a labor union, to take on the problems. Like labor unions, toxic waste activists face a jobs versus environment blackmail situation. Further, corporations have an easier case against environmental causation because they can easily claim that there were many other potential sources of exposure for residents and that the residents' own behavior put them at risk.

Many people can grasp the significance of corporate and governmental denial when they look at the issue of radiation exposures due to government activities. The legacy of the "atomic veterans" is particularly crucial: many soldiers and sailors were deliberately exposed to radiation in the 1940s and 1950s to test exposure levels that would produce radiation-induced illnesses. Workers at nuclear weapons facilities and nearby residents faced high levels of accidental and deliberate releases. These victims fought long for recognition of their diseases and encountered deep government secrecy and implacable denial.[9]

Given such a history, it is not surprising that today many people distrust official corporate and government statements on risks and hazards. This history makes it easier for people to believe that there are many environment health effects for which both business and government are responsible. The public responses to the toxic crises of our time—whether Love Canal–like contamination episodes or increases in cancers and in immunological and neurological disorders—represent not just the responses of individual organizations to specific problems. Rather, they represent a broad and growing public distrust of how science and government have operated and a belief that our society has become so full of risks and hazards that it is dominated by them—what Ulrich Beck terms the "risk society."[10]

WHY PICK ASTHMA, BREAST CANCER, AND GULF WAR–RELATED ILLNESSES?

To explore these issues, I examine three cases of disease with known or suspected environmental causation—asthma, breast cancer, and Gulf War illnesses. In addition to those cases, I discuss many other contested illnesses throughout the book, though in far less detail. Out of so many diseases with potential environmental causation, why pick these three? In this section, I provide brief highlights of the answers to this question and go into depth on these matters in later chapters.

PROMINENCE IN THE PUBLIC EYE

There is much debate over identification, environmental causation, treatment, and the proper place of lay involvement. Asthma incidence has increased dramatically over the past decade, and there is widespread debate on the contribution of air contamination, especially small particles as yet unregulated by the EPA. Discussions about more stringent regulation call into question the whole history of air pollution regulation, provoke industry opposition based on cost considerations, and raise scientific concern about the research methods. Breast cancer has been linked to environmental causes, especially xenoestrogens that act as hormone disrupters.[11] The growing evidence supporting this endocrine disrupter hypothesis has created much public and scientific discussion and is especially important given the ubiquity of those chemicals. People are shocked that the lifetime incidence for breast cancer is one in eight.

Gulf War–related illnesses (GWRIs) were very visible for a while, but are less so now because of the highly political nature of the government's reluctance to treat fairly the soldiers who fought in a widely popular war. The Agent Orange controversy following the Vietnam War, during which veterans experienced increased risk of various diseases due to dioxin exposure, failed to provide a guide for how to deal with military-related environmental disease because the government was also then very reluctant to acknowledge Agent Orange's health effects. For the Iraq War starting in 2003, high rates of psychiatric disorders have been reported, making this issue once again a significant public-health concern.[12]

PUBLIC SCRUTINY OF RESEARCH AND REGULATION

Funding, hypothesis formulation, research design, interpretation, and pressure for regulation are highly visible features of these contested illnesses. Pressures from laypersons and professionals have led directly to specific state and federal action to fund research. Asthma awareness advocates and activists, along with public-health professionals and air pollution scientists, have made air quality regulation a focus of their efforts. They have fought to reduce emissions through stronger amendments to the Clean Air Act, the 1963 federal legislation aimed at reducing pollution, as well as through many local air pollution regulations. Breast cancer activists won state and federal funding in Massachusetts and New York for specific studies of a possible link between the environment and breast cancer. Along with other environmental activists, they have pushed for local, state, and federal adoption of the precautionary principle, which places the burden of proof for health effects of chemicals on the producers rather than on the consumers and declares that proof of safety should exist before chemicals are utilized in order to curtail or stop the use of potentially dangerous chemicals, even in the absence of definitive studies. The landmark 2003 San Francisco resolution directs all city agencies to use the precautionary principle to guide their activities.[13] Gulf War veterans formed advocacy groups that obtained congressional support for investigations of their ailments when the Department of Defense (DOD) proved unhelpful in addressing their concerns. The research on these contested illnesses has been discussed widely both inside and outside of science, with a large amount of mass media coverage. Of course, these outcomes are positive, which does not always happen. Sometimes lay efforts do not lead to regulatory change or other desired outcomes, and the dispute over causation continues. Or, even if groups get funding for studies, they do not find results that help them, as we will see with the Long Island breast cancer studies.

PROMINENCE OF SOCIAL MOVEMENTS

As suggested by the previous point, social movements are central in bringing these contested illnesses to light and in pursuing research and treatment. Asthma activism has become a core element of many *environmental justice* groups, who organize around unequal distribution of environmental hazards, mostly involving racial inequality. Environmental justice organizing has highlighted how an illness such as asthma, which disproportionately impacts poor people of color living in urban areas, is driven largely by disparities in housing quality, access to health care and treatment (e.g., access to maintenance medications as opposed to a trip to the emergency room when suffering an attack), and exposures to environmental sources of pollution from point and mobile sources. In this way, environmental justice highlights the political economy of illness and how discrimination shapes environmental health disparities such as the incidence of asthma.

Pellow noted this type of approach in his environmental inequality model, which emphasizes the interaction of three methodological and analytical points: (1) the need to view environmental inequality as a sociohistorical process rather than as a discrete event; (2) the need to understand that environmental inequality involves a multiplicity of stakeholders with "shifting interests and allegiances" rather than a simple victim-perpetrator dyad; and (3) the need to view environmental inequalities as a cyclical process of production and consumption.[14] Pulido views environmental justice efforts as "subaltern struggles" in which activists are in "direct opposition of prevailing powers" and challenge "the entrenched and all-encompassing ways in which power relations are constituted and experienced."[15] The key theoretical insight offered by Pulido's approach is that subaltern struggles over environmental issues are never solely about the environment. These struggles call into question forms of structured inequality and directly challenged institutionalized forms of domination. Subaltern groups must address a variety of social structures and institutions that contribute to environmental and social inequality because of their position within the socioeconomic structure. Because these groups are at the bottom of the socioeconomic structure, their goals and actions are always aimed at reconfiguring the distribution of power and resources in society. That is why their efforts to deal with asthma are at the same time efforts to gain more economic resources and political power for their communities. Agyeman takes this point further, arguing that environmental justice groups have developed a "just sustainability paradigm" that proactively combines environmental justice and sustainable development into a new hybrid. In this model, "environmental quality and economic and social health are inextricably linked."[16] Thus, it is not surprising that the social movement component of asthma awareness advocacy and activism has made asthma into one of the key health issues around which minority populations organize.

Breast cancer activism has been one of the most significant health social movements of our time, involving people across a wide range of politics and achieving notable successes in increased breast cancer research funding. The environmental breast cancer movement (EBCM), a subset of this larger movement, has become an important avenue for pursuing environmental factors in breast cancer, as well as for more broadly integrating environmental politics and health movements.[17] It has gone beyond breast cancer to take up many other environmental health concerns and in that process has been a major player in advocating for the precautionary principle as an overall social perspective, mainly through pushing for tighter control of endocrine disrupters. Because the movement focuses on endocrine disrupters as possible causes of breast cancer, it has helped to identify many potentially dangerous chemicals and to study exposures that have previously been unknown or poorly known.[18] This focus on prevention has also affected the direction of the overall breast cancer movement.

Gulf War veterans organized to demand recognition of their symptoms, greater disability compensation, and more research on environmental causation. Though they did not achieve the breadth of organization and the continued mobilization gained by asthma and breast cancer activists, they did force the issue to public and governmental attention, winning some important gains. Their continued pressure kept attention on GWRIs, even in the absence of positive data on health effects, and compelled the military to revise sharply upward its estimates of how many soldiers were exposed to toxins, especially by finally admitting to widespread exposure from the Khamisiyah weapons depot. Various government reports were also critical of the delivery of care, of poor DOD and Department of Veterans Affairs (VA) collaboration, and of the lack of adequate research overall.

PUSHING FOR DEMOCRATIZATION OF SCIENCE

Activists in the areas of all three conditions have increased democratization of science by raising issues of medical science and environmental research as part of public debates, rather than allowing them to remain behind the closed doors of science and government. Asthma activists in environmental justice organizations have become major players in pushing for the growing federal funding of *community-based participatory research*, where citizens and scientists work together on environmental health research. Breast cancer activists have succeeded in getting movement members to be part of state and federal review panels, where they play a role in deciding what research to fund. The lay-professional collaboration in these contested illnesses has helped expand the overall push for democratic participation in science, making citizen-science alliances increasingly part of the social fabric of science.

CONTROVERSY ON DISEASE DEFINITION AND CAUSES

In the study of *contested illnesses,* it is clear that not every disease or condition will have the same level of scientific consensus about environmental causation. The scientific base for GWRIs is the weakest of the three conditions on which I focus here. For many physicians and researchers, GWRIs are themselves suspect as an illness or disease category. This suspicion contributes to the difficulties Gulf War activists have had in achieving biomedical recognition and government action. Breast cancer, in contrast, is highly recognizable, with little or no dispute about whether or not such a condition exists. Research on links between breast cancer and the environment is mixed, however, with only a small degree of support for environmental causation, but with enough positive findings and suggestive directions to help generate more research funding and to expand public awareness. For asthma, the science base for the role of air particulates, ozone, and other criteria pollutants in creating and exacerbating the disease is very strong, making it more possible to garner scientific support, though the social movement is weaker than that for breast cancer, and it has been less successful in achieving funding increases and public awareness. There is some degree of scientific skepticism about whether all diagnosed asthma cases are in fact asthma. We learn from the comparison of these three conditions that social movements are as crucial as the science base when it comes to achieving results.

Controversy reads differently for different parties. Activists use the creation of controversy to advocate precautions because they are challenging the taken-for-granted assumptions about the safety of chemicals and certain products. But for many people in government or industry, pointing to controversy is equal to saying that something is not falsifiable and therefore that no action should be taken. Within such controversies, there is a large imbalance in the ability of either side to generate evidence. Corporations have extensive resources in the form of staff scientists, lawyers, and managers, many of whom have the ear of government regulators. Advocacy groups and social movement organizations often do not have scientists on staff, and when they do, they have perhaps one or only a few. For them, it is typically an uphill battle to remove or regulate something already in production and use, and there is a natural governmental and corporate resistance to such action. In this sense, corporate interests frame advocate-defined controversy as an assault on progress. These points about controversy should be kept in mind whenever we examine specific disputes over environmental causation and the remedies people seek.

EXCELLENT RESEARCH SITES

Although the reasons stated so far for why I have chosen to focus on asthma, breast cancer, and GWRIs make much sense, it was also necessary to find good research sites where I could observe the way that activists and scientists collaborated. I was fortunate to have exceptional groups near me, and as I mentioned in the preface, I had connections to them via the local environmental health networks, and they warmly welcomed me, my faculty colleagues, and my students.

ALTERNATIVES FOR COMMUNITY AND ENVIRONMENT Alternatives for Community and Environment (ACE), an environmental justice group in Boston founded in 1993 by two lawyers, provided me with a fine example of how an environmental justice group organized around environmental causes of asthma. ACE had originated in a struggle against the siting of an asphalt plant and had become widely respected as the major environmental justice organization in the area. In assessing local people's desires for action, ACE understands asthma as key and is able to work on many levels: teaching in schools, organizing in communities, getting government agencies to install air monitors, collaborating with academics in research, working for transit system reform. It includes a mixture of activists, professionals, and grassroots neighborhood people, covering many racial and ethnic groups. ACE works with other local and regional environmental justice organizations and has a very broad perspective on how its local struggles fit into a bigger picture. It is a great example of how nonscientific organizations in practice can learn much science, engage in scientific research, and push for policy changes stemming from that research.

WEST HARLEM ENVIRONMENTAL ACTION West Harlem Environmental Action (WE ACT) in New York City was not one of my original research sites because I was focusing on the Boston area. But it became clear that WE ACT was doing many of the same things as ACE, and I liked the idea of seeing this kind of work in two different locations. This type of "multisited ethnography" provides a more holistic and complex picture of social movements, especially in showing that material circumstances are creating the same social outcomes in different locations.[19] Building on struggles around a garbage transfer station, WE ACT had generated enormous credibility in its community and was able to organize effective campaigns on transportation and asthma. It has a tighter connection to collaborative work with universities and is one of the pioneer environmental justice groups in the National Institute of Environmental Health Sciences (NIEHS) program. It plays an important national role in conferences that help shape the community-based participatory research model of environmental justice and is very engaged in local and regional networks as well. WE ACT also has a strong approach to doing and applying scientific research through various collaborative projects with Columbia University.

SILENT SPRING INSTITUTE Silent Spring Institute in Newton, Massachusetts, offered a perfect location to study environmental causes of breast cancer. Originated by a breast cancer activist organization (Massachusetts Breast Cancer Coalition) in 1994, Silent Spring Institute is the only research center in the country dedicated to researching breast cancer and the environment. Silent Spring Institute comprises scientists, activists, public-health advocates, and elected officials, and it exemplifies the notion of a research organization dedicated to an activist-scientist collaboration. It has also engaged in a nationally watched study of breast cancer on Cape Cod, in which it analyzed toxic exposures, pioneered new household toxics-sampling measures and geographical information system (GIS) approaches, and characterized many toxics for the first time. Silent Spring Institute is committed to working with laypeople, to sharing research data openly, and to being part of national effort on environmental breast cancer. It is also, along with its parent organization, the Massachusetts Breast Cancer Coalition, active in advancing the local and state environmental community through the Precautionary Principle Project.

BOSTON ENVIRONMENTAL HAZARD CENTER The Boston Environmental Hazard Center (BEHC), a collaborative project of the VA Medical Center in Boston and the Boston University School of Public Health, was a valuable location to study Gulf War illnesses. It was one of only four such centers in the country and was committed to strong involvement of Gulf War veteran activists. The Boston University connection was in the Department of Environmental Health, long a national model for how epidemiologists and other scholars put their skills in the service of community environmental groups, many of which otherwise have no resources. Epidemiologists there have been national leaders in helping communities to evaluate whether they should conduct health surveys or seek such surveys from the government and in assisting them in some cases where they did do health studies.

TOXIC USE REDUCTION INSTITUTE, PRECAUTIONARY PRINCIPLE PROJECT, AND ALLIANCE FOR A HEALTHY TOMORROW Because I was concerned with approaches to reduce toxic exposures that can cause so many diseases, I also studied the Toxic Use Reduction Institute (TURI), a state agency in Lowell, Massachusetts. TURI was formed through the 1989 Toxics Use Reduction Act and guided by environmental activists. It has been the national model for state efforts in this field. TURI represents a general environmental health practice that derives from public concern, and its activities can directly impact asthma and breast cancer incidence through reduction of exposure to toxins. At the time that I was setting up connections to various environmental justice groups, individuals involved in both TURI and other departments and centers at the University of Massachusetts–Lowell were beginning their efforts to launch the Precautionary Principle Project, and because breast cancer and other environmental health activists were eagerly pursuing the precautionary principle, it was logical to include this group in my study, along with its successor, the Alliance for a Healthy Tomorrow, both of which play a national role in guiding environmental health theory and practice.

All of these groups are exceptional in many ways. They have a solid organizational structure that allows them to remain committed firmly to their goals, and at the same time they are very flexible in adapting to new directions. They all are very dedicated to serving the needs of the affected and potentially affected public and to doing so via much public involvement. They have much respect in their local communities, but also have gained the regional and national limelight. They cooperate with an extensive number of other organizations, some within their specific focus (e.g., one environmental justice group cooperating with other environmental justice groups) and others outside their immediate focus (e.g., breast cancer groups cooperating with toxic waste groups). They connect scientific research and social activism. Put together, such attributes make these organizations trendsetters in environmental health.

CONCEPTS AND THEORETICAL FRAMEWORKS FOR EXAMINING ENVIRONMENTAL HEALTH DISPUTES

In this section, I present the *dominant epidemiological paradigm* (DEP), by which I mean the generally accepted perspective on diseases held by many parties. I then examine how environmental health activists critique the DEP and put forth an alternative *public paradigm*. In doing so, they engage in *popular epidemiology* (the lay identification of disease clusters or toxics exposure or both) and in other forms of citizen-science alliances, many of which employ community-based participatory research. I show how an activist-oriented scientific framework of critical epidemiology and other sources of professional activism help in this process.

The term *paradigm* is no longer fashionable in science studies, and a huge critical literature has developed around this shift. A central concern is the assumption that research fields are characterized by a single main conceptual framework and set of methods. The paradigm concept has tended to be replaced by concepts such as consensus knowledge. However, within health research communities and policy circles, the term *paradigm* continues to be widely used, including the idea of a dominant paradigm, and to some degree the Kuhnian idea does describe fairly well the way that consensus knowledge achieves a stranglehold over research funding and acceptable research problems. As a result, I follow my informants and use this term, with the understanding that in this context it does not imply all that has come to be associated with Kuhn's use of it.[20]

THE DEP

When we see the enormous obstacles faced by citizens and scientists in showing the dangers of well-established toxics such as lead, mercury, radiation, and tobacco, we realize how much goes into challenging the status quo of science and government policy and public understanding—in other words, the DEP. This DEP is a set of beliefs and practices about a disease and its causation embedded within science, government, and public life. It includes established institutions entrusted with the diagnosis, treatment, and care of disease sufferers, as well as journals, media, universities, medical philanthropies, and government officials. Many structures and institutions contribute to a generally accepted view of disease, but people do not immediately see them.

Furthermore, the DEP is both a model and a process. It is a model in that it helps us understand the complexity of disease discovery. It is a process in that it delineates a variety of locations of action. Actors can enter the DEP process at different locations and take action on one or more of the components. A DEP develops over time, becoming the product both of general factors such as the rise of professional authority, medical dominance, and "scientization" and of specific factors such as the rise of organizations, agencies, and research traditions.

The rise of professional authority, most notably in medicine, has placed experts in control of many areas previously in the purview of laypeople. Over time, lay experience and knowledge became devalued; traditional professions and occupations were curtailed or eradicated; and alternative methods of looking at the world and carrying out scientific and medical action were diminished.[21] In the world of medicine, a new model of medical education, geared solely to allopathic medicine, pushed out competitors and reduced the vitality of multiple approaches. For example, midwives were sharply attacked and their profession greatly reduced, despite their widespread success, a success that is recently being understood in light of the poor infant mortality rate in the United States in comparison to the rate in other advanced industrial nations that use midwives extensively.

"Scientization" is a logical outcome of professional authority and medical dominance. It refers to the variety of ways that science and technocratic decision making have become an increasingly dominant force in shaping public attitudes, media coverage, and most particularly social policy and regulation. The quest for "better science" in policymaking has become a powerful authority used to support dominant political and socioeconomic systems. Through this scientization of decision making, industry exerts considerable control over debates regarding the costs, benefits, and potential risks of new technologies and industrial production by deploying scientific experts who work to ensure that battles over policymaking remain scientific, "objective," and effectively separated from their social context. This process has several outcomes. First, questions are posed to science that are virtually impossible to answer scientifically due to data uncertainties or the infeasibility of carrying out a study.[22] Second, the process inappropriately frames the "transcientific" issues—the political and moral questions—as necessarily scientific, thus limiting public participation in decision making and keeping the decisions under experts' control.[23] Third, the scientization of decision making delegitimizes the importance of those questions that may not be conducive to scientific analysis. All of these processes can exclude the public from important policy debates and diminish public capacity to participate in the production of scientific knowledge itself.[24]

Despite the power of science, throughout the book I show that science itself is not the defining characteristic of the policy outcomes. The science itself sometimes seems less relevant than who has power; at other times, it is not. In the chapter comparing asthma, breast cancer, and Gulf War illnesses (chapter 5), I show how the strength of the social movement can be more important than the accepted science base, as demonstrated when we look at the successes of a social movement in a given disease area. Government and industry often cast the arguments in terms of narrow questions of sampling, significance levels, confounding variables, and other such features, and they use "controversy" as a means of dismissing community groups' claims. This approach often pushes groups to engage in science, either on their own or in collaboration with scientists. Even if social movement activism can sometimes trump the science base, activists generally prefer that science backs up their claims because it makes their organizing, policy, and education tasks much easier.

A DEP is also produced by the rise of organizations, agencies, and research traditions. For example, the American Cancer Society (ACS) became the nation's largest health voluntary (also called a *medical philanthropy* or *health charity*, an organization that works in varying degrees with health providers and government on issues of disease education, research, and treatment), uncritically supporting a biomedical model that avoids any attention to environmental or occupational factors, and even failing to take a strong stand on tobacco's role on lung cancer at an early point when the evidence for that connection was clear. On the government side, the National Cancer Institute (NCI) became the premier institute within the National Institutes of Health (NIH), also eschewing any focus on environmental factors. The ACS and NCI, along with other NIH institutes, private foundations, and medical societies, created what many scholars have called the "cancer establishment," focusing on genetics and individual lifestyle choices and putting very little support into prevention.[25] This approach has led to a growing focus on genetic determinism, which fails to look adequately at interactions with the surrounding world and even adheres to the primacy of genetic explanations when genetic factors do not account for the bulk of cancers (as shown later for breast cancer). Health promotion messages from the cancer establishment emphasize the actions individual people can take to avoid cancer, making them solely responsible. The problem with the individualist model is not that some foods and substances are unhealthy, but that it seeks only to change people's personal behavior and not to regulate the industries and institutions that produce, sell, use, and dispose of items that are unhealthy.

As a result of these factors, the DEP is typically characterized by a hegemonic outlook on disease that emphasizes individual behavioral factors rather than environmental and social factors as keys to disease prevention. Such an individualistic focus is common because it seems more straightforward to change individual behaviors rather than to reorganize social institutions or to promote fundamental changes in industrial production and government regulation. Such individualist approaches also carry a moralistic undercurrent that holds individuals responsible for their health status, despite population-level data that demonstrate the importance of social structural factors in determining health and disease in populations. These approaches are frequently termed *lifestyle* approaches because they deal with factors that are apparent choices, such as smoking, diet, alcohol use, and late first parity. A lifestyle approach fails to see, however, that personal behaviors are largely shaped by social structures; for example, unhealthy school food and the absence of full-service supermarkets in poor urban areas set up children to consume fast food. This emphasis on individual behavior saves money for corporations and the government through minimizing structural responsibility.

Although the DEP is hegemonic in terms of the many components that combine into a powerful process, many lay challenges to it, coupled with internal dissention within scientific and governmental institutions, have surfaced. In some cases, these challenges are characterized by a form of critical epidemiology in which laypeople join with epidemiologists in citizen-science alliances to do a combination of research and policy advocacy. These challenges to the DEP, when cohesive and rooted in social movements, provide a public paradigm (to be shortly discussed) that offers an alternative conception of epidemiology.

Observing health social movements involved in environmental health shows that there is contestation with far more actors than just state authority. In particular, there is contestation with science and medicine, the media, medical philanthropies or health voluntaries, research organizations, professional bodies, journals, and educational institutions. This complex of actors is what makes it necessary to view these contestations through the lens of the DEP. Current scientific and public debates regarding potential links between toxics and contested illnesses epitomize the multifaceted ways in which struggles to shape the DEP can play out. The controversy over environmental causation of disease is the subject of debate among social movement activists, scientists, and policymakers. Deliberation within and between each of these groups has helped shape public awareness, scientific research, medical practice, and government regulation.

As the diagram in figure 1.1 illustrates, the DEP comprises many components, and social movement challengers must address many, if not all, of these components in order to effect change. The diagram begins with three prediscovery elements that—through government, science, and public life—shape medical practices and the experience of the medical system. By "prediscovery," I mean the set of beliefs and practices that exist before the discovery of a disease cluster or the exposure of a toxic substance or the awareness of a potential environmental cause of disease. The "postdiscovery" period occurs after a debate on the legitimacy of the DEP, in which organized forces have mobilized to engage with as many of the DEP's components as possible, including direct challenges to public understanding, scientific knowledge, and policy.

Once people organize to change the way a disease or condition is defined or treated, they confront the hegemonic components of the DEP as they work

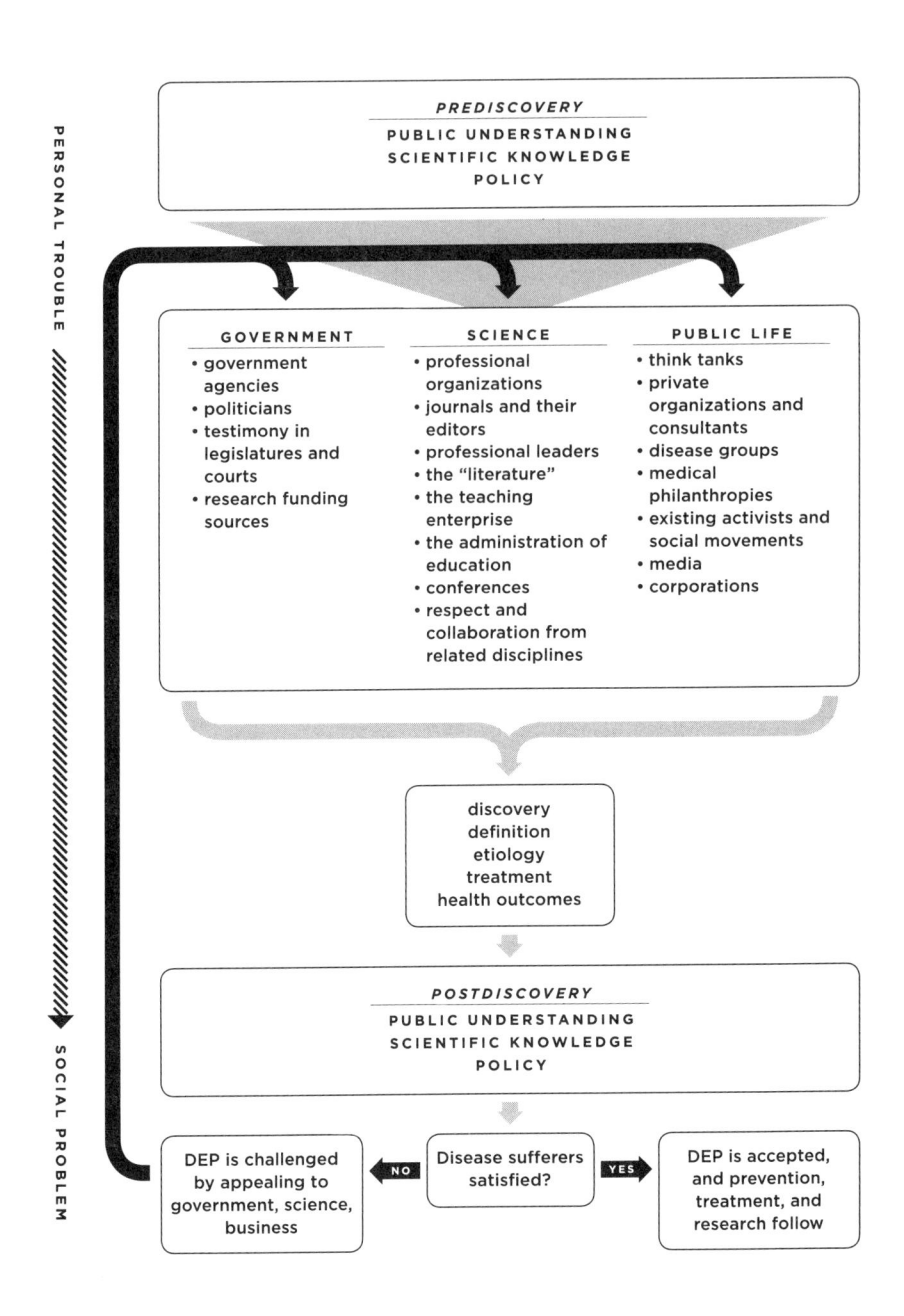

FIGURE 1.1 The dominant epidemiological paradigm's (DEP's) process of disease discovery, definition, etiology, treatment, and outcomes.

to alter key medical processes: disease discovery, illness definition and etiology, treatment protocols, and evaluation of health outcomes. If as a result of mobilized campaigns a new or modified paradigm emerges that satisfies social movement organizations and their publics, the paradigm will be accepted. If the end result is unsatisfying to some or all the actors, then mobilization continues, again challenging various practices within the government, scientific, and private sectors. Mobilization against a DEP is central to health social movements because these movements seek much more than increased access to health services; they seek to change fundamental conceptions of the nature of disease, its causation, and its links to a range of other societal elements. In the process, they pursue the development shown vertically on the left margin of figure 1.1: C. Wright Mills's notion of how individual troubles are transformed into social problems.

PUBLIC PARADIGMS

In opposition to a DEP, environmental health activists put forth a *public paradigm*. My notion of a public paradigm builds on Krimsky's concept of a public hypothesis that is formed when the public feels it has a stake in scientific study, debates, and consequent outcomes and therefore demands to participate in them.[26] Krimsky coined the term *public hypothesis* to describe the work of scientists and activists who have developed and pursued a hypothesis concerning the importance of endocrine disrupters as causes of developmental and reproductive abnormalities, perhaps also of breast cancer.

A public paradigm has not necessarily been fully developed for each disease or condition. For breast cancer, the endocrine disrupter hypothesis serves as such a paradigm, and, for asthma, the growing focus on outdoor air pollution does so. More generally, the public paradigm represents an expanded pattern of public action and attitudes not just for a specific scientific argument, but also for current scientific and societal processes. It is a paradigm not for one disease or condition, but for an overall approach to environmental causation. The precautionary principle is the best example of a public paradigm. In many cases, activists base their movements on the precautionary principle. So powerful is this principle as an organizing configuration that it orients many people to a dramatically new perspective. Environmental justice has also attained public paradigm status as an approach that links race and class stratification with a large set of characteristics of social organization, including production, distribution, and disposal of chemicals and other substances; housing; urban development; land use; water rights; and transportation. A new public paradigm leads people to alter fundamentally how they view many daily aspects of their lives, and it poses deep levels of criticism as a route to an alternative model of social life that provides greater health and democracy. Creating a new public paradigm necessitates participation in science and at least symbolic influence over its formation. To create a public paradigm, it is necessary to have extensive public involvement in science, as discussed later.

There can be more than one public paradigm, and these paradigms can potentially conflict with each other. One example of such conflict involves the significance of endocrine disrupters. EBCM activists have a public paradigm that views these endocrine disrupters as a likely cause of breast cancer. Other key environmental health activists and scientists, however, consider endocrine disrupters to be well known for reproductive and developmental effects, but not for breast cancer. Indeed, they worry that breast cancer activists' push for an environmental causation model centered on endocrine disrupters will harm the overall progress of the environmental health movement because there is already much mainstream criticism of breast cancer activists for focusing on endocrine disrupters.

The lower left of figure 1.1 shows a box labeled "DEP is challenged by appealing to government, science, business." The action to forge an alternative public paradigm is the substance of that box. A shorthand version of this model of disputes is shown in figure 1.2, whereby two parallel tracks of lay and professional action lead to a relatively unified challenge. In living with an actual or potential disease, lay people undergo an *illness experience* in which they relate to their symptoms and disease in various ways: self-concept, work, life, relationships, and attribution of responsibility for their situation. Medical sociologists typically think of illness experience as occurring in people with a disease, condition, or syndrome. My expanded environmental health definition views illness behavior as also occurring in the face of potential disease from known hazards. We have strong evidence for such behavior—for example, mental health effects from being in proximity to an environmental crisis—even in the absence of direct physical disease.[27] Tying together illness experience and awareness of local hazards can then lead people to a *social discovery* of a disease cluster, a heightened incidence, or a higher than expected exposure. As individuals share this illness experience with others, it becomes a *collective illness experience* for people either in their locality, as with a toxic waste crisis, or among a nongeographically bounded population such as women with breast cancer. When organizers frame this collective illness experience in terms of corporate and governmental responsibility and suggest a political route to seek redress, the illness becomes a *politicized illness experience* that may spawn a health social movement. That new worldview of politicized illness experience and health social movements then allows people to engage in popular

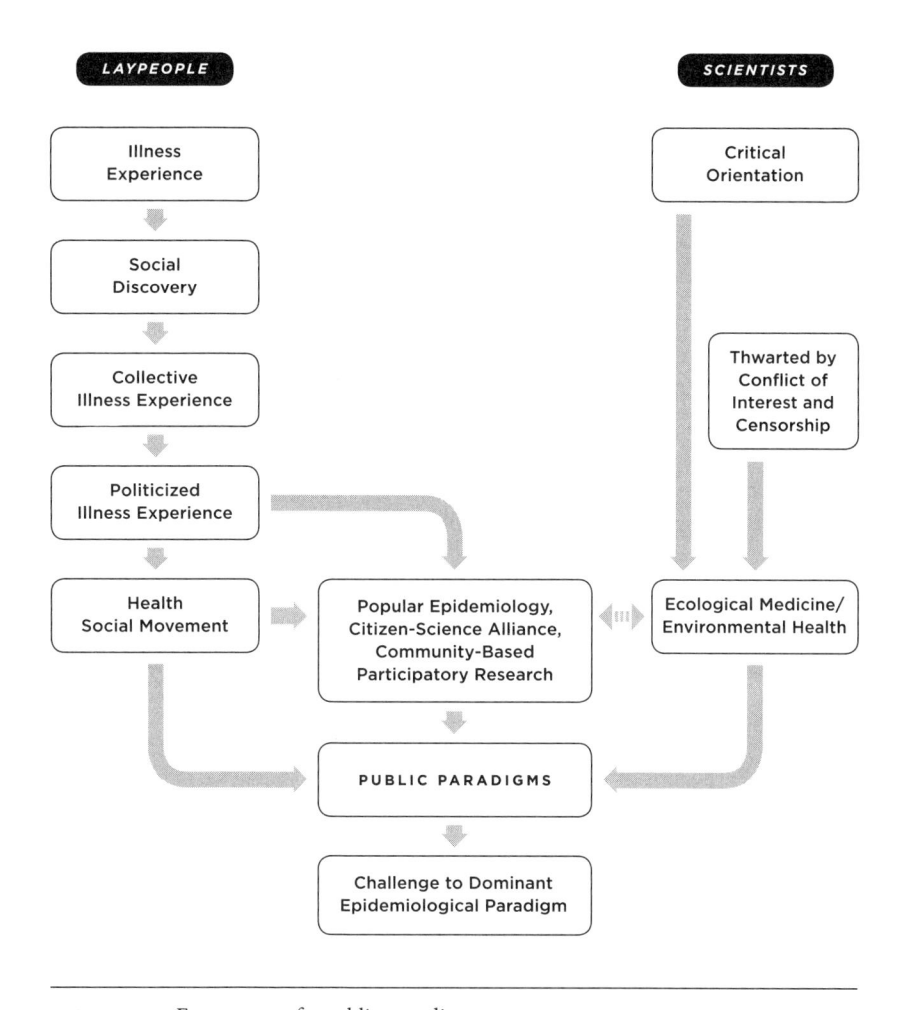

FIGURE 1.2 Emergence of a public paradigm.

epidemiology, citizen-science alliances, or other forms of community-based participatory research.

On a separate track, scientists develop a challenge to existing science. Some scientists come from a *critical orientation* that predisposes them to a critical perspective and to community-professional cooperation. Other scientists, without such inclinations, try to carry out the best science possible and are thwarted by *conflict of interest* and *censorship* from business and government. They are also thwarted by traditional scientific canons that have no tolerance for innovative environmental health or ecological medicine perspectives. From either the prior critical background or the newer critical background, scientists find their way to a critical epidemiology or "advocacy science" perspective that leads them to collaborate with concerned citizens and lay activists.[28]

The convergence of these two tracks—lay and scientific—takes shape as a public paradigm to challenge the DEP. It is not always a completely unified perspective, and as I show later, conflicts remain because laypeople and professionals still have different backgrounds and points of view. Nor do lay-science collaborations always occur.

HEALTH SOCIAL MOVEMENTS AND EMBODIED HEALTH MOVEMENTS

So far I have said much about the importance of social movements. Now I want to define these movements more precisely as they are related to health issues. *Health social movements* are collective challenges to medical policy, public-health policy and politics, belief systems, and research and practice that include an array of formal and informal organizations, supporters, networks of cooperation, and media. Health social movements make many challenges to political power, professional authority, and personal and collective identity. They address (a) access to or provision of health care services; (b) disease, illness experience, disability, and contested illness; and (c) health inequality and inequity based on race, ethnicity, gender, class, or sexuality. Based on these categories, I developed a preliminary typology representing ideal types of health social movements; however, the goals and activities of some movements may fit into more than one of these categories.[29]

Health access movements seek equitable access to health care and improved provision of health care services. Such movements seek things such as national health care reform, increased ability to pick specialists, improved Social Security drug benefits, and extension of health insurance to uninsured people. *Embodied health movements* address disease, disability, or illness experience by challenging science on etiology, diagnosis, treatment, and prevention. They focus on contested illnesses that either are unexplained by current medical knowledge or have disputed environmental explanations. As a result, these groups organize to achieve medical recognition, treatment, and research. In addition, some established embodied health movements may include constituents who are not ill, but who perceive themselves as vulnerable to the disease; many EBCM activists fit this characterization and join other women who do have the disease. Among these movements are the breast cancer movement, the AIDS movement, and the tobacco control movement.

Constituency-based health movements address health inequality and health inequity centered on race, ethnicity, gender, class, and sexuality differences.

These groups address disproportionate outcomes and oversight by the scientific community and weak science. They include the women's health movement, the gay and lesbian health movement, and the environmental justice movement.

These categories of the typology are ideal types, and the range of organizational agendas within any movement will not always fit neatly into one category and may frequently overlap into other categories. For example, the women's health movement can be seen as a constituency-based movement, but at the same time it contains elements of both access health social movements (e.g., in seeking more services for women) and embodied health movements (e.g., in challenging assumptions about psychiatric diagnoses for premenstrual symptoms). Nevertheless, by virtue of having a large categorical constituency, the women's health movement directly raises issues of sex differences and gender discrimination, and it represents a large population with specific interests, so that the constituency nature is significant. For another example, the members of environmental justice organizations typically center their actions on their own illnesses or their fear of becoming ill. At the same time, they address the disproportionate burden of polluting facilities and health effects carried by communities of color. As a result, these organizations share features of both embodied and constituency-based health movements. They also have features of access-based movements because activists within them seek better access to treatment and health insurance.

The health social movements that involve asthma, breast cancer, and GWRIs are embodied health movements, so it is necessary to provide more groundwork on the latter. They are defined by three characteristics. Though many other types of health social movements have one or even two of these characteristics, embodied movements are unique in possessing all three. First, they introduce the biological body to social movements in central ways, especially in terms of the embodied experience of people who have the disease. The use of embodiment in social movement formation and strategizing can also be seen in the disability movement and women's health movement.[30] Second, they typically include challenges to existing medical and scientific knowledge and practice. Such challenges also characterize the environmental movement, antinuclear movements, and other movements, though, as I discuss later, they do so in different ways. Third, embodied health movements often involve activists collaborating with scientists and health professionals in pursuing treatment, prevention, research, and expanded funding. Although the simultaneous possession of these three characteristics makes these movements somewhat unique, they are nevertheless much like other social movements in that they depend on the emergence of a collective identity as a mobilizing force. In the case of illness, people's first approach is to work within existing social institutions. When these institutions, namely science and medicine, fail to offer disease accounts that are consistent with these individuals' own experiences of illness, or when science and medicine offer accounts of disease that they are unwilling to accept, they may adopt an identity as aggrieved illness sufferers and move on to collective action.[31]

Such an identity emerges primarily from the biological disease process occurring inside the person's body. The body is often also implicated in other social movements, especially identity-based movements. But identity-based movements emerge because a particular ascribed identity causes a group of people to see their bodies through the lens of social stigma and discrimination, as with the women's and lesbian and gay rights movements. With embodied health movements, in contrast, the disease process, rather than the ascribed characteristics, results in the development of a particular disease identity. This identity represents the intersection of social constructions of illness and the personal illness experience of a biological disease process.

Embodied experience constrains the options available to a movement once it is mobilized. Illness sufferers can work either within or against their target, in this case the system of the production and application of scientific and medical knowledge. They are less free, depending on the severity of their condition, simply to exit the system. Though some illness sufferers seek alternative or complementary therapies, many others either need or seek immediate care and are forced to pursue solutions within a system they perceive as failing their health needs. Most important, ill people have the unique familiarity of living with the disease process, its personal illness experience, its interpersonal effects, and its social ramifications. Their friends and family, who share some of the same experiences, may also engage in collective action. These personal experiences give people with the disease or condition a lived perspective that is unavailable to others. It also lends moral credibility to the mobilized group in the public sphere and scientific world.

Challenges to existing medical and scientific knowledge and practice are a second unique characteristic of embodied health movements, whether they are working within or challenging the system. After all, if the aims of such movements are to bring the social construction of the disease more in line with the actual disease process, and if disease processes are understood through scientific investigation, then the movements become inextricably linked to the production of scientific knowledge and to changes in practice. Just as embodied health movements are not the only movements that involve the physical body, they also are not the only movements to confront science and scientific knowledge and practice. Environmental groups, for example, often confront scientific justifications for risk management strategies, determination of endangered species, global warming, or resource use by drawing on their own scientific evidence for alternative courses of action. However, many environmental disputes can also center on nature and the value placed on it by opposing interests. In these cases, some environmental groups can abandon scientific arguments, appealing instead to the public's desire to protect open spaces for psychological or spiritual reasons or to preserve resources for enjoyment by future generations. However, it is not necessarily embodied health movements' challenge to science that sets them apart from other movements, but *how* they go about doing it. Embodied health movement activists often judge science based on their own intimate, firsthand knowledge of their bodies and illness.[32]

Furthermore, many such activists must simultaneously challenge *and* collaborate with science. They do not typically have the luxury of ignoring the science, which may be providing them with medications and other treatments. Although they may appeal to people's sense of justice or shared values, they nevertheless remain to a large extent dependent on scientific understanding and continued innovation if they hope to receive effective treatment and eventually recover. As Epstein points out, when little was known about AIDS, activists had to engage the scientific enterprise in order to prod medicine and government to act quickly enough and with adequate knowledge.[33] Even embodied health movements that focus on already understood and treatable diseases are dependent on science. Although they may not have to push for more research, they typically must point to scientific evidence of causation in order to demand public policies for prevention. For example, asthma activists who demand better transportation planning for inner cities and who seek better quality affordable housing do so knowing that the scientific evidence linking outdoor and indoor air quality to asthma attacks supports them.

Embodied health movements' dependence on science leads to the third characteristic: activist collaboration with scientists and health professionals in pursuing treatment, prevention, research, and expanded funding. Lay activists in these movements often strive to gain a place at the scientific table so that their personal illness experiences can help shape research design, as Epstein points out in his study of AIDS activists.[34] Even if activists do not get to participate in the research enterprise, they often realize that their movement's success will be defined in terms of scientific advances or in terms of transformation of scientific processes. Part of the dispute over science involves a disease group's dependence on medical and scientific allies to help them press for increased funding for research and to raise money to enable them to run support groups and get insurance coverage. The more scientists can testify to those needs, the stronger patients' and advocates' claims are. All these points indicate that science is an inextricable part of embodied health movements, thus placing them in a fundamentally different relationship to science than many other social movements.

POLITICIZED ILLNESS EXPERIENCE The social movements discussed so far have a profound effect on how people experience illnesses that may have environmental causes or triggers. For many people and organizations, the experience of illness has been transformed from an individual disease into a social movement focused on health and social inequalities. This process is what I term a *politicized illness experience,* whereby community-based organizations help people make direct links between their experience of a disease and the social determinants of their health. Medical sociologists study the illness experience in many ways. They look at the personal experience of illness and symptoms and how individuals adapt to their illnesses in order to function in everyday life. Beyond the experience of symptoms and subsequent adaptation, they also look at how illness shapes personal identity. Finally, they examine how individuals search for a cause of their illness and how they subsequently attribute responsibility for the illness.[35]

Collective identity around a disease and set of potential causes becomes politicized through activism. Both individuals and groups link social and physical realities, thereby developing a set of shared grievances that they attribute to discrimination, structural dislocation, shared values, or other social constructions. Through the process of collective framing, these people and organizations transform the personal experience of illness into a collective identity that is focused on discovering and eliminating the social causes of a disease. This collective framing leads to the politicized illness experience. Although my concept of the politicized illness experience is new, it fits well with existing studies by medical sociologists and medical anthropologists on community-based approaches to environmental hazards and catastrophes and on collective approaches to illness experience, as with breast cancer.[36] People with a disease are casting off a passive patient role, countering the stigma of their disease, fighting against treatments that cause unnecessary pain or disfigurement or that involve dangerous side effects, and developing alternate explanations for the causes. In these ways, they challenge the DEP and create a new public paradigm.

For a moving example, we can see how Maren Klawiter illustrates the way the breast cancer movement has transformed the "regime of breast cancer" so that "collective identities, emotional vocabularies, popular images, public policies, institutionalized practices, social scripts, and authoritative discourses" give women with breast cancer today a fundamentally changed experience from decades ago. By tracing the experience of one woman with breast cancer and its recurrence over two decades, Klawiter demonstrates how this politicization alters the very experience of breast cancer: Clara Larson moves from an original period of isolation and subservience to medical authority to a new era of identification and solidarity with others, an attendant activism, and personal control of health care decisions. This useful example shows how the transformation of illness experience can be used as a measure of social movement success.[37]

The concepts defined here provide a framework for a comprehensive approach to looking at health and the environment. In discussing each disease case, I revisit them: the DEP, public paradigms, citizen-science alliances, community-based participatory research, critical epidemiology, health social movements, and the politicized illness experience.

BOUNDARY MOVEMENTS People who seek prevention and remedies for environmentally induced diseases have involved themselves in manifold challenges to the usual science or medical search for knowledge and practice. They have crossed over many of the usual boundaries, which is why embodied health movements are *boundary movements*.[38] That is, in pushing the limits of what is defined as normal scientific practice and in bridging previous social movements, they represent hybrid movements that blur the boundaries between lay and expert forms of knowledge and between activists and the state.

Boundary movements are a combination of social movements and their constituent organizations. In some cases, this configuration might include some or all of the following: individual activists, outside supporters, scientists, academics, legislators, government officials, government agencies (usually parts of them), and foundations. So many components blur traditional distinctions, such as those between movement and nonmovement actors as well as between laypeople and professionals. In particular, five characteristics define boundary movements.

First, they attempt to reconstruct the lines that demarcate science from nonscience. They push science in new directions and participate in scientific processes as a means of bringing previously unaddressed issues and concerns to clinical and bench scientists.

Second, boundary movements blur the boundary between experts and laypeople. Some activists informally become experts by using the Internet and other resources to arm themselves with medical and scientific knowledge that can be employed in conflicts with their medical care providers. Others gain a more legitimate form of expertise by working with scientists and medical experts to gain a better level of understanding of the science underlying their disease. Through this process, boundary organizations gain power and authority by obscuring the line between the expert and layperson.

Third, boundary movements often have state allies. This connection usually involves only sectors of state agencies; for example, the tobacco control office of a public-health department might be part of an antitobacco movement in tandem with a nonprofit organization or a political group.[39] Studies of social movements show that states are usually the targets of activism, and, indeed, movements still seek action by the state. But states are not monolithic, and many social movement successes stem from alliances with officials, staff, and scientists of state agencies. There are more possibilities for challengers in some administrations than in others, and conservative regimes have an antiregulatory stance that goes against the grain of reform. Still, we see in the interstices of the state varying opportunities for collaboration with activists. Activists sometimes employ legacies from prior eras when reform was more acceptable—for example, laws that require citizen input—or units of agencies that retain their reform goals. At other times, they find agency people who resist the political control of science by their superiors or by the White House. In yet other situations, they readily make their logical case to authorities, who view it as sensible.

Fourth, boundary movements transcend the traditional conceptions (i.e., boundaries) of what is or is not a social movement. They move fluidly between lay and expert identities and across various organizational forms. Their fluidity allows them to move in and out of organizations and institutions in ways that traditional social movement activists do not. Raising money to fund their own research exemplifies how they are boundary organizations because doing so blurs the boundary between previously distinct and autonomous institutions: science and civil society.

Fifth, boundary movements use "boundary objects," or objects that overlap different social worlds and are malleable enough to be used by different parties.[40] For instance, a mammography machine is a diagnostic tool in science, but for black activists it is a symbol of unequal health care access and for EBCM activists a symbol of overemphasis on mammography as well as a false claim that mammography is prevention.

Boundary movements work in the cultural and analytical spaces between existing social movements in order to negotiate the meaning of science and to challenge the definitions of acceptable scientific practices and products. Many individuals can identify with and participate in such a movement without having to be part of a specific organization. Finally, a boundary movement crosses two or more social movements, while blurring the boundaries of those separate movements.

The citizen-science alliances I mentioned earlier exemplify boundary movement activity. My conception of boundary movements is akin to Ray's "fields of movements," in which social action stems from collective efforts of groups of social movements.[41] A field-of-movements approach describes more than just a lineage of social movements. Fields of movement capture lineage, but they also point to the way movements intersect with one another. Because ideas, values, and action strategies move from one movement to another, people find it easier to participate in more than one group and movement. Social movements are in this sense identified less as discrete organizational entities and more as communities that involve actors, organizations, and tactics not only from other social movements, but from science, academia, and government.

The interaction among boundary movement groups is made possible by the fluidity with which these groups can move back and forth between organizational cultures and between the roles of activists and experts. This fluidity allows professionals to play varying roles over time, occasionally becoming part of the movement as either members or "advocacy scientists," other times being somewhat detached scientists, and other times being uninvolved observers.[42] Fluidity is similarly represented in Epstein's notion of "analytical blurrings," which holds that we can no longer adhere to binary sets, such as insider/outsider and lay/expert, because of the fluidity of such shifts among both individual participants and organizations.[43]

With this background on health social movements, it is now time to examine how they act to merge the efforts of ordinary people and experts to understand and act on toxic exposures. I look at forms of collaboration between laypeople and professionals, first through popular epidemiology, citizen-science alliances, and community-based participatory research, and then through critical epidemiology and other professional inputs.

POPULAR EPIDEMIOLOGY, CITIZEN-SCIENCE ALLIANCES, AND COMMUNITY-BASED PARTICIPATORY RESEARCH

Studies of environmental factors have played a pivotal role in science and policy by prompting public involvement in research through *citizen-science alliances:* lay-professional collaborations in which citizens and scientists work together on issues identified by laypeople.[44] These citizen-science alliances both contribute to new knowledge as well as challenge and sometimes change scientific norms by valuing illness sufferers' embodied knowledge. These alliances between researchers and community groups may be formal or informal. Formal alliances often occur in university-based "science shops" (most well known in the Netherlands) and in research consulting organizations (such as the Loka Institute in Amherst, Massachusetts, or John Snow, Inc., in Boston), or through government-funded programs that set research agendas based on lay demands (such as Silent Spring Institute, initially funded by the Massachusetts legislature, which studies potential environmental factors in breast cancer). More informal citizen-science alliances have been built in response to contaminated communities, most notably in Love Canal and Woburn. Furthermore, these alliances may be initiated by citizens or professionals or created through a joint-affinity model where lay and researcher interests are aligned.

My notion of the citizen-science alliance stems from my earlier idea of popular epidemiology, which I introduced in the preface when I explained the Woburn childhood leukemia case. *Popular epidemiology* refers more specifically to grassroots efforts begun by citizens, involving such activities as "lay mapping," wherein residents construct local maps of disease clusters and go door to door to recruit neighbors to their efforts. Laypeople often recruit scientists to help them determine possible causes, sometimes through health studies, though in other cases they simply use their identification of diseases or hazards as a way to seek regulatory or legal action or to organize politically. In a citizen-science alliance, lay involvement does not necessarily initiate the collaboration with professionals. Further, regardless of whether laypeople or professionals initiate the relationship, the citizen-science alliance is a more institutionalized approach, and it does not require that laypeople be actually participating in all phases of the work. Therefore, popular epidemiology is a subset of citizen-science alliances.

Another distinction between popular epidemiology and citizen-science alliances, at least in some cases, is that popular epidemiology may be more applicable to individual cases where citizens deal with a specific crisis, as with the leukemia cluster in Woburn; citizen-science alliances are formed at a later stage, where a broader collectivity of activists joins with scientists to deal with a range of cases—for example, the overall involvement of asthma and breast cancer activists with sympathetic scientists. These alliances typically take shape as national-level interactions. Citizen-science alliances have been boosted by governmental support for community-based participatory research, most notably through the NIEHS's Environmental Justice and Community-Based Participatory Research programs, which fund dozens of projects on issues such as asthma, lead, migrant worker health, Native Americans' exposure to uranium tailings, and Alaska Natives' exposure to military base toxics.

Citizen involvement in public-health programs began in the 1960s with community mental health centers, neighborhood health centers, and rural health centers. Since the 1990s there has been a resurgence of lay involvement, increasingly aimed at participation in health research. Foundations and government agencies have positively responded to national reports such as *Healthy People 2000* and *Healthy People 2010,* both of which set community-based programs as priority areas by funding health research that incorporates public involvement.[45] Furthermore, there has been an increase in federal funding for environmental health research with community-participation components, though some of this "research" is merely participation in review panels rather than full-scale community-based participatory research. The NIEHS, the NCI, the DOD, the EPA, the Centers for Disease Control and Prevention (CDC), the Department of Housing and Urban Development (HUD), and the Agency for Toxic Substances and Disease Registry support programs with a range of public involvement. In the early 1990s, the NIEHS initiated the Translational Research Program to encourage collaboration between community members and scientists through "a methodology that promotes active community involvement in the processes that shape research and intervention strategies, as well as in the conduct of research studies."[46]

Community-based participatory research goes beyond the traditional models of public involvement by requiring active participation at every stage in the research process. However, it is increasingly used mistakenly as a term to describe research that involves a community in any capacity even though members of these communities may have little actual input in the study. The NIEHS has continued to be the primary federal agency supporting community-based participatory research, especially in the environmental justice area and often focusing on asthma. Shobha Srinivasan was a key director of this effort, and her program has been responsible for funding many dozens of environmental justice collaborations; further, by legitimizing such work, the program has paved the way for other collaborative projects to form. In 2002, the National Institute of Occupational Safety and Health, with Adele Childress as program officer, began partnering with the NIEHS to coordinate joint reviews of community-based participatory research projects in environmental justice, a logical development given the many similarities in how occupational and environmental health deal with hazards and with corporate resistance and problems with governmental regulation. It is important to note that participatory research is an overall *approach* to lay-professional joint efforts, whereas citizen-science alliances are the *phenomena* of actually doing the collaborative work.

Community-based participatory research is not confined solely to researcher-community interactions. It is a term that has become increasingly codified as the dominant language for citizen participation, at least in federal funding circles. It is an approach that is inclusive of all affected parties and all potential end-users of the research, including community-based organizations, public-health practitioners, and local health and social services agencies. Both participatory and translational research have been advanced by the NIEHS and other funding agencies within the area of environmental health. The main idea here is that public involvement in environmental issues generally and in environmental health research specifically can take several forms, and the variation in type reflects differences in practice, discipline-specific terminology, and philosophy—that is, the extent to which "expertise" or "science" pass unchallenged and to which "experts" dictate roles, relationships, and agendas.[47] Popular epidemiology, citizen-science alliances, and community-based participatory research are all forms of such public involvement.

The most egalitarian community-academic collaboration occurs when there is a grassroots movement because this origin makes it more likely that community groups will initiate or play key roles in the process, as opposed to merely being invited to sit on advisory boards. More comprehensive citizen involvement in research often occurs as the social problem becomes more public and the social movement gains strength and momentum. This change takes place because the increasingly public nature of discussion on the problem, amplified by a cogent critique of the failure of science to address local concerns, has the potential to highlight the constraints of traditional scientific approaches. In this process, activists become increasingly knowledgeable about the disease and begin to push for involvement in defining and researching it.

Laypeople and scientists who challenge traditional approaches put forth alternative hypotheses that in many cases are relatively parallel. These hypotheses can become unified into a public paradigm through the work of citizen-science alliances. This work has gone on for a long time, as with cases such as the United Farm Workers and their scientist allies' fight against pesticides and, later, Love Canal and other toxic waste sites. More recently, community-based participatory research has provided a vehicle to make this connection more feasible through federal funding, journal publication, and the resultant legitimation of such work. (In the following section, I say more about the professionals' origins and practices.)

As demonstrated in the three disease cases focused on in this book, citizen-science alliances vary considerably. Alliances for asthma and breast cancer have been driven by activists and hence have had a strong amount of lay input, whereas activists for GWRIs were less successful in pushing for alliances and hence had little influence on the conduct of research. Also, for GWRIs, the alliances developed relatively late in the controversy, by which point many veterans had already lost trust in the ability of science and government to find answers. For asthma and breast cancer, however, scientific research still holds more promise.

CRITICAL EPIDEMIOLOGY AND OTHER SOURCES OF PROFESSIONAL ACTIVISM

Professional allies are needed for a citizen-science alliance. To create such alliances, lay groups do not simply go out and recruit *any* scientists to help in their efforts. They need to find scientists who are already sympathetic. These scientists are largely found among what I term *critical epidemiologists.* A primary objective for critical epidemiology is to understand how epidemiology is an organized system of human endeavor to construct knowledge about the distributions and determinants of population health. Critical epidemiology examines how the historical context of epidemiologic investigations, in particularly the dominant political ideology, influences what we know and do not know about disease etiology, health services, and prevention strategies.[48] Critical epidemiology is fundamentally about applying the discipline's epistemology for social justice. This process includes taking a social-structural and health inequalities approach to epidemiology, working with laypeople, and seeing critical epidemiologists as tied to social movements and thus as a challenge to the DEP.

One example of critical epidemiology is when environmental epidemiologists and related scientists work with community groups to search for potential environmental causation of diseases, especially when traditional funding sources and regulatory and public-health bodies strongly oppose such connections. This approach can be seen in the work of numerous scientists at Woburn, Love Canal, and other contaminated communities, where they have joined in the citizens' popular epidemiology efforts. Another example of critical epidemiology is when epidemiologists teach lay activists enough science so that the latter can engage with scientists and officials. The National Breast Cancer Coalition's (NBCC) Project LEAD (Leadership, Education, and Advocacy Development) is one such effort, where lay activists participate in expert-led workshops on breast cancer in order to prepare themselves to serve as lay members of state and federal review panels. In a few specialized applied settings, entire research organizations, such as Silent Spring Institute, merge their scientific inquiry with direct participation in social action.[49]

We also see critical epidemiology at work in more academic settings, where scholars challenge "risk-factor" epidemiology with a social-structural perspective that emphasizes the role of "fundamental causes" such as race, class, and sex.[50] This challenge also involves written critiques of mainstream approaches and represents a turn away from fetishizing the most modern mathematical models and toward practicing "shoe-leather epidemiology" and making a commitment to the social justice that initiated the discipline. Wing argues that epidemiologists should

1. Ask not what is good or bad for health overall, but for what sectors of the population.
2. Look for connections between many diseases and exposures, rather than looking at merely single exposure-disease pairs.
3. Examine unintended consequences of interventions.
4. Utilize people's personal illness narratives.
5. Include in research reporting the explicit discussion of assumptions, values, and the social construction of scientific knowledge.
6. Recognize that the problem of controlling confounding factors comes from a reductionist approach that looks only for individual relations rather than a larger set of social relations. Hence, what are nuisance factors in traditional epidemiology become essential context in a new ecological epidemiology.
7. Have humility about scientific research and commit to supporting broad efforts to reform society and health.[51]

Although few epidemiologists will go so far as to adopt all of these changes, a growing number are retooling their work to move in that direction. Advocacy scientists have increasingly extended their personal responsibility and commitment to their professional work. Krimsky, in his narrative of the emergence of the environmental endocrine disrupter hypothesis, witnessed several scientists' becoming visible activists for the hypothesis despite gaps in their knowledge and the subsequent risks to their image and professional careers. In Krimsky's words, advocacy scientists "view their role as bifurcated between advancing the scientific knowledge base and communicating to the public, the media, and policymakers."[52] Critical epidemiologists are definitely one of the groups of scientists that make up this broad array of advocacy scientists.

In some cases, professionals realize at an early point that they have strong links to individual laypeople who are dealing with environmental hazards and diseases, and they steer their careers in that direction. This trajectory is much more likely for professionals who come from a background of activism prior to or concurrent with their professional training. It would certainly include people who have formed a critical epidemiology perspective, but there are not that many such people. There is a larger set of prior activists who do not have such a well-worked-out political approach to their science, but who nevertheless feel committed to serving the interests of those people most affected by environmental threats. Some of these scientists feel a tug of contradiction between their clear critical viewpoints on general social and political phenomena, on the one hand, and the pressure to conform to established scientific routines, on the other.[53] Some are politicized by the larger social movements of their era and form radical caucuses and alternative science organizations to balance their commitment to both science and politics.[54]

Other professionals neither come from a political background nor have a specific goal of working with environmental activists. They are simply trying to do their best science, fulfilling their desire to figure out scientific puzzles and advance human knowledge. They are often thwarted by business and government. Business wants to minimize or even hide dangers and goes very far in efforts to control the normal range of scientific research, and government either sides directly with business or puts forth other reasons for failure to regulate or outlaw hazardous substances (I take up such corporate and governmental actions in later chapters, starting with the next chapter on breast cancer). Besides dealing with corporate leaders' conflicts of interest and direct malfeasance, scientists may realize that corporations have many scientific resources at their disposal and can push data and policy as they see fit. Scientists who are developing an awareness of such structures come to realize that a fundamental inequality affects people and communities who do not have such tremendous resources on their side. Scientists who face such obstacles to making their best scientific effort may therefore have an affinity with laypeople's critiques and will find citizen mobilizations to be the best vehicle for challenging those who hamper science. In this case, it is the activists who show the scientists how traditional models of research and policymaking have failed to protect people. The activists then work to convince the scientists to go beyond their routine academic habits and stances and to take more radical action such as supporting popular epidemiology efforts and entering into citizen-science alliances.

Once an alliance or collaboration is made, however, it does not mean that all future work will go through without conflicts. As shown in chapter 3 on asthma, laypeople and scientists sometimes converge and other times diverge. They may unite in a broad concern for tracing environmental factors in disease or in showing that there are hazardous exposures, but they may disagree over the degree of scientific confirmation needed to make strong or definitive statements about environmental factors.

DATA AND METHODS

My contacts with the groups I was researching, including some past working relationships, facilitated entrée. Research involved participant observation and focused interviews at the primary research sites and then at other related sites. As mentioned earlier, the primary sites were the environmental justice group ACE; Silent Spring Institute, which studies environmental causes of breast cancer; the BEHC, which examined Gulf War illnesses; TURI; and the Precautionary Principle Project (Alliance for a Healthy Tomorrow). For breast cancer, the other related sites included various groups in Long Island, New York, where a major federally funded project was investigating breast cancer and the environment, and several groups in the San Francisco Bay Area, where there is considerable activism and research on breast cancer and the environment. For asthma, I also studied WE ACT, an environmental justice organization that is very similar to ACE in its asthma focus, but that maintains more ongoing collaboration with public-health researchers.

I conducted interviews with activists and researchers involved in these organizations, as well as with other activists, researchers, and government officials in these fields: 37 interviews for breast cancer, 18 for asthma, 23 for GWRIs, and 18 for toxics reduction, for a total of 96 interviews. I also collected and analyzed policy-related government documents and printed materials from each organization.

I also conducted ethnographic observations to examine the everyday workings of some of the groups. For ACE, this process involved observing its work in teaching asthma education programs for various public-school and after-school programs, as well as workshops and meetings the group organized for adult audiences. For Silent Spring Institute, it involved observing the group's public-education programs, one of its scientific advisory board meetings, and a peer-review session held by the Department of Public Health (DPH) to determine the future of their funding. For the BEHC, this process involved attending monthly meetings of the center staff as well as periodic meetings of the science advisory committee and the veterans' advisory committee, other VA Medical Center meetings that involved the group, and two annual national conferences of federally funded Gulf War researchers. For toxics reduction, observations included workshops and conferences held by the various toxics reduction groups. I ended up with 10 observations for breast cancer, 12 for asthma, 13 for GWRIs, and 7 for toxics reduction, a total of 42 observations.

I also examined the main scientific papers on environmental factors in asthma, breast cancer, and GWRIs in order to trace the lines of development and dispute over environmental factors. As the project developed, I realized that my examination of selected sites and movements was tapping into a larger phenomenon, which led to the later focus on health social movements. So I began to study a wide array of movements, which adds to the book's broader subject matter. As a result, my fieldwork has expanded to various other funded research projects and a number of environmental activism efforts, and those experiences have added far more data.

My approach is congruent with Marcus's notion of "multisited ethnography." An individual research site, although capable of rich description and analysis, is insufficient to convey larger trends in an increasingly complex and interdependent world. Hence, the ethnographer must trace a cultural formation across diverse sites, while simultaneously developing the interaction between the macrosocial context and those specific sites. For Marcus, any ethnography of a single cultural formation is by extension a study of the larger system in which that single formation is embedded. What knowledge the researcher gains of the microlevel affects his or her understanding of the macrolevel, and vice versa. Further, this "mobile ethnography" enables the researcher to have a more emergent and complex view of the local site than would be possible by merely studying that single site.[55] As Burawoy remarks, in a postmodern world where there are many local connections to the world system, it is necessary to engage in "welding ethnohistory to ethnography, combining dwelling with movement." For the multisited ethnographer to have done his or her job, Burawoy argues, he or she must have "delved into external forces," "explored connections between sites," and "uncovered and distilled imaginations from daily life."[56] Rapp speaks of this multisited ethnography as an "endeavor to break the connection of space, place, and culture" because there are no clear boundaries between the research sites, the people who populate them, and the places from which those people come.[57]

Indeed, this multisited approach well describes my method. Throughout this book, I trace interconnected locales that make up environmental and health social movements that are not tied together in a formal organizational form and whose boundaries are continually in flux. I view this approach in terms of boundary movements that traverse a wide range of actors and institutions, and that continually cross between social movements and across the margins of lay, professional, and governmental worlds.

As I wrote in the preface and in this chapter, the collaboration between academics and community groups in research, education, and organizing is central to my approach, and I revisit this idea in the concluding chapter. Activists-scholars and advocacy scientists are increasingly playing a role in environmental health and other forms of research. Through trusted coparticipation, this approach enriches the knowledge we can gain, and it provides to often underserved communities the resources of academic and scientific institutions that are supported in large part by public funds. It also allows for people like myself to convey the realities of toxic exposures in a way that a more detached method could not. I share the participants' words and emotions in order to show how science and policy issues are played out in people's lived experience.

2

BREAST CANCER
A POWERFUL MOVEMENT AND A STRUGGLE FOR SCIENCE

IN THIS chapter, I show how the movement formed by women with breast cancer, their friends, and families has become a major force in American society and in particular has transformed many facets of treatment, patients' rights, social and cultural sensitivities, and research directions. I examine the EBCM that has grown out of the larger breast cancer movement and other health social movements, how it has focused on potential environmental causes, and how it has changed how breast cancer is researched and publicly perceived. This chapter covers the scientific controversy over environmental factors and breast cancer and goes into detail about how activists have participated in that controversy in three ways: (1) moved debates about causation "upstream" in order to address the cause; (2) shifted emphasis from individual to societal level factors; and (3) pushed for lay involvement in research to raise new questions, change research methods used, and revise standards of proof. Innovative methods of knowledge construction receive particular attention here. Through studying environmental breast cancer activism in Long Island, the San Francisco Bay Area, Boston, and Cape Cod, I provide vivid portrayals of how these challenges are played out in diverse settings and with varying allies. We will see how the EBCM has influenced science and policy by participating in the development of a new public paradigm about environmental causation, especially with regard to endocrine disrupters.

SOCIAL AND SCIENTIFIC CONSTRUCTION OF THE BREAST CANCER EPIDEMIC

Breast cancer rates have been increasing steadily for at least fifty years, to the point where in the United States a woman is diagnosed with breast cancer every three minutes and another woman will die of breast cancer every twelve minutes. The lifetime likelihood of a woman's getting breast cancer is one in eight. Around 211,240 new cases of invasive breast cancer and 58,490 cases of in situ breast cancer are diagnosed annually, and it is the leading cause of death for women ages thirty-four to fifty-four.[1] In 2005, an estimated 40,870 individuals died from breast cancer.[2] Cancer surveillance data for the United States shows that from 1973 to 1999 the incidence of invasive breast cancer rose by nearly 28 percent and that the trend was slightly greater, around 35 percent, for African American women and for women age fifty and older.[3] There has also been a marked increase in ductal cell carcinoma in situ since the early 1980s, comprising more than 85 percent of breast cancers diagnosed annually in the United States.[4] Breast cancer rates have increased globally, and this type of cancer has become the most common cancer among women throughout the world.

Since the early 1990s, a part of the breast cancer movement has advanced a new public paradigm centered on environmental causes of breast cancer. This public paradigm represents an expanded pattern of public action and attitudes not just for a specific scientific argument, but also for the broader perspective underlying current scientific and societal processes. Activists base their new paradigm on the precautionary principle, which places the burden of proof for health effects of chemicals on the producers rather than on the consumers and believes that proof of safety should exist before chemicals are utilized. The EBCM works toward four goals: (1) to broaden public awareness of potential environmental causes of breast cancer; (2) to increase research into environmental causes of breast cancer; (3) to create policy that might prevent environmental causes of breast cancer; and (4) to increase activist participation in research.

Activists believe that science has not appropriately served their needs, at least in part because it is a system controlled by people other than disease sufferers. They focus on the importance of the political context, such as the conservative role of the ACS and the rest of the cancer establishment. They also point to the economic power of the corporations, such as pharmaceutical company Astra-Zeneca's control of Breast Cancer Awareness Month. They point out that some firms that donate money in Shop for the Cure efforts spend a huge amount advertising their participation, yet have a low cap of what they actually donate. Shop for the Cure promotions include automobile companies donating money for each test drive, and stores and manufacturers donating money for each routine item bought. These amounts are typically very small, and the cost to participate often obviates the donation. For example, a yogurt lid costs the consumer thirty-nine cents to mail but unknown amounts for clerical processing in order for the company to donate one dollar. Furthermore, activists argue, some of those companies are producing products that are dangerous to health, perhaps even causing breast cancer. By drawing these connections, activists conceptualize the scientific as political. Consequently, they seek new approaches that insert individual-level knowledge into scientific and political processes in order to affect population-level changes. This valuing of the personal perspective, along with the belief that activism plays a critical role, creates a unified philosophy for all movement actors. Movement actors exemplify boundary movement activity by crossing the often rigid boundaries of the scientific world as lay actors. They use the personal experience of breast cancer and their individual knowledge of potential environmental exposures to provide themselves with credibility in science.

Due to public attention in the past twenty years largely produced by the broader breast cancer movement, the popular conception of breast cancer has changed radically from a private occurrence to a collective and politicized experience. The breast cancer movement has addressed issues of care for patients, including better insurance coverage; knowledge about treatment options, especially in regard to mastectomies, lumpectomies, and radiation; risks of hormone replacement therapy and Tamoxifen trials for prevention; knowledge of the limits of mammography; personal and social support for those affected by the disease; and increased research funding. The movement's other successes include the production of a breast cancer stamp, whose additional cost above normal postage is given to governmental research institutions to further breast cancer research; National Breast Cancer Awareness Month, during which fund-raising walks and runs involve tens of thousands of people every year; and the Shop for the Cure campaign, in which merchants and credit card companies give a portion of the proceeds to breast cancer foundations. The general breast cancer movement's success can also be seen in the increase of breast cancer research dollars from $90 million in 1990 to $600 million in 1999 and in the ability to win federal legislation, such as the Breast and Cervical Cancer Prevention and Treatment Act of 2000.[5] That act, spearheaded by the NBCC, gives states the opportunity to extend Medicaid eligibility to women diagnosed with those forms of cancer. It fills in the gap left by the Breast and Cervical Cancer Mortality Prevention Act of 1990, which provided only screening. Two years after the bill's passage, all states and the District of Columbia implemented the option or were in the process of doing so.[6]

Breast cancer became a powerful public issue in the 1970s and 1980s as women publicly presented their personal stories. Instead of a private experience that could be overcome with a positive attitude and a supportive family, breast cancer became politically relevant, especially in terms of options for treatment. As women gained more control over treatment options, an advocacy movement emerged, focused on increasing research funding and finding a cure. Although activists criticized the medical control of their bodies (e.g., in the use of radical mastectomy), they did not always challenge the biomedical model that focused breast cancer research on methods for treatment and promoted individualistic methods for cancer prevention. As the breast cancer movement became more powerful and the environmental movement put more attention on health effects of toxics and endocrine disrupters, a new EBCM that combined the two movements emerged to challenge the biomedical model.

This EBCM has reframed the successes of the broader breast cancer movement in order to focus on potential environmental causes and to change how breast cancer is researched and publicly perceived. Some of those general movement successes are criticized by the EBCM—or by what one activist leader terms the "political breast cancer movement."[7] For example, for years people took for granted the position of the ACS, NCI, and other parts of the "cancer establishment" that "mammography is the best form of prevention." EBCM activists counter that belief by saying that if a tumor is detected, prevention has obviously failed because the tumor now exists. This stance is assisted by the growing scientific awareness that mammography is not very effective in women younger than fifty. Activists have even mounted a campaign to have breast cancer stamp revenues shifted to the NIEHS from the NCI, where research on environmental factors is rarely supported. Activists also challenge a range of corporate influence in breast cancer advocacy.

Even for activist leaders who are not in groups that see themselves as part of the EBCM, there is much interest in environmental factors. In 2004, Silent Spring Institute interviewed fifty-six leaders of general breast cancer organizations in the United States and Canada. The study showed a substantial interest in the environment among leaders of grassroots organizations. Out of those interviewed, 45 percent said that questions about environmental factors had come up for their organization, and 23 percent said that they were actively addressing local environmental issues. Approximately three-quarters rated as "very important" any research about workplace chemicals, air pollution, pesticides, household chemicals, drinking water, and endocrine-disrupting compounds. Grassroots leaders said that nearly half of the breast cancer research dollar should be spent on environmental research.[8]

There has been much scientific debate concerning the causes and treatment of breast cancer. The dominant scientific paradigms used in studying breast cancer have been shaped by the biomedical model and have consequently focused on individual-level factors such as diet, exercise, age at first parity, and genetic makeup.[9] However, these studies have also shown that such factors account for a limited number of cases. The discovery of the BRCA-1 and BRCA-2 gene mutations led to much attention to genetic causes even though those mutations account for only some 5 to 10 percent of all cases.[10] Activists also point to the fact that the genome does not change rapidly enough to account for the increase in breast cancer incidence, which in 1964 was one in twenty for women who lived to be age eighty and is now (in 2006) one in eight. For the EBCM, environmental factors are a crucial part of the complex set of the causes of the disease.

SCIENTIFIC CONTROVERSY OVER ENVIRONMENTAL FACTORS IN BREAST CANCER

In 1962, Rachel Carson's groundbreaking book *Silent Spring* drew attention to the detrimental effects of pesticides on wildlife and human health, shaping the modern environmental movement and catalyzing a policy shift in environmental regulation. Carson's eloquent writing about what she saw as the "human price" of industrialization ended an era of public ignorance and created collective motivation for advancing environmental health research. Although Carson's legacy is often considered in terms of wildlife, her book very much emphasized understanding environmental links to health and disease. Carson died from breast cancer, though she did not speak about it for fear of being seen as an unreliable advocate.

The debates and controversies that characterize the state of breast cancer science today provide a backdrop to traditions of research that are in the process of changing. Large amounts of research dollars have historically been devoted to analyzing individual and behavioral risk factors, such as childbearing, lifestyle, and diet, but this approach has yielded fewer results than hoped for. Despite major investments in cancer research, disappointing statistics for breast cancer and other cancers raise concern that the "war on cancer" has not yielded promising results for understanding causation and lowering incidence. This situation has led a small but significant sector of the public and scientific community, along with some policymakers, to advocate for new research attention to potential environmental causes. The rising challenge to traditional breast cancer research and the ensuing disputes over the validity of environmental causation theories highlight the underlying belief system of the cancer industry, or the DEP that I introduced in the first chapter.

```
          DOING SCIENCE
         ↗              ↘
ACTING ON          INTERPRETING
 SCIENCE              SCIENCE
         ↖_____↙
```

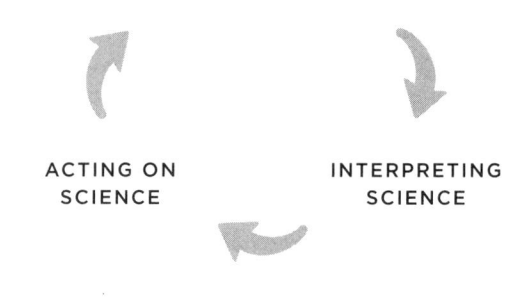

FIGURE 2.1 The doing-interpreting-acting model for science.

Controversies over scientific research in breast cancer take place in three arenas that constitute the process of fact making and knowledge production: doing scientific research, interpreting science, and acting on science, as shown in figure 2.1.[11] This knowledge production structure is not linear, but rather cyclical, in which the process of acting on science leads to the further doing of science. These three processes—doing, interpreting, and acting on science—set the stage for understanding struggles against the DEP of breast cancer etiology.

This paradigm struggle occurs in a three-dimensional manner along three axes, as shown in figure 2.2: (1) upstream versus downstream approaches to prevention (upstream approaches try to prevent the occurrence of the disease by preventing exposure); (2) individual versus environmental risk factors in disease causation; and (3) the degree of community involvement in research and data collection. These three interconnected paradigm struggles involve scientists, environmental breast cancer activists, the media, and government representatives—all of whom play shifting and fluid roles in these conflicts. I view these actors as the fundamental players in the game to shape the DEP and the direction of breast cancer research toward a more participatory and environmentally focused endeavor.

THE DEP

The DEP models how the three primary sectors—government, science, and public life—contribute to public understanding, scientific knowledge, and social policy. In delineating the DEP in terms of breast cancer and the environment, the NCI, the CDC, the EPA, Congress, the DOD, and state departments of public health appear as the major players in the governmental

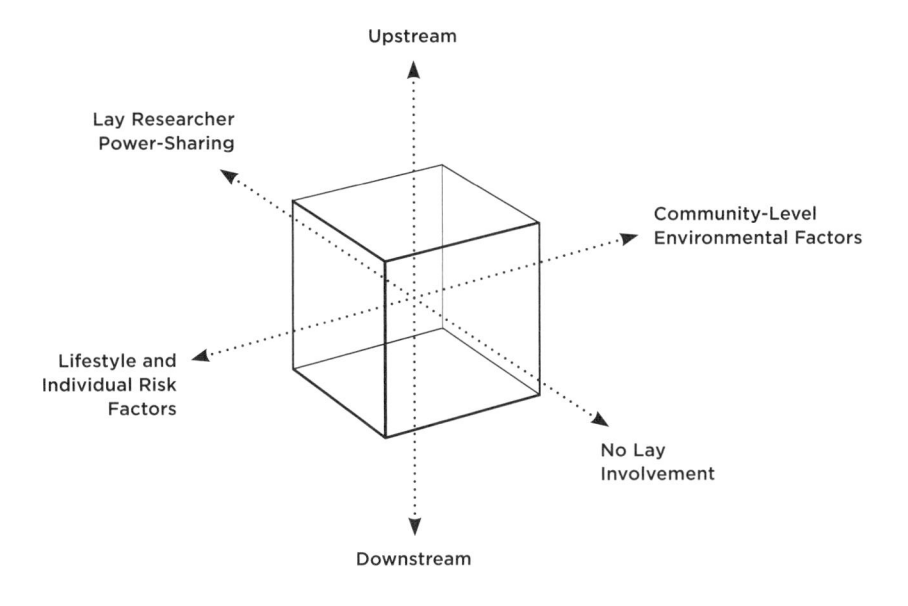

FIGURE 2.2 Dimensional model of disputes over breast cancer research (adapted from a figure by Todd DeMelle).

sector. When we examine public life, we see that health voluntaries (especially the ACS) and media sources shape the dialogue pertaining to cancer causation; corporations push for leniency in regulation of chemicals and produce massive amounts of research to support that action; and social movement groups attempt to stimulate public awareness about environmental causes. Environmental causation is contested and debated in scientific circles through journals, professional organizations, educational institutions, and internal scientific discussions.

Breast cancer science is more than the field and the lab; it includes the regulatory apparatus, the policymakers, social movements, and health voluntaries because those actors help define the parameters of scientific knowledge and research and shape the practical applications of that science. Nevertheless, the DEP is by no means monolithic; not all scientists, officials, and other actors are always in agreement on all aspects of a disease paradigm. Interestingly, the challenges to the DEP that I speak of are found primarily in the United States because of the strength of the breast cancer movement, especially its environmental wing, and hence this analysis reflects a uniquely U.S. phenomenon.

Public challenges, coupled with internal dissension within scientific and governmental institutions, can create opportunities for major shifts in how public-health research is conducted. These challenges are made both by activists through popular epidemiology, an approach that uses lay observations as a central component of examining environmentally induced diseases, and by professionals through critical epidemiology, which examines disease on a population level rather than on an individual level, analyzes the importance of race, class, and gender stratification, and highlights conflicts of interest, including the role played by and the power of pharmaceutical firms. These challenges to the DEP are fundamental issues in modern science, in particular the way that activists and affected people work to shape research methods and influence the direction of scientific inquiry.

The debates and controversies that characterize breast cancer epidemiology serve as a backdrop to current changes in broader research traditions. As with breast cancer, large amounts of research dollars for other diseases have been devoted to analyzing individual and behavioral risk factors, but this line of inquiry has yielded equivocal answers.[12] Therefore, researchers are increasingly exploring whether and how environmental factors play into human disease. These paradigm struggles are often precipitated by external political events, including incremental developments (e.g., an accumulated body of research), major milestone events (e.g., the release of a state cancer database), and the converging of political movements in new ways, such as between the environmental and women's health movements. These struggles are also precipitated by dissent within the scientific community itself, led by a critical mass of researchers who have become dissatisfied with the status quo, who feel that the dominant research programs are not effectively explaining disease etiology, and who want to open new lines of scientific investigation.

Many scientists—including epidemiologists, toxicologists, geneticists, immunologists, and molecular biologists—puzzle over what causes breast cancer and how to prevent and treat it. Some pose alternatives to the mainstream models. To examine disputes, I make an assumption in my conceptualization of "traditional" approaches as being those scientific approaches that focus on individual-level characteristics and factors, such as age at first parity and diet. Although it appears that individuals have control over such factors, those factors are indeed structurally shaped. For example, school lunch programs start children off at an early age on unhealthy, fat-laden diets. Traditional approaches to science and to consequent policy are further characterized by a focus on individual versus population-level factors (a population-level approach asks why entire populations or groups within populations have differing morbidity and mortality, such as with the increase in breast cancer incidence in Japanese women who migrate to the United States), and by research on individual chemicals rather than on synergistic effects of multiple chemicals. Some scientists might make such conceptual leaps on their own, but activists have been key movers in developing alternative directions. From the activist perspective, the DEP's focus on individual-level problems benefits powerful actors such as government and corporations by placing the burden on isolated people and avoiding the economic and political costs of major social transformations.

THE ARENAS OF KNOWLEDGE PRODUCTION: DOING, INTERPRETING, AND ACTING ON SCIENCE

As noted in the previous section, the three arenas of scientific knowledge production are interrelated and occur simultaneously: doing scientific research, interpreting science, and acting on science (i.e., making policy recommendations). Paradigm struggles take place in all three, as portrayed in figure 2.1. These realms of scientific knowledge production set the stage for understanding movement along the dimensional axes.

DOING SCIENTIFIC RESEARCH

Doing scientific research involves how scientists choose particular topics and questions, how they proceed with their investigations, and how they view their relations with funding, research, and support organizations. Doing research also includes how organizations shape research protocols and allocate funding.

Scientific research is circumscribed by trends in theoretical approaches, where disciplines are historically limited by strict disciplinary boundaries and "scientific bandwagons."[13] In breast cancer research, genetics and lifestyle bandwagons predominantly shape the research agenda.[14] This situation is understandable because scientists who follow the dominant approach have a greater ability to obtain research funding, scientific prestige, and career advancement. On their own, scientists might not be able to adopt alternative scientific approaches, but activist pressure on funding and research institutions enables them to do so. For example, the breast cancer movement pressured the DOD to include activists on review panels, an idea that was very novel at the time but is now more common.[15] Breast cancer organizations participated in NIEHS workshops in 2002 that resulted in a program to fund research centers working on breast cancer and the environment. These centers, first established in 2003, collaborate in their focus on the hypothesis that there are periods of vulnerability in the development of the mammary gland when exposures to environmental agents may impact the breast in ways that can influence breast cancer risk in adulthood. The centers will use basic science techniques in laboratory animals and cell cultures in order to characterize the molecular basis of the mammary gland over the lifespan and to determine how this development may be affected by exposure to environmental agents. They will also do an epidemiological study that will examine the environmental and genetic determinants of puberty by prospectively following several cohorts of young girls to determine how hormonal changes, obesity, diet, family history, psychosocial stressors, environmental exposures, and genetic polymorphisms, among other factors, may interact to control mammary gland development and other landmarks of puberty.[16] Many activists feel the centers have a limited range. One positive innovation is that these centers are mandated to provide community outreach through a "translational core" that makes scientific information into usable public knowledge.

The DEP validates only science that is constructed by experts. Its mainstream approach does not include affected persons in the scientific process and therefore misses an embodied account, or what Haraway calls "situated knowledge."[17] The latter idea resonates with Hartsock's "standpoint" theory, which states that accurate knowledge must derive from affected communities. For Hartsock, knowledge from affected groups is not merely another voice in a relativistic world. Instead, knowledge is a "counterhegemonic force" introduced by those usually excluded from science.[18] Science is not routinely subject to public accountability. Consequently, as Hubbard argues, the pretense of objectivity "obscures the political role science and technology play in underwriting the existing distribution of power in society."[19] Such an approach to science implies different methodologies, such as lay involvement in research, which many breast cancer activists advocate.[20] Through the citizen-science alliance, environmental breast cancer activists have recruited and collaborated with scientists willing to investigate environmental factors and to legitimize lay perspectives in research. Activists have also made explicit the role of politics in research.

The NIEHS has strongly supported research on environmental factors; however, it is one of the smallest and least influential NIH institutes. Conversely, the well-endowed and most powerful institute, the NCI, invests relatively little on research that examines environmental links to cancer. As evidenced in its documents, NIH leaders often consider lifestyle factors, such as diet, to be "environmental," so they claim to be conducting more environmental research than they actually are; activists are especially critical of this claim.

In the absence of significant funding for research on environmental factors, the breast cancer movement has had to press state and federal legislatures for special bills. The Long Island Breast Cancer Study Project was funded by an earmarked federal bill, and the Silent Spring Institute's Cape Cod Study was supported by a state bill pushed by the Massachusetts Breast Cancer Coalition and other breast cancer activists (these two major studies are discussed later). Activists were also able to work with federal legislators to put considerable breast cancer funding in the Congressionally Directed Medical Research Program, which is run by the DOD.

INTERPRETING SCIENCE

Interpreting science involves how scientists make sense of data. Two normative standards are critical to the interpretation of science: standards of proof and weight of evidence. Standards of proof are a major element of what Collins terms "interpretive flexibility" because they are the norms with which research results are interpreted.[21] The notion of interpretive flexibility informs us that different conclusions can be drawn from the same data. Multiple scientific truths can coexist or lead to disputes over what constitutes sound methodology and proof of causation. There are also disputes over what studies should be included in deliberations about the weight of evidence regarding the carcinogenicity of certain exposures and risk factors. Results can be interpreted in isolation or in combination with experimental evidence from other fields, such as toxicology. Although a discipline-specific body of evidence might suggest that certain environmental exposures contribute to increased breast cancer rates, novel hypotheses might require the consideration of multiple types of evidence. For example, Theo Colborn, Dianne Dumanoski, and John Peterson Myers' groundbreaking book *Our Stolen Future* synthesized research findings from human, environmental, in vivo, and in vitro toxicology studies to argue that endocrine-mimicking chemicals have effects on multiple species in multiple contexts.[22] In addition, occupational studies more frequently support a correlation between breast cancer risk and chemical exposure, but are often not considered in tandem with general population studies.

Standards of proof are representative of the dominant conceptions of knowledge that try to make science neutral and value free through accepted methodology, a universal reflection of reality, and control by a scientific community that can delineate its own work from personal interests.[23] These standards of proof include strength of association, significance level, study design (prospective cohort versus case control), temporality, biological plausibility, and timing of exposure in biological developmental cycles. The accepted standards of proof in science have important implications. Mainstream science has a tendency to err on the side of uncertainty by not drawing conclusions until there is almost absolute certainty. Mainstream scientists worry that relaxing any component of scientific rigor will reduce the credibility of all science, and they therefore believe that scientific standards should remain strict, regardless of circumstances. Reliance on these standards of proof affects whether peer review allows researchers to get government or private funding for further scientific study and for publication of results. Furthermore, it also affects whether government takes action in creating new regulations or in strengthening existing ones.

Controversies over standards of proof often forestall scientific closure, despite overwhelming evidence.[24] Data can lead scientists to different conclusions because debates over the meaning and implications of scientific data are not absolute, but change based on social circumstances. For example, regulatory levels of contaminants such as lead, radiation, and mercury have been reduced over time in part due to scientific evidence demonstrating adverse health effects at increasingly lower levels, but also due to public pressure on policymakers and regulators to act on the existing evidence.[25]

In breast cancer research, there is equivocal data for an environmental hypothesis. Some studies show health effects of chemicals, whereas others do not. But the dominant interpretation focuses on the lack of demonstrated health effects, and those who take traditional approaches believe the weight of evidence does not confirm the role of environmental factors. Disputes over the weight of evidence involve determining how much evidence is enough to warrant conclusions and whether the body of research literature shows reliable patterns across many studies. Perhaps the biggest issue is whether data from animal studies can be used as good predictors of carcinogenicity in humans. These disputes also involve questions about whether the research findings come from appropriate channels, typically meaning peer-reviewed articles in major journals.

Interpretive flexibility is also an issue, especially in the debate on whether "gray literature" (unrefereed material that is not published in peer-reviewed journals) should be considered valid research. For example, as a National Academy of Science conference pointed out, much useful research conducted by public-health departments is never published because of government reticence, outright censorship, or lack of interest and time to publish in academic forums.[26] The National Research Council followed up that claim with another volume that detailed methods of using the gray literature, including unpublished reports from the EPA, the National Toxicology Program, and manufacturers' test data.[27] These reports are the only scientific studies on some of these chemicals and their cancer risk. This source is important given the absence of exposure data on many chemicals, which has been emphasized by environmental breast cancer researchers. Much of the gray literature published by government agencies is peer reviewed and hence actually does conform to scientific canons of valid knowledge, even if it never gets published in journals. Indeed, the U.S. Department of Energy (DOE) organized a 2000 conference of government agencies to set up an Internet-based gray literature network, GrayLIT Network, to facilitate use of such government reports from the DOE, the DOD, the EPA, and the National Aeronautics and Space Administration.[28]

In thinking about the weight of the evidence, Silent Spring researchers conclude: "Given the relatively modest relative risks associated with the recognized breast cancer risk factors, an integrated research agenda for study of environmental pollutants in both laboratory and human settings has great potential. Even if the relative risks of environmental factors are modest, discovery of a risk that can be modified would save thousands of lives."[29]

ACTING ON SCIENCE

Once research is conducted and interpreted, it must be acted on because it will not automatically diffuse into practices such as regulating chemicals or reducing toxic emissions. This process involves scientists calculating the weight of evidence, deciding how to approach scientific uncertainty about the subject at hand, and creating policy that takes these factors into account. Bruno Latour's actor-network theory helps explain how scientists and advocates join together to spread their hypotheses, approaches, and findings and to push action in policy directives.[30] Actor-network theory holds that the acceptance of scientific discovery is not a linear progression of pure scientific deliberation. Rather, it is accomplished largely through scientists' recruitment of support from other scientists, media, government officials, and the general public. In this path, "advocacy scientists" or "dissident scientists" play a key role in mobilizing support by going outside traditional science into the realm of public advocacy.[31] In addition, social movement activists press for greater research on environmental factors by recruiting researchers sympathetic to their research agenda, as shown later.

The citizen-science alliance specifically exemplifies the "analytical blurrings" that take place through boundary movements. Expert and lay roles are altered, and typically rigid scientific practices take new forms as a result of these blurrings. For example, one woman spoke about breast cancer activists' attempts to find "a place at the decision-making table" that would bring out "not just our intuition and our native intelligence and our common sense and our perspectives as the patient [and] the affected community, but also really learning the science and the medicine of breast cancer." In this way, activists' role and knowledge base change as a form of analytical blurrings through the citizen-science alliance.

A crucial part of acting on science is for scientists and regulators to recognize the extent of our knowledge gaps. Research on breast cancer and the environment is impeded by what activists call "toxic ignorance."[32] More than 85,000 chemicals are registered for commercial use in the United States, but only a small portion of them, less than 1,000, have been tested for carcinogenicity, and even fewer have been fully and comprehensively tested for noncancer outcomes. Many toxins, such as dioxin, are formed as accidental by-products of production or incineration, and in many cases scientists and regulators have not even identified them.[33]

Moreover, the *Third National Report on Human Exposures to Environmental Chemicals*, released in 2005 by the CDC, examined the body burden of environmental contaminants, including pesticides, metals, dioxins, PCBs, and phthalates (plasticizers found in children's toys, cosmetics, and medical devices), in a sample of the U.S. population. The study found 148 chemicals, including known or suspected carcinogens, in the blood and urine of study subjects, up from 116 in the second report, with the notable addition of several PCBs, certain phthalate metabolites, the pyrethroid insecticide permethrin, and the organochlorine pesticides aldrin, endrin, and dieldrin. Although the CDC hopes to get several sets of data before addressing trends in chemical exposures, the current data are useful not only in establishing the effectiveness of current public-health efforts and in identifying which chemicals need to be further addressed, but also—by dividing the data into groups by age, gender, and race—in identifying disparities in these exposures.[34] The CDC will continue to increase the number of chemicals monitored, including 309 and 473 chemicals in 2007 and 2009, respectively.[35] Despite evidence of ubiquitous exposures to many chemicals, little is known about the long-term cancer risks associated with low-level chronic exposures. Activists and scientists have called for increased human and environmental monitoring to generate crucial information about the origins and potential long-term effects of chemicals. Indeed, Scandinavian countries' ability to implement the precautionary principle on many policy decisions is possible largely because they have high-quality monitoring databases with information on problems such as contaminants in breast milk. Chapter 7 addresses more fully the important new development of human exposure measurement (often called *body burden research,* or *biomonitoring*), which involves the assessment of the presence and concentration of chemicals in humans, by measuring the parent chemical, its metabolite, or reaction product in human blood, urine, breast milk, saliva, breath, or other tissue, including efforts by lay advocacy groups. Activists and scientists have also argued for the adoption of pollution prevention strategies that would decrease environmental emissions and population exposures.

Acting on science involves two key decisions: whether to act and how to act, both of which are determined in large part by state and federal legislatures and agencies. But scientists also make those decisions, such as when they consider whether to press for regulatory policy. Decisions on whether to act are not based primarily on the weight of the evidence, but rather on political and economic considerations; for example, policymakers might consider the economic cost of substituting one chemical for another. Even when a decision to act is made, there is leeway in how to intervene. In terms of breast cancer, this choice involves decisions about whether to pursue an individual responsibility approach (lifestyle and childbearing choices) or a social responsibility approach (restricting or removing chemicals).

Some scientists have a growing dissatisfaction with the ability of mainstream science to answer difficult questions and with the resultant lack of action. They pose questions about scientists' social and moral responsibilities to affected people and communities, as well as to the larger society. They also address the role played by funding sources (especially corporate ones). They question the veil of objectivity that separates science from social policy, obscuring the societal pressures that shape science. Some scientists raise questions about whether it is their responsibility to point out elevated disease rates and whether prepublication release of research is acceptable if it benefits public health. These challenges increasingly emphasize the precautionary principle's basic assumption that "[w]hen an activity raises threats of harm to human health or the environment, precautionary measures should be taken even if some cause and effect relationships are not fully established scientifically. In this context the proponent of an activity, rather than the public, should bear the burden of proof."[36] When scientists develop such perspectives, they are ripe for a citizen-science alliance.

THREE AXES OF CHALLENGES TO BREAST CANCER RESEARCH

As mentioned in the previous chapter, a paradigm refers to an established worldview that shapes what problems scientists are encouraged to study and how to study them, but excludes those theories, hypotheses, and observations that do not match this worldview. Major alterations in science occur when a critical mass of researchers begin to question the validity of a dominant paradigm that no longer adequately explains empirical observations. Here, I refer to the DEP that drives routine science. In addition to efforts within science, debates on the efficacy of the dominant paradigm take place within social movements, media, and congressional hearings, and they overflow from external debates into internal scientific circles. Paradigm shifts in environmental health almost always hinge on activists' efforts and typically involve the public paradigms where scientific innovations are discussed openly among many parties.

The dimensional model shown in figure 2.2 helps us examine how struggles for paradigm shifts occur. These axes represent continua involving various combinations of research, prevention, policy, and activism. Axis 1 (upstream/downstream) reflects the level of prevention; it is based on the familiar notion of downstream, or tertiary, approaches being after the fact and upstream, or primary, approaches being the most preventive (for example, research on treatment efficacy is downstream research, compared to upstream research on the relationship between dioxin exposure and breast cancer). Axis 2 represents a continuum of research foci from individual risk factors on one end to community-level hazards on the other. Finally, axis 3 reflects the degree of public participation in research, ranging from no lay involvement to research efforts that are mostly controlled by lay groups.

AXIS 1: UPSTREAM/DOWNSTREAM

Axis 1 addresses whether the research focuses on treatment, intervention, or prevention. These approaches are metaphorically referred to as upstream or downstream approaches.[37] Primary prevention (upstream) emphasizes disease prevention in populations. In an environmental health context, this emphasis implies strategies aimed at preventing human exposure to toxics through pollution prevention and toxics use reduction. Secondary prevention (midstream) aims to provide screening, early detection of disease, and prompt intervention for people at risk for disease. Tertiary prevention (downstream) minimizes the impact of disease in people who are already quite sick. For breast cancer, a walk upstream implies a radical shift in research and intervention away from tertiary approaches such as treatment efficacy (surgery, radiation, and pharmacological treatments) and secondary prevention such as screening and early detection (through mammography,[38] breast self-exam, and biopsy) and toward minimizing exposures to risk factors and toxic substances that may be linked to disease. When intervention and research strategies emphasize community and population exposures to toxics, they raise the formidable methodological challenges that are inherent to environmental epidemiology. Specific methodological barriers to studying links between chemical exposures and breast cancer include adequately measuring chronic low-level chemical exposures, measuring long latency periods for diseases such as cancer, and controlling for other covariates such as diet.

These scientific challenges are coupled with the political and economic controversies associated with reorienting research and intervention toward a focus on the impact of chemical exposures on human health. The 2001 Bill Moyers PBS special *Trade Secrets* exposed how chemical manufacturers concealed the polyvinyl chloride (PVC) contamination of employees and demonstrated how chemicals have "far more rights than people" because of industry's economic and political clout and its overwhelming influence on the regulatory process.[39] Similarly, describing the history of lead poisoning in children, Rosner and Markowitz argue that health and safety standards sometimes affect industrial policy, but only as a response to public outcry and national concern.[40] Indeed, government reluctance to regulate the industry's production decisions proactively and many research scientists' skepticism regarding environmental links to cancer make agencies such as the NCI reluctant to fund environmental causation research. In the absence of solid funding, professional support, and other resources, environmental breast cancer researchers see themselves as breaking traditional scientific norms in their pursuit of alternative methods and advancing new theories of disease causation. As I discuss in the section on axis 2 (individual lifestyle and environmental factors), these researchers are challenging the lifestyle-based DEP held by many segments of science, government, and the public.

THE ENDOCRINE DISRUPTER HYPOTHESIS The endocrine disrupter hypothesis has played perhaps the most significant role in pushing research interests upstream. The hypothesis "asserts that a diverse group of industrial and agricultural chemicals in contact with humans and wildlife have the capacity to mimic or obstruct hormone function—not simply disrupting the endocrine system like foreign matter in a watchworks, but fooling it into accepting new instructions that distort the normal development of the organism."[41] Although the link between endocrine disrupters and breast cancer remains tentative, many researchers and activists looking at environmental causes of breast cancer feel that the hypothesis supports the direction of their own work. Based on this increasing knowledge about the health effects of many endocrine-mimicking chemicals, a growing number of scientists have moved to an upstream approach. As one stated, "We need to understand exposure to chemicals that are of interest for research on breast cancer and endocrine-related health outcomes. So this was the first step in learning more about exposure and that these chemicals are chemicals that need to be studied further and considered for regulatory [policy]."

One scientist from Silent Spring Institute, which studies environmental factors in breast cancer, began to develop new scientific methods in order to capture the effects of endocrine disrupters:

> There are some significant methods developed in the first phase of the research that I consider to be central to our scientific contribution. One is the creation of ... environmental sampling tools. We tested wastewater, groundwater, and drinking water for endocrine disrupters. There weren't any chemical methods just on the shelf that you could use, so we had to adapt them. We also used E-screen [estrogen screen test], which was developed by Tufts medical school before the study started, but the extraction method that allows for the bioassay to be used for environmental samples was developed in the Cape study.... So we became the first team to identify estrogenic activity in groundwater. Endocrine disrupters had been previously found in surface water.

These new methods are based on the idea that despite the lack of clear evidence that endocrine disrupters cause breast cancer, the endocrine disrupter hypothesis warrants more investigation. Endocrine disrupters are working their way into people's bodies, though it is often unclear how. The high body burden of chemicals is increasingly the subject of study, and there is evidence of a connection to hormonally related disease. Research and regulatory policy have flowed from this awareness, including the EPA's formation of the Endocrine Disrupter Screening and Testing Advisory Committee in 1996, the European Union's policy to require toxicity information on all major chemicals in commerce and industry, a growing number of specialized research conferences organized by the NIEHS and universities, and increasing numbers of publications in scientific journals on the issue.[42] Even if the endocrine disrupter hypothesis is not confirmed with regard to breast cancer, it has led to widely accepted knowledge concerning developmental and reproductive disorders.[43] The endocrine disrupter debate has made a huge impact on toxicology and epidemiology and has cast its stamp on public policymaking.[44] Such debates are mechanisms for advancing science in general, even if they do not confirm all or most assumptions presently under study. Indeed, the tradition of scientific curiosity shows countless such examples where novel ideas catalyze new approaches in related areas.

ISSUES OF METHODS AND MEASUREMENT Scientists' approaches are affected by methodological challenges. As previously mentioned, environmental research faces the same obstacles as traditional epidemiology: difficulty with measurement, limited information about exposure, the multiplicity of variables that must be considered, and the newly emerging set of methods that must be adjusted to account for the causes being analyzed.[45] Although these factors reduce researchers' proclivity to focus on environmental causation, they also provide an opportunity to reshape scientific norms and practices to fit emerging knowledge about environmental factors.

Past studies of environmental causes of breast cancer have focused only on certain chemicals: dichloro diphenyl trichloroethane (DDT), dichloro diphenyl dichloroethylene (DDE), PCBs, and other organochlorines. There is little reason for this focus other than that these chemicals are easily detectable and found in women's blood or tissue.[46] Although the prevalence of these chemicals offers a large enough study population to create statistically significant findings, some researchers explain that this supposed benefit of studying organochlorines is actually a drawback: "We are dealing with compounds where exposure is ubiquitous; some of them cannot be measured ten, twenty, thirty years later, and our ability to find differences between populations is really marred by that." Because there is such omnipresent exposure in women, specifically in their breast milk, it is impossible to find a control group against which a population can be studied.[47] In addition, little is known about which specific organochlorines may be related to incidence. Therefore, the complexity within this one group of chemicals is unaccounted for across the scientific literature. The source from which the chemical is collected (i.e., blood serum or tissue) may also be a factor that influences findings; some chemicals may produce carcinogenic mutations but disappear from the assayed tissues.

Studies of DDT, DDE, PCBs, and other organochlorines actually represent a small proportion of research on breast cancer. The vast majority of research emphasizes tertiary prevention, which means treatment. Breast cancer research has been a prominent part of NIH spending. Although breast cancer activists applaud the importance of this research, they also question whether the money is being invested wisely. As one scientist remarked, "Almost from the beginning when epidemiologists in an organized and systematic way started looking at the epidemiology of breast cancer, they focused on a very narrow range of factors, almost all of them were related to women's reproductive systems, age at menarche, menopause, parity, lactation, then it branched out into body mass, host factors, family history, and genetic factors. Breast cancer and lung cancer, between the two of them, are probably the most widely studied, and I would say that maybe of the thousands of studies that have been done up to the early 1990s, almost none of them looked at environmental factors."

The NBCC has urged Congress and the executive branch to consider changes in the grant mechanisms and research structure at the NCI. Activists are calling for external monitoring of the NIH and believe the public should design and participate in an oversight process that will track how the money is being spent and whether it is being allocated wisely.[48]

In summary, current and traditional approaches to prevention tend to fall along the downstream end of axis 1. The endocrine disrupter hypothesis challenges the DEP and pushes the research focus upstream, though findings based on this hypothesis are often inconclusive and heavily criticized (more is to be said on this subject in the next section). Support for the endocrine disrupter hypothesis has led to significant amounts of primary prevention of toxic exposures in Europe, for example, by the banning of phthalates from children's toys and pacifiers and most recently in the United States by some cosmetic firms' halting phthalate use. (I am not implying a connection between phthalates and breast cancer; indeed, the stronger links are between phthalates and reproductive disorders, especially in males. The point is that there can be rather quick regulatory action on endocrine disrupters.) Nevertheless, various methodological challenges make the upstream approach much more difficult to pursue.

The endocrine disrupter hypothesis is not the only environmental hypothesis that can push research upstream to a more preventive model. Some toxins produce direct damage to chromosomes and intercellular signaling mechanisms. More attention gets focused on endocrine disrupters because that hypothesis poses a new challenge to science, whereas the more traditional examples of direct damage are more in line with acceptable approaches.

Movement along axis 1 involves disputes over doing science, interpreting science, and acting on science. As noted, there are great challenges to conducting environmental science that takes a more upstream approach. This type of science faces problems with measurement, exposure timing and recording, and lack of a control group. In part because of these difficulties, there is far less funding for research on environmental factors than for research on lifestyle and genetics. Despite these problems, we can observe an increase in research based on the endocrine disrupter hypothesis, one of the main scientific hypotheses that is moving scientific investigation upstream. Struggles over interpreting science are very intense, with a general reluctance among scientists to accept the evidence from animal studies and human reproductive health effects as applicable to human carcinogenesis, accompanied by a belief that the weight of evidence favors a lack of connection to environmental causation. One of the hindrances to moving farther upstream is that gray literature is not included in the body of evidence used to interpret scientific support for or against an environmental hypothesis.

Possibly the greatest obstacle to moving science farther upstream is the trepidation concerning how a finding would then need to be acted upon. If science takes a more upstream approach and finds that chemicals need to be better regulated, industry may face major financial difficulties. Therefore, it is in the best interest of industry to take a more downstream approach that focuses on curing diseases rather than on preventing them. Not only does this approach avoid the costs of prevention, but pharmaceutical companies profit from selling chemotherapy drugs and the broader medical industry profits on treatment. In sum, acting on science involves the choice of focusing either upstream on prevention or downstream on curative approaches. Acting is only rudimentary for this area of breast cancer and the environment, yet there has been some upstream regulatory action, less in the United States than in Europe.

AXIS 2: INDIVIDUAL RISK FACTOR AND COMMUNITY-LEVEL ENVIRONMENTAL HAZARDS

Whereas axis 1 has implications for intervention, axis 2 examines the focus of research—the individual level (characteristics of individuals) or the community level (aggregate characteristics of geographical areas). For axis 1, I discussed the emphasis on xenoestrogens (endocrine disrupters) as part of the upstream approach because the identification of chemical causes leads to a primary preventive approach of reducing or eliminating such exposure. Here, in axis 2, I revisit chemical factors, this time looking at individual versus collective units of analysis. In addition, whereas axis 1 represents the general trends of epidemiological research, axis 2 relates these general struggles to the specific case of breast cancer research. Axis 1 is concerned with how a number of social sectors (e.g., industry, government regulators) can prevent exposure in the first place, thereby reducing disease. Axis 2 and axis 3 are concerned with the internal methods and data collection of medical research, including the degree to which lay involvement plays a role.

The traditional and dominant approach to disease research focuses on individual risk factors, whereas environmental breast cancer researchers and activists pursue factors that are difficult for individuals to modify and that require community-level action. Underlying science and policy are what Tesh calls "hidden arguments," or political ideology about what constitutes legitimate sources of knowledge, and moral arguments about human nature and what constitutes a good society.[49] These hidden arguments, or conceptual frameworks, inform which questions get asked, which do not, and how researchers go about investigating them. Furthermore, these frameworks theorize about the causes of disease, so they influence how researchers conceptualize and operationalize the determinants of health.[50] Finally, theoretical frameworks also inform how researchers prioritize possible models for disease prevention. This formulation is, quite noticeably, similar to the idea of the DEP. By analyzing the hidden arguments and bringing them into the open, people challenge the DEP and often work out a public paradigm.

The biomedical model is the dominant framework that informs how most health researchers are trained to conduct research in the health sciences, although the tenets of this model are rarely specified.[51] The central question of research informed by the biomedical model is: How do humans become ill? Disease, according to this conceptual framework, is purely a biologic phenomenon that can be understood through positivist, value-free research. Further, this model assumes that diseases are best addressed through treatment and through mitigation against individual-level risk factors for disease. However, the other defining assumption of the biomedical model is that it does not address societal or structural determinants of health and therefore cannot explain why the distributions of health and illness tend to correspond to society's economic and social structure.[52] Biomedical models emphasize biological and physiological phenomena, whereas lifestyle models emphasize people's behaviors. These two approaches show up together in research studies because both focus on the characteristics of individuals. The primary distinction between them is that according to the lifestyle model, prevention or intervention is achieved through individual behavior change rather than administered by a medical professional.

In contrast to frameworks that conceptualize disease causality and prevention at the individual level, models focusing on the political economy of health or the social production of disease ask how economic and political relations affect health. This focus includes an emphasis on the role of environmental factors in disease. Under the assumptions of this framework, disease prevention is achieved not through medical treatment or through individual behavior change, but through changes in industrial production practices. Theories that center on the political-economic and social production of disease look at the dynamic relationship between macrolevel structures and individual bodies: the individual is always seen in relation to the political economy and social world.[53] According to Geoffrey Rose's distinction between studying sick individuals versus sick populations, an individual risk factor approach seeks to answer the question, "Why do some individuals have breast cancer?" Conversely, a population-based line of inquiry asks, "Why do some groups of women have breast cancer, but in other populations it is rare?"[54] Although the former approach may help us understand why there is variation in disease among individuals, it misses the fundamental public-health and epidemiological challenge of elucidating those determinants that explain population differences in disease incidence. In terms of intervention, the population-based approach is more radical because it implies the need for mass environmental control methods or the alteration of socioeconomic norms that give rise to widespread hazardous exposures and of collective behaviors that enhance certain communities' vulnerability to disease.[55]

TRADITIONAL APPROACH TO INDIVIDUAL RISK FACTORS In a period of growing genetic determinism, genetic makeup has been a major research focus, and knowledge claims based on genetic explanations of disease are considered especially credible.[56] The discovery of the BRCA-1 and BRCA-2 mutations led to much attention to genetic causes, even though, as noted previously, it has since been recognized that genetic causes account only for some 5 to 10 percent of all cases.[57] A recent study of forty thousand sets of twins found that the environment, not genetics, is the dominant cause of cancer.[58]

Activists and some scientists have criticized the way that genetic approaches focus on only individual risk factors. Such an approach may account for some individual susceptibility, but it cannot explain why breast cancer rates have increased in a period that is too short for genetic changes to occur in the population.[59] If genetics research focuses on individual risk factors, it will likely result in campaigns for people to alter their individual behavior, based on why some women are more susceptible to endocrine disrupters. Such research is not unimportant, but nonmainstream scientists argue that it should not be the sole focus of searches for the cause of breast cancer. For example, one researcher pointed to other areas of study that reveal promising leads: "We have evidence that synthetic hormones affect breast cancer risk. I also think that the wildlife studies are very compelling as a whole. . . . The effects of endocrine disrupters on wildlife, I think, are a very important body of support for further investigation of environmental causes of breast cancer. And then there are laboratory studies that have identified mammary carcinogens in animals. Cell studies have shown synthetic chemicals make human breast cancer cells grow in a lab. That's a signal to me that it should be a priority to study them in humans."

The interpretation of genetic factors may impact medical practices. Knowledge of genetic factors might be used to focus on a variety of practices, ranging from late intervention to prevention. Differing approaches to genetic risks may also lead some scientists to ask why certain populations are more vulnerable now than they were previously. The increasing ability to discover more vulnerable populations might be a tool for making the case for tighter regulation, rather than advocating such drastic practices as the use of preventive Tamoxifen (a drug originally used only to treat estrogen-positive breast cancers) or prophylactic mastectomy. Alternative approaches might examine the interaction of genetic risk and environmental exposure. Another direction of genetic research that some scientists find promising is toxicogenetics, in which researchers study biomarkers of toxic exposure and effect at the genetic level.

Similarly, diet and reproductive behaviors are widely implicated in breast cancer, and it would be possible to ask the same population-based questions as with genetics: Why are some subpopulations more vulnerable, and what can be done to reduce that vulnerability (especially with dietary factors)? If these behaviors are risk factors, they are certainly more than *individual* risk factors.

Some lifestyle factors are indeed important; there is strong evidence regarding the effects of alcohol and physical activity and persuasive research on tobacco smoke. But medical and public-health efforts focus primarily on these factors rather than on population-level factors and on "chemical trespass" of toxins into people's bodies without their direct use of products containing those toxins. Individual risk factor approaches continue to dominate for several reasons. First, they fit well with traditional biomedical conceptions of disease causation, which emphasize individual-level variables. Second, leading cancer researchers, institutes, and agencies have placed most of their efforts on this perspective largely because it is easier to design and test individual risk factors than environmental factors, and hence such research is funded more easily. Third, there is currently only weak evidence for environmental causation. Fourth, research into environmental causation places responsibility on the business sector and on the government agencies that fail to regulate that sector adequately. As one scientist commented, "It's much more acceptable to talk about people's diet and smoking than to try to pin things on industrial chemicals."

Based on interviews with breast cancer researchers and my review of the research, I conceptualize the existence of two main approaches to research on environmental factors. *Traditional* research focuses mainly on individual risk factors and occasionally adds environmental variables. It is centered in the DEP for breast cancer. *Innovative* approaches put primary emphasis on environmental factors, which have otherwise been largely ignored. Proponents of an innovative approach argue that more environmental research should be conducted because it offers more opportunity for prevention and because traditional approaches have led to little knowledge of causation. They also argue that genetic studies do not show enough genetic causation to warrant the attention given to it, and they point to findings from immigration studies that show rapid increase in breast cancer when Asian women come to the United States, which is likely due to dietary changes such as a reduction in soy and omega-3 consumption.

MIXED RESULTS AND CONFLICT IN THE SEARCH FOR ENVIRONMENTAL FACTORS Proponents of an innovative approach argue that genetic explanation cannot account for the jump in lifetime risk of a woman's developing breast cancer from one in twenty (by age eighty) in 1950 to current estimates of one in eight. They also point to the immigration-related data that show the breast cancer risk for Asian immigrant women increasing 80 percent in the first generation and the rate for their daughters approaching that for U.S.-born women.[60]

The body of environmental breast cancer research, which has been equivocal, is far smaller than the literature on genetics and individual risk factor epidemiology. Some studies have shown a correlation between environmental toxins and breast cancer incidence, but others have not supported this conclusion. Although Hunter and colleagues' findings in the Nurses' Health Study and the Long Island results (discussed later) are a setback to the search for environmental factors, the positive findings for polycyclic aromatic hydrocarbons (PAHs) in the Long Island study have interestingly been downplayed, with most media and scientific focus on DDT, DDE, and PCBs.[61] However, many researchers emphasize that lack of evidence does not disprove an argument for environmental causation. Rather, they say that movement in this direction has been slowed by a constant critique from mainstream researchers.

In the first five years of the new millennium, a growing number of studies on environmental factors showed more promising avenues, even while realizing that the older emphasis on DDT and DDE is not likely to be fruitful.[62] Wolff and colleagues were among the earliest to show a direct correlation between increased risk of breast cancer and DDE, the chemical breakdown product of DDT, a pesticide commonly used worldwide. Since then (1993), an increasing number of studies have focused on this group of organochlorine pesticides used mainly since World War II. Wolff and colleagues' landmark study was the first large-scale study to draw major scientific attention because of its conclusion that there is a relationship between DDE and breast cancer incidence.[63] But, as noted previously, the later work by Hunter and colleagues cast doubt on these conclusions and posed a major obstacle to proponents of environmental causation.[64]

Other research has focused on chemicals more generally. Some studies are retrospective studies using health data, employing ecological design rather than individual body burden. Examples of such research are Griffith and colleagues' and Najem and colleagues' studies of increased breast cancer rates in women living in the vicinity of hazardous waste sites.[65] Most of the significant data come from body burden studies. Earlier work on PCBs found no support for PCBs' playing a role in breast cancer etiology.[66] But more recent studies examined subgroups and found that a combination of menopausal status and gene variations led to increased breast cancer risk.[67]

The massive dioxin contamination from the 1976 Seveso, Italy, factory explosion has led to a twofold increase in breast cancer incidence for a tenfold increase in dioxin in the blood. Because the women in that cohort are still relatively young, higher breast cancer rates may occur as they age.[68] Recent attention to organochlorine pesticides other than DDT and DDE have linked dieldrin, hexachlorobenzene, betahexachlorhexane, trans-Nonachlor, and oxychlordane to breast cancer in at least one study for each compound, and this avenue of research is still new.[69]

In evaluating research on pesticides and breast cancer risk, Snedeker notes that although there is a lack of support for a correlation in Western women of European descent, the most common population being studied, further studies of other populations should be conducted.[70] Krieger found differences in morbidity and mortality across racial groups as they age.[71] Millikan and colleagues, in their study of black and white women, first conclude that strong support for a correlation between breast cancer risk and exposure to pesticides does not exist, but then note correlations among subgroups.[72]

One point of controversy in these studies is whether the recency of exposure affects risk. Studies in developing nations where DDT is still used are a way to test this question. Research in Mexico and Colombia has shown a strong correlation between DDT serum levels and breast cancer risk, but other evidence gathered from women in Vietnam, Brazil, and Mexico has shown negative results.[73] Hence, there is mixed evidence on this point as well.

Other chemicals have been implicated in various types of exposures. A case-control study of women who were accidentally exposed to perchloroethylene leaching from drinking-water pipes found higher breast cancer risk.[74] In an ecological study based on the EPA's Toxic Release Inventory, women in Texas had higher breast cancer incidence linked to ten of twelve contaminants.[75]

The mainstream scientific community has been very skeptical of environmental causation of breast cancer, and proponents of the connection have received criticism from their peers and from others outside of the scientific community. One environmental breast cancer researcher commented: "It is hard to get people's attention when you are from Silent Spring [Institute] and doing research about environmental causes of breast cancer. It's hard to break through and get people's attention, scientists' attention. You go through a kind of hazing where you have to establish talking about what you're doing, your credibility with each new member. You don't have on your name tag, Harvard University. You really have to go through a process of demonstrating your credibility."

Some studies contradict findings that link environmental factors and breast cancer, and editorials occasionally condemn environmental causation hypotheses.[76] In addition, critiques of environmental breast cancer research from more conservative scientists call it "junk science," arguing that environmentalists exaggerate or fabricate data that support the role of environmental factors in disease and in essence conduct a politically motivated assault on modern science.[77] Hunter and colleagues' study was accompanied by a deprecatory critique written by Stephen Safe, a researcher funded by chemical firms, who has been a major attacker of environmental breast cancer research.[78] He began his editorial by using the term *chemophobia* to imply that those interested in chemicals were merely paranoid and referred to environmental breast cancer research as "paparazzi science." Because his article was published in the influential *New England Journal of Medicine,* it was relatively influential. At the same time, that the *New England Journal* would publish such a diatribe indicates the extent to which established science and medicine will go to discredit the environmental hypothesis.

Similarly, the scientific and media response to the August 2002 publication of the results of the Long Island Breast Cancer Study Project was exceedingly sharp and appeared to be more of an attack than a scientific discussion.[79] Critics pointed out that the negative results from the Long Island study show both that environmental causation is not a useful research direction and that public participation is dangerous for science. Scientists from well-known institutions who opposed this backlash were unable to get op-eds or even their letters to the editor published, including James Huff, a leading scientist at the National Toxicology Program, and Lorenzo Tomatis, former head of the International Agency for Research on Cancer.

But even apart from the sharp attacks, the scientific findings from Long Island do put a powerful constraint on supporters of environmental causation. Many people had hoped for positive results, even in the light of activists' criticism that the study examined only a small number of the many chemicals they believed should have been included. One activist pointed to the methods she felt were problematic: "Epidemiology is not really a very good science for noninfectious disease. Particularly a noninfectious disease that is clearly environmental in origin because in order to arrive at some significant statistical understanding of the outbreak of a disease, you need a control group. How can you have a control group? So all you get when you use epidemiology is a statistically insignificant result. So that epidemiology becomes an apologetic when it's used with cancer, and that's exactly what happened in Long Island. No statistical significance was found. On top of which they looked only at chemicals in the blood."

Occupational studies have found some support for environmental causation.[80] Hansen found that for 7,802 relatively young women employed in industries using organic solvents (e.g., textiles, chemicals, paper and printing, metal products, and wood and furniture), increasing duration of employment was associated with greater risk of breast cancer.[81] Among those employed ten years or more in a solvent-intense industry, the adjusted odds ratio for developing breast cancer was 2.0, meaning on average the risk of cancer doubles after ten years of occupational exposure to solvents. A recent study showed that in agricultural and manufacturing employment where women were exposed to chemicals, there was an increased breast cancer risk.[82] A growing number of other recent studies are adding even more occupational linkages.[83]

Occupational studies may capture information about chemicals that do not bioaccumulate—in other words, that dissipate quickly—and hence are missed by body burden studies (biomonitoring to assess chemicals in human fluids and tissues). Although occupational studies were conducted earlier and have been more supportive of environmental factors, they have tended to receive less attention in the scientific community. The low level of interaction between occupational and nonoccupational studies is an indication of the fragmentation in current breast cancer research and must be viewed as a major obstacle, even though that fragmentation is in part caused by some supporters of the environmental hypothesis.

Epidemiologists and toxicologists who study environmental factors in breast cancer understand that some of the issues and obstacles mentioned previously will directly affect their work. My interviews found that some scientists who were unable to detect environmental causation were disappointed, saying things such as, "I wanted to find something." They had a plausible hypothesis, were motivated by a desire to serve the public health, and expected that they might find positive results. As one scientist remarked, "The more people who are interested in something, I think the better chance is that you're going to find out something about it."

But, failing to do so, they have not given up on such efforts. They have turned their efforts to new approaches and measurements, as with Silent Spring Institute's research (discussed later). Indeed, Silent Spring Institute's review of studies for the 2000–2005 period shows a strengthened approach to environmental factors research, with investigators studying more chemicals, more types of exposure, and more parts of the life course.[84] Some have sought to advance a critical epidemiology perspective that challenges canons of traditional epidemiology (also discussed later on). Others have understood the difficulties in environmental epidemiology to be so substantial that they invoke the precautionary principle as a defensive public-health approach to reduce or ban chemicals even in the absence of definitive knowledge of environmental causation.

In summary, aided by a broad social awareness of the disease, environmental breast cancer research has been used to construct a more political-economic framework (e.g., one that includes the role of the economy in the increase in breast cancer and has a role for laypeople in research) and to highlight new areas of legitimate scientific discourse (especially the endocrine disrupter hypothesis). My analysis is driven by an attempt to understand the obstacles to researching environmental causes of illness and, by extension, broader challenges to constructing new knowledge. Research at the environmental end of this second axis, however, remains a weak part of the overall field.

AXIS 3: LAY INVOLVEMENT IN RESEARCH

Axis 3 captures the extent to which research involves lay activists. At one end of the continuum, scientists work independently of lay input, and laypersons may serve as participants or subjects without contributing to research question formulation, methods, data analysis, or dissemination of data. The opposite end of the continuum is anchored by cases with active lay involvement, where laypersons direct research efforts in collaboration with scientists. Here, "nonscientists" and lay knowledge are equally important components to the research and knowledge production processes. In some situations, expert-produced science has not adequately explained changes in health status or has contradicted lay experience. As a result, laypeople and activists have generated new science, while minimizing or erasing lay-expert boundaries and transporting science into public spheres for use by a broad array of people. They have accomplished these results by seeking out independent scientific advisors, conducting health surveys, creating what I call "lay maps" of disease clusters, and interjecting lay knowledge into traditional, scientific ways of generating explanations. Lay participation can also take on other forms that lie between these poles, depending on the unique constellation of actors, resources, sociopolitical factors, and underlying philosophies about science.

The inherent uncertainty in breast cancer epidemiology makes citizen involvement particularly important. Lay activists, especially those who focus on environmental factors, push science in new directions, forge novel collaborations, and generate new methodologies. They do not accept data that contradict their lived experience, and they continually refocus researchers and funders to acknowledge the significance of that personal experience. They do not see science as a privileged domain, but as a tool to achieve their goals. Despite institutional barriers that may bar them from certain types of participation, they continue to push for involvement in several ways. Breast cancer activists have challenged researchers to explore environmental hypotheses, secured funding, worked to redirect research questions, critiqued outcomes, shaped public acceptance of projects outcomes, and called for additional research with retooled hypotheses and methods.[85] As one activist noted, "Power isn't only knowledge; it isn't only that I know how to run GIS statistical analysis. That's not all power is. Power is perspective; it is sensibility, and I even think power is intuition, too. . . . Maybe you need to use the scientific process, but you need to broaden the approach, broaden what you would ask, think of a new angle. There needs to be a lot more flexibility around doing that than there has been. Part of it is bringing new ideas and ideas to the table that scientists may not think about."

Many activists, with technical assistance from supportive epidemiologists, have achieved proficiency in the scientific literature to engage scientists effectively, to navigate scientific arenas, and to gain entrée into formal settings where policy decisions are made. The NBCC's Project LEAD is one example where lay activists participate in expert-led workshops on many scientific issues in breast cancer to prepare themselves for service on state and federal review panels. Since 1995, NBCC has trained more than seven hundred advocates.[86]

The citizen-science alliance has generally been an effective vehicle for activists to engage more fully in scientific endeavors. Such an alliance is exemplified in the efforts of the Environmental Working Group (EWG), which teams up with scientists to conduct "impact research" where results are leveraged using savvy media and communication strategies to promote changes in environmental regulation and policymaking. An advocacy and watchdog organization founded in 1993, the EWG assesses corporate and governmental accountability, conducts research on toxics, works for stricter environmental regulation, and provides public education on environmental hazards. It launched a national media campaign to publicize results outlined in the CDC's *National Report on Human Exposure to Environmental Chemicals* in advance of the official release. In 2003, the EWG released on-line a study entitled *Body Burden: Pollution in People,* which involved a scientific collaboration with the Mount Sinai School of Medicine in New York and Commonweal.[87] Results were also published in a peer-reviewed public-health journal.[88] Researchers found an average of 91 industrial compounds, pollutants, and other chemicals in the blood and urine of nine high-profile volunteers, including television journalist Bill Moyers and Andrea Martin, recently deceased founder and director of the Breast Cancer Fund. A total of 167 chemicals were found in the group. None of the people tested worked with chemicals on the job or lived near industrial facilities. The EWG's decision to collaborate on a body burden study was strategic from a scientific and political point of view; indeed, results from body burden testing have often resulted in swift action by government and corporate leaders. Following a medical study showing high mercury levels in the blood of patients whose diets were high in mercury-contaminated fish, the state of California recently sued several grocery chains to force them to put labels on these products. When a chemical used in Scotchgard stain repellant was found in virtually all Americans, 3M Company was compelled to change the formula, even without data on health effects.[89]

Silent Spring Institute is a classic example of the citizen-science alliance approach. Institute scientists are both similar to and different from many mainstream scientists. They work in an alternative institutional site that is not part of a university, think tank, or government agency, although they collaborate with researchers from universities. Their mission statement contains an affirmation of the citizen-science alliance as part of their work. However, although Silent Spring Institute scientists are appreciative of lay input, they are staunch in their unwillingness to alter their standards of evidence. They believe that standards of technical excellence are necessary for good science and for establishing their legitimacy in the scientific world. To reach these goals, they believe, required "a lab of our own," as Executive Director Julia Brody said in a 2004 talk.[90]

That sentiment of "a lab of our own" captures the desires of a whole movement of activists and scientists who want a societywide lab of their own in which to examine environmental factors in breast cancer—not just physical settings for this kind of research, but committed action on the part of the regulatory apparatus and research-funding network, a more inviting atmosphere from journal editors and reviewers, a firmer place in university curricula, and more balanced news coverage. In other words, a "lab of our own" can be seen as a paradigm challenge to the existing DEP in all its institutional components.

Silent Spring Institute sits in the middle of paradigm struggles over doing science in that its researchers support challenges that activists pose to traditional standards of proof and yet consistently maintain academic rigor. In working toward these ends, Silent Spring Institute uses a wide variety of innovative methods in order to create a well-rounded understanding of the multiple possible factors that may be related to breast cancer incidence. They include GIS applications, survey methods, and novel epidemiologic and toxicologic practices.

One innovation is exemplified by the development of a scientific "shopping trip" in which researchers collected samples of air and dust from various locations in stores and homes to examine them for endocrine-disrupting and carcinogenic chemicals. Researchers initiated these new methodologies because of the need to collect information about exposures to women in multiple locations and over time. This endeavor was also a change from examining chemicals in the lab, as in traditional research, to focusing on the actual experience of women in their day-to-day lives. They collected samples from places women typically frequent, including sites not usually considered in epidemiological research, such as stores and dry cleaners. Through these techniques, they were able to characterize chemicals previously unstudied and to conclude that hormonally active agents are common and important indoor exposures that require further research.[91] Then, in sampling air, dust, and urine in Cape Cod homes, they extended their gaze to a wide variety of chemicals. The result was that Silent Spring Institute researchers characterized many substances for the first time, leading the way for other household and body burden studies.[92] Institute scientists also determined that length of time living on the Cape was statistically significantly associated with breast cancer risk. This finding did two things: it provided evidence that higher rates were not due to people's migrating to the Cape, to established risk factors, or to mammography test rates; and it suggested that an unknown risk factor, perhaps an environmental factor, affects the region.[93]

Silent Spring Institute's commitment maintains public channels of communication (including keeping an office on Cape Cod until the Massachusetts DPH's withdrawal of funds led to the closing of that office), engages in outreach via educational sessions, and provides extensive scientific information to the general public. One example of the latter is the *Cape Cod Breast Cancer and Environment Study Atlas* (http://www.silentspring.org/newweb/atlas/index.html), which contains Cape-specific information on breast cancer incidence, a history of pesticide use for agricultural land and tree pests in the area, drinking-water quality, census data such as income and education, and land-use information, including the location of waste disposal sites and the dramatic transition from forested land to residential housing. Silent Spring Institute believes that scientific data should be circulated widely to provide the fullest possible access to information for people actually and potentially affected by breast cancer.

Another form of collaboration can be seen in the Carolina Breast Cancer Study, where researchers brought study participants and breast cancer advocates into the research process. Those new partners pressed the scientists to provide a more detailed description of the study, to communicate results to participants, to make local physicians more aware of the study in order to maximize recruitment, and to provide participants with more knowledge on how their confidentiality was protected. Advocates and participants also suggested hypotheses for investigation and helped revise the research protocol in terms of notifying people of data. Upon learning of limited education on breast cancer and poor access to follow-up care, the researchers sought and received funding from the Susan G. Komen Breast Cancer Foundation to provide a resource directory for women with breast cancer; they distributed more than four thousand copies in their first year.[94]

Through citizen-science alliances, activists can alter scientific agendas by pressing scientists to think about the research questions they pose and the immediate relevance of their questions to community health in general and to women's health in particular. Citizen-science alliances create new knowledge by using novel research designs and methodology (e.g., body burden testing and GIS mapping), informed by the personal experiences of women with breast cancer. In doing so, these alliances have the potential to influence permanently the methods through which future studies are conducted because of new ideas introduced by people usually excluded from the scientific realm. Citizen-science alliances are also often framed in ways where study results can be effectively leveraged to expedite policy action. In this way, the shift toward increased lay-activist collaborative participation in scientific research has influenced or paralleled shifts in breast cancer epidemiology previously mapped along axes 1 and 2.

The heightened interest in and the controversial public interpretation of the Long Island results demonstrate that boundary-crossing activities are important to both the public and science. At the same time, boundary crossing makes a social movement vulnerable to criticism and defeat in multiple venues. In citizen-science alliances, it threatens traditional roles played by laypeople and scientists and hence has ramifications for the power held by these actors.

Ultimately, though, the significant improvements in lay involvement in review panels and research participation have not been matched by the inclusion of laypeople in the policy apparatuses where regulatory decisions are made. Therefore, activists and advocates have become part of only two of our science arenas—doing science and interpreting science—but have not been integrated into the third, acting on science. Activists obviously do act to seek implementation of policies they believe are useful; for instance, a coalition including the Breast Cancer Fund sued the EPA to speed up its endocrine disrupter analysis. This example demonstrates that activists need to take exceptional measures to become part of the policy practice.

Now I turn aside from the scientific struggles to focus more directly on the activist groups that have formed such a powerful new social movement. To do this, I examine activism in three locales—Cape Cod and Boston, Long Island, and the San Francisco Bay Area—and then I show how these social movement organizations and related research groups have shaped a growing EBCM. I chose these places because they were where the most significant action was taking place around environmental causation of breast cancer and where well-defined social movement organizations and research centers were located.

THE EBCM IN THREE LOCALES

Creating a new public paradigm necessitates participation in science and at least symbolic influence over its formation. Such influence requires a unique type of social movement—what I discussed in chapter 1 as *boundary movements*. The fluidity and analytical blurrings in these boundary movements explain how movement actors in different locations have a variety of relationships with the state, other movements, and experts, while maintaining their own unifying movement philosophy. As I show, a number of Long Island activists found it easy to work with mainstream politicians, whereas Bay Area activists were more likely to link up with AIDS, women's health, and toxics activists, and Boston-area activists found allies among those involved in precautionary principle organizing. Local context shapes the way each area approaches the problem, even though they all are part of one overall social movement. Such localized "cultures of action" involve assorted groupings of movement and nonmovement actors in changing alliances.[95]

Boundary movements also lead us to examine the role of *boundary objects* that traverse the human/nonhuman boundary.[96] These material and nonmaterial objects transect a variety of contexts, maintaining enough similarity in each to create coherence across circumstances even while being used distinctly in each one. Hence, we see the importance of *nonhuman actors* serving as boundary objects: mammography machines, genetic testing for breast cancer, patents on the BRCA-1 and BRCA-2 sequences, pharmaceutical firms, Breast Cancer Awareness Month, and Avon's Breast Cancer Walk.[97]

The three areas I examine—the San Francisco Bay Area, the Boston and Cape Cod area in Massachusetts, and Long Island, New York—have higher incidences of breast cancer than the rest of the United States, and public attention has led to numerous studies on these areas.[98] This activism was initiated in Long Island, developed soon thereafter in Massachusetts, and most recently spread to the San Francisco Bay Area. A combination of local and national contextual factors influenced the development of the movement, research, and policy in each location. Local factors include media response, the political climate and political connections, preexisting social movements, the response of local government, and potential funding sources. A general environmental framework unites the three locales into a national movement whose cohesion is still in its formative stage, yet which has already affected local and national policy. The citizen-science alliance central to this social movement has altered scientific processes and legitimacy and has served as a beacon to other health movements.

BACKGROUND ON THE LOCALES

After providing detailed background information on the three major locales of movement formation, I then focus on the overarching framework of the EBCM, including its links with other social movements and its creation of citizen-science alliances.

LONG ISLAND, NEW YORK In the early 1990s, women in western Long Island began noticing that they were "surrounded" by breast cancer. Women with breast cancer and women who had never had the disease began to map breast cancer cases in their community. At first, only a few women were involved in lay mapping, or working with a map onto which they located cases. However, after one group was formed that developed mapping for its individual community, it assisted similar groups that formed later in other communities. Groups interested in environmental causes of breast cancer, such as the Huntington, West Islip, and Southampton Breast Cancer Coalitions, were formed around 1992 with the help of preexisting breast cancer support groups that were sometimes administered through local hospitals.

Long Island activists have a widely dispersed network of organizations and a configuration of research projects, and they employ small-scale utilization of GIS in multiple ways. Although they conduct GIS research, they also provide support services to women with breast cancer. Some conflicts between groups have existed alongside a relative level of cooperation. One umbrella organization, the Long Island Breast Cancer Network, unites the diverse set of groups but is not used as a forum under which to make statements or give opinions. Outside of this forum, projects between organizations overlap only sporadically due to their different areas of coverage and different agendas.

This formation process was also facilitated by support from scientists at a local university. Activists sought scientific study from these researchers to support their hypothesis that more women on Long Island were getting breast cancer than were women in other places. In 1993, they gathered backing from other well-known scientists and held a conference on the issue at which the CDC, the EPA, and the NCI were major presences. This alliance enabled the passage of the federal bill that planned and funded the $32 million Long Island Breast Cancer Study Project through the NCI.

Long Island activists were noteworthy in their ability to forge strong connections with Republican politicians, especially Senator Alphonse D'Amato, considering that Republicans usually stay clear of such environmental positions. These connections were largely responsible for a government public hearing on breast cancer. Involved scientists and activists claim that the attention activists brought to the issue, the political and scientific connections they made, and their individual testimony in the Senate were the reasons for the success in funding the research. The Long Island organizations' relationships with local and state government are much more supportive than at other locales, probably because personal relationships with political representatives were nurtured over a long time.

Another example of the political effectiveness of the organizations on Long Island is the work of the Southampton Breast Cancer Coalition. Although largely focused on helping local women, these activists serve on research advisory boards and the New York State Department of Health's Research Science Advisory Board, which decides on allocation of funds generated by the breast cancer research check-off box on state tax returns. Leadership by a local governmental representative has helped the Southampton Breast Cancer Coalition gain major support from the local government, including a GIS technician to conduct research from an office in the town hall.

Activists in the area have benefited from much positive media attention. Many organization leaders have been on major television news, specials, and talk shows such as *Oprah, Eye to Eye,* and *Primetime Live,* and breast cancer events have been routinely covered in print and broadcast media. The results of this coverage have been mixed, however, as discussed later with regard to the Long Island study.

Long Island groups were not linked to environmental groups for some time, perhaps due to their members' Republican connections, but that situation changed as the breast cancer groups developed. The West Islip group, the first to form, is now fighting against pesticide use on golf courses. The Huntington Breast Cancer Action Coalition is currently focused primarily on environmental causation, as shown by their pushing of organic horticulture and their newsletter's promotion of displaying pink flags in yards grown without chemicals.

Currently, many research projects are under way or have recently been completed on Long Island. Two major projects stand out as having had the most impact and the most community involvement. The first is the Long Island Breast Cancer Study Project. Ten subsidiary projects exist under that umbrella, including studies ranging from the investigation of potential impacts of overhead power lines to the search for chemical contaminants in tissue samples. Activists from all over the Long Island area have sat on the advisory boards of many of the studies being conducted. The second most significant set of studies is the one utilizing GIS to map breast cancer cases and environmental factors in local areas, where new technologies have elaborated the original lay mapping.

The first results of the Long Island Breast Cancer Study Project were released in the summer of 2002 and garnered wide press coverage. The study found little to no support for a hypothesis that PCBs and organochlorines correlated with increased breast cancer risk, but some evidence of higher risk from PAHs, which are common by-products of fuel emissions, tobacco smoke, and grilled meat and fish. For women in the highest quartile of exposure to PAHs, there was a 50 percent increase in breast cancer risk.[99] Although the researchers called for further examination of a link between chemicals and breast cancer and asserted that their results were far from conclusive, the press interpreted the report as proof that there was no link between pollution and breast cancer and that collaborations between activists and scientists are detrimental to scientific advancement, thus discrediting both activists and those researchers who were looking at environmental causation. Activists and scientists felt that this presentation was unfair and would potentially discourage funders from pursuing similar projects in the future. Activists pointed out that they sought the examination of a wider range of chemicals than were eventually selected for research, an announcement that brought to light some tension between laypeople and scientists. Contrary to much of the scientific and press attitude, Deborah Winn of the NCI, the chief federal scientist who oversaw the whole Long Island research, believes that lay involvement was indeed positive and that the Long Island study helped facilitate lay involvement in other studies.[100]

The two papers reporting the Long Island study called for further research but were not as specific as they should have been. Activists and researchers note the need for more measures at the time of exposure, studies of early exposure, examination of gene-environment interaction, and research into other chemicals.

MASSACHUSETTS Long Island activists only slightly preceded Massachusetts activists, who gained state support in 1994. Cape Cod, Massachusetts, has a rate of breast cancer about 20 percent above the state average, with nine of the Cape's fifteen towns having statistically significant elevated levels. After several years of activism based on this problem, the Massachusetts Breast Cancer Coalition founded Silent Spring Institute in 1994 as a research organization focused on studying environmental causes of breast cancer with a citizen-science alliance approach. It also sought to educate the public about environmental causes of breast cancer and was successful in getting a bill through the state legislature to provide $3.6 million for a three-year study.

As with much environmental epidemiology, things take longer than planned. In this case, the DPH always controlled the funds and on several occasions withheld them and tried to cancel part of the study. In 2002, all state funding for Silent Spring ended when the government incurred a massive budget cut, but the institute has been able to transition successfully to support from private donors, foundations, and federal grants.

The Silent Spring study is still ongoing. The first phase, completed in 1997, had three main accomplishments: (1) development of the GIS system that enables researchers to map diagnosed women with environmental data; (2) historical study of pesticide use and drinking-water quality on the Cape; and (3) establishment of new field methods to study environmental estrogens and discovery of estrogenic activity in groundwater that supplies drinking water.[101] In the second phase, Silent Spring Institute is pursuing the testing of a very large array of chemicals. The classes of chemicals it has measured include many that are considered emerging contaminants—chemicals for which few monitoring data are currently available. In particular, the data include the first reported measurements in residential environments for more than thirty of the compounds the study detected—in many cases among the first reported measurements anywhere. Key findings include identification of diethyl phthalate, dibutyl phthalate, 4-nonylphenol, 4-tert-butyl phenol, o-phenyl phenol, and methyl paraben as important indoor air contaminants; measurements of brominated flame retardants in air and dust samples; and twenty-three different pesticides in air and twenty-seven in dust.[102]

Phase two also involves interviewing 2,100 Cape women to identify individual risk factors (family history, age at birth of first child, amount of physical exercise, exposure to toxic products), testing to determine possible estrogen mimics in households and estimating environmental exposure for each interviewee, based on GIS mapping.

The Massachusetts Breast Cancer Coalition is more focused on environmental causation and the related political economy critique of industry than is its national organization, the NBCC. It has gotten the NBCC to add support for environmental research to its mission and agenda. The Massachusetts Breast Cancer Coalition has considerable clout because of its role in pursuing environmental factors through its support of Silent Spring Institute and in this way has been key to the EBCM. Although the coalition and Silent Spring Institute are now separate entities, many people are involved in both, and they view themselves as sister organizations.

The Women's Community Cancer Project is the other main movement group in Massachusetts, though it focuses on a feminist analysis of all cancers. It has engaged in public protest and education, including a large-scale mural in Harvard Square. The more radical nature of this activism includes a critique of the "cancer industry" (the ACS, the NCI, leading researchers, major cancer research centers, leading philanthropists and foundations), a focus on corporate responsibility, and organization against direct marketing of cancer drugs. A December 2000 statement by the project noted the organization's dissatisfaction with "the unwillingness of policymakers to place public health ahead of profits . . . the role of pharmaceutical companies in determining the course of cancer research, and . . . the race and gender discrimination found throughout the medical establishment."

The Women's Community Cancer Project is not involved in research, but it shares with the Massachusetts Breast Cancer Coalition and Silent Spring Institute an involvement in the Alliance for a Healthy Tomorrow, which grew out of the Precautionary Principle Project. That work includes forums presented to college, community, environmental, and health groups in order to educate laypeople, government representatives and staff, and scientists about the precautionary principle, including its relationship to diseases other than breast cancer. These forums are a part of the tactics employed to push the precautionary principle in state and local policy, in local and statewide action to restrict pesticide spraying, and in the legislative bills Safer Alternatives to Toxics Program, Safer Alternatives to Mercury Products, and Safer Cleaning Products.

SAN FRANCISCO BAY AREA Activism in the Bay Area began in the early 1990s, with a 1992 victory in which state cancer research review committees were mandated to include laypeople. This initial work was significantly enhanced when the Northern California Cancer Center published statistics showing that the Bay Area has some of the highest rates of breast cancer in the world. Breast cancer groups responded strongly to the publication and dissemination of these data, and the movement flourished in three distinctive geographic areas: San Francisco, Marin County just north of San Francisco, and the East Bay. The major activist groups located in San Francisco—Breast Cancer Action and the Breast Cancer Fund—are focused primarily on the nation as a whole. The Breast Cancer Fund raises money to support its mission to change detection, treatment, and prevention policy so that it recognizes and focuses on environmental causes. Breast Cancer Action focuses on education and direct action, such as demonstrations, coalition work with other political organizations, and critiquing "pink" marketing campaigns (corporate advertising that promises contributions to breast cancer research and treatment, but yields little and often promotes potentially harmful products). They are active in opposing pharmaceutical companies' direct marketing of cancer drugs. Bay View/Hunter's Point Community Advocates is also located in San Francisco, but it is focused only on the South San Francisco neighborhoods after which it is named. In Marin, there are Marin County Breast Cancer Watch and more recently Marin Cancer Project, both of which focus exclusively on the local area. The Women's Cancer Resource Center is the main organization in the East Bay.

Bay Area activists have been enriched by the perspectives of the Bay View/Hunters Point Community Advocates, an environmental justice organization that grew out of the earlier Southeast Alliance for Environmental Justice and that has been concerned with abnormally high rates of breast cancer among black women in their area, especially because such women typically have a lower incidence than white women. These activists have criticized the lack of racial identity recognition in epidemiological research and the institutional failings of risk assessment.[103] Even though this group is not specifically a breast cancer organization, it has been involved with groups such as Breast Cancer Action, thus contributing a more multiracial perspective and helping to broaden breast cancer activism to include environmental justice.

Marin County Breast Cancer Watch, which includes the precautionary principle as part of its mission statement, was one of the first Bay Area breast cancer organizations and is the only one that has a citizen-science component to it. The organization originally received funding to conduct a study of adolescent risk factors for breast cancer and housed several researchers part time in order to do so. It gained state funding to begin another study more specifically focused on potential environmental causes. In 2003, it received one of the seven-year NIEHS Breast Cancer Centers of Excellence grants, in collaboration with the University of California at San Francisco, Kaiser Permanente Health Care, Lawrence Berkeley National Laboratory, and the San Francisco and Marin County health departments.

These organizations' efforts were preceded by other cancer activism. Founded in 1986, the Women's Cancer Resource Center was one of the first women's cancer organizations and was a precursor to the movement in the Bay Area. It set precedents in providing services to women with cancer and now links its activism with the environmental movement as well. Its mission is to educate the public about cancer and to support women with cancer by providing educational workshops, a resource library, support groups, and a hotline.

Bay Area groups have different levels of focus. Whereas the movement in San Francisco contains both locally oriented and nationally oriented groups, Marin activism is only local. The different local and national levels of focus as well as the slightly differentiated geographic locations of these groups often keep them from competing for the same resources. These factors and the general lack of large-scale government research funding in most of the Bay Area, excluding Marin, has also reduced competition between groups.

Unlike Long Island groups, Bay Area activists are less concerned with their relationships with the state and do not seek state funds, though they use such connections to generate public attention and were very active in pushing San Francisco to apply the precautionary principle. They have been very involved in the Campaign for Safe Cosmetics to get endocrine-disrupting chemicals out of personal care products. Most recently, they have been engaged in legislation to mandate biomonitoring, which became law in 2006. In 1992, Breast Cancer Action and the Breast Cancer Fund joined with the National Alliance of Breast Cancer Organizations to push for lay involvement on research proposals funded by the California Breast Cancer Research Program. This lay involvement proved to be a model for the well-regarded citizen participation in the DOD's breast cancer research program and has included some projects on environmental factors. Breast Cancer Fund has also advocated for stricter standards on chemicals regulated under the federal Food Quality Protection Act. They work closely with the NIEHS and the Food and Drug Administration (FDA), the latter collaboration largely around clinical trial protocols. Breast Cancer Action has collaborated with the San Francisco DPH Breast and Cervical Cancer Services and with various community and health organizations to organize town hall meetings on breast cancer in low-income communities of color. Representative Nancy Pelosi (D-Calif.) has been very supportive of the organizations by pushing for funding for breast cancer research and by recognizing these organizations' efforts in her public speeches. Other local representatives have also offered symbolic support by creating days of recognition for the organizations' events. Marin activists at times faced challenges by the local community, such as being prohibited from holding meetings in public buildings, but have now received a large amount of government funding to research breast cancer incidence in the area.

Although other local groups also focus on issues related to universal access to care, activists in the Bay Area were the only ones who brought it up on a regular basis. A recent change in federal policy attests to the effectiveness of this activism. Federal funding supported access to mammography, but it did not provide aid to women who needed financial assistance to have breast cancer treatment. Bay Area activists pushed for new legislation to provide that support.

Bay Area activists have been distinct in their focus on public action and awareness, in addition to their fund-raising for research. They rely largely on street theater, protests, and graphic displays such as a series of posters and billboards of young women with mastectomies, shown nude from the waist up. Although such radicalism may initially have caused some difficulty in gaining public legitimacy and support, the strong social movement climate in the region has offered a basis for acceptance, and the breast cancer movement has ultimately gained that legitimacy. The Marin group's radical nature is manifest in its holding of events that emphasize environmental causes of breast cancer, highlight community-based participatory research, and feature speakers from environmental justice and environmental health activist groups such as West County Toxics Coalition and Commonweal. Despite these radical tactics, Marin Breast Cancer Watch has received little resistance within the community, perhaps in part because of the widespread concern over the alarmingly high rates of breast cancer in the area. The Breast Cancer Fund also raises money to fund breast cancer research that might not otherwise be funded by mainstream sources.

Activists in this locale have been intricately connected to environmental activism. In 1995, Breast Cancer Action became a founding member of the Toxics Link Coalition and helped organize the first Cancer Industry Tour. One head of a breast cancer group in the Bay Area described her organization as having a stronger connection with environmental groups than with general breast cancer advocacy. There is even crossover in leadership between breast cancer and environmental organizations. As I discuss in chapter 6 on precautionary approaches, environmental breast cancer groups have been key founders of organizations based on the precautionary principle and have been centrally involved with the Campaign for Safe Cosmetics. Some Bay Area groups have had a strong environmental component from the beginning. For other groups, a transition occurred as they became more concerned about environmental causation, as noted by an activist in Breast Cancer Action: "We adopted a policy in 1998 about not taking money from known environmental polluters or pharmaceutical companies or anybody profiting on cancer. In the process of doing that, we realized that our alliances were not going to be with breast cancer organizations. . . . In conjunction with that, we adopted a policy about when we would join a coalition, and we can't join a coalition which violates that policy. It makes it a little tricky, but not in environmental work. So we began seeing our alliances more in the environmental field and in the broader women's health field."

EBCM activists in all locales are concerned with cooptation of the broader breast cancer movement. Ferguson and Kasper note that "breast cancer is a growth industry, with Race for the Cure runs and walks in most major U.S. cities, the constant entry of new drugs and clinical trials to combat the disease, whole bookshelves devoted to the topic at local bookstores, and a cornucopia of tee-shirts, hats, pins, and pink ribbons."[104] Zones believes that many corporations are getting good public relations out of donations to breast cancer efforts and have even named breast cancer a "dream cause . . . [because it is] the feminist issue without the politics."[105] These activists argue that companies are exploiting breast cancer for profit, with the collusion of some mainstream breast cancer groups. As EBCM activists see the situation, corporations spend little on their own campaigns, relying on the public to put up the money by buying products or by signing up for road races and walks. Overhead costs run as high as 40 percent, which is deemed more than acceptable by advocates of proper charity procedures, and corporations boast how "pink" campaigns increase sales and profits.[106] This "subsidized philanthropy," as King terms it, diffuses or neutralizes political unrest at the same time as it enriches corporations.[107] This attitude and related actions exemplify the split between the EBCM and the general breast cancer movement in the Bay Area. Environmental breast cancer activists have sometimes stood at the finish line of the Race for the Cure and handed out their own literature about environmental causation. Similarly, they were willing to join in a coalition to support the Breast Cancer Summit led by the mayor of San Francisco, yet also picketed outside during the gathering to criticize lack of attention to environmental causation.[108] In this process, Myhre argues, they were acting as "outsiders within."[109] Although the movement has benefited from general breast cancer activism, in some ways there is a tenuous relationship between the two. Most recently, groups such as Breast Cancer Action have started the "Think Before You Pink" coalition of organizations that point out the flaws and biases in the many mainstream, corporate pink campaigns.

Radical activists in the Bay Area also point to pharmaceutical companies, medical supply manufacturers, genetic testing firms, and mammography producers that are making a great deal of money, often by scaring women. These more radical activists point with dismay to the methods these actors employ, such as the ad that a gene testing company put in the *Hartford Jewish*

Ledger (because Ashkenazi Jewish women are more likely to have the BRCA-1 mutation) that said, "If you carry damaged breast cancer genes and you live long enough, you are almost guaranteed to develop breast cancer." These activists simultaneously claim that pharmaceutical producers and doctors often downplay side effects, including uterine cancer from Tamoxifen. This conception of companies exploiting the breast cancer situation for profit is supported by the fact that even the NCI notes that there are two to three times the number of mammography machines necessary. Activists, as a part of their critique of the larger breast cancer movement, point out that Imperial Chemical Industries, the parent company of Zeneca (later merged with Astra to become Astra-Zeneca), initiated Breast Cancer Awareness Month and retains authority to approve or disapprove all printed materials used by participating groups. Tying the political economy critique together with their belief in environmental causation, activists note that Astra-Zeneca at the same time produces pesticides and herbicides that may be causing breast cancer.[110] It goes without saying that Astra-Zeneca gains much publicity for its widely sold drug Tamoxifen as a result of its sponsorship for Breast Cancer Awareness Month.

THE FRAMEWORK FOR ACTIVISM—PERSONAL, SCIENTIFIC, AND POLITICAL

This overview of movement activism in the three major locales demonstrates the importance of the personal experience as a political and scientific tool for the movement as a whole. Activists' knowledge of self-exposure to environmental toxics led them first to forming a new movement and second to challenging scientific perceptions of breast cancer.

Environmental breast cancer activists work out of a framework centered in the classic feminist stance "the personal is political." This stance is the foundation of activist initiative in that the politicization of individual women's experiences transforms a personal problem into a public agenda. One activist described the initiation of her organization in these terms: "I refused to accept the fact that one out of every eleven women will die of breast cancer. I remember reading Gloria Steinem said the day the revolution starts is when one person looks back in someone's eyes and says no, I refuse to budge, I refuse to accept that."

Activists believe clinical and support services are crucial, but not enough. For example, the traditional breast cancer movement response is to provide support groups that give emotional sustenance, whereas the EBCM emphasizes a model of political action that facilitates empowerment in the political realm, replacing the primary focus on individual responsibility with a focus on the responsibility of corporations, government institutions, and science. This individual empowerment was demonstrated by a scientist who had experienced breast cancer: "I was diagnosed with breast cancer, and soon after that I went to a rally . . . where they asked women [who had breast cancer] to come down out of the stands and stand in a chalked area that was supposedly to be the shape of Massachusetts, and so many people came out of the stands, I couldn't believe it. Young women, old women, black women, white women. And I was flabbergasted. . . . When I went to medical school breast cancer was rare, . . . I don't have a family history. I suddenly have breast cancer. I'm looking around at hundreds of women who have breast cancer, and this isn't rare."

All interviewees, both scientists and activists, reported a strong belief that activism and social movements play a critical role in advancing research, educating the public, and changing policy about environmental causes of breast cancer. One scientist said that activists "have been the catalysts. They're the drivers. The environmental movement, the women's movement, specifically the breast cancer activist movement, they're the ones that are driving the establishment of research about environmental factors."

Activists feel that social movements have been especially instrumental to the development of research on environmental causation. One long-time activist emphasized the difference between the activist and scientific perspective: "Changes that have to happen around breast cancer and all the things that influence it are going to come from the ground up. They're not going to come from the top down. . . . If you think about it, activists were the first to argue for this, and some of the scientific community has been sort of dragged along."

Activists also have clearly articulated their wish to advance changes in attitudes and practices outside of the breast cancer issue, thus demonstrating the new public paradigm they propound: "I think the role of social movements is to push the 'establishment' . . . whether it's the scientific community, federal government, drug companies, or whoever. . . . We as activists are pushing people to broaden the scope of what they consider to be environmental problems."

For movement activists, empowerment and politicization often involve utilizing knowledge of individual environmental exposures to inform science. In this process, both social movement actors and scientists are boundary actors crossing into a new world to advance social movement claims. So the classic feminist perspective of "the personal is political" is here altered to "the personal is scientific, and the scientific is political." This claim articulates the collective politicized illness experience also seen among asthma activists. Indeed, the broader breast cancer movement has been one of the most powerful manifestations of collective illness experience, and the EBCM has made it into a highly politicized form of that experience.

LINKS WITH OTHER SOCIAL MOVEMENTS

The "personal is political" framework brings together multiple social movement perspectives in order to highlight the political and economic issues in breast cancer causation, research, and treatment; to challenge traditional epidemiological models; and to draw attention to unrecognized geographic disease patterns. The EBCM has taken shape as a merging of actors and agendas from the environmental movement and the breast cancer advocacy movement. Moreover, it is rooted in the larger feminist, women's health, and AIDS movements. This hybridization of movement activity is representative of the boundary crossing that movement actors have accomplished without diminishing the distinct nature of their own agendas.

MAINSTREAM BREAST CANCER MOVEMENT By the time the EBCM had begun in the early 1990s, the national breast cancer movement had already achieved major successes in several different venues. In fact, the first breast cancer movement activity took place in 1952 in the form of Reach for Recovery, a self-help organization.[111] Breast cancer activism took further shape in the 1970s and 1980s as a part of the general women's health movement. In each one of the locales I analyze, general breast cancer organizations preceded the development of the breast cancer movement, and EBCM activists often emerged from the general movement, including the NBCC. The EBCM's origins in the general breast cancer movement has provided simultaneously a strong base from which to recruit, but also some tension between general movement actors over the EBCM's more political environmentalism.

The overall environmental movement provided legitimacy for environmental causation theories and offered an environmental activist network that could provide support. As environmental activists became more sophisticated in focusing on health effects, they provided a logical basis for connecting to breast cancer activism. The awareness created by a national grassroots environmental movement provided a basis from which the public could understand potential environmental causation and from which government could recognize a constituency of voters. There is also some overlap in leadership and agendas among local environmental organizations and EBCM actors.

WOMEN'S HEALTH MOVEMENT The EBCM also emerged from the developing women's health movement, which has worked to increase funding for research on women's illnesses, to educate women about their bodies, to include women in clinical trials, to criticize the medicalization of women's experiences, and to fight for self-determination of health care options.[112] The EBCM benefited from the feminist movement in terms of increased public attention to women's issues, existing mobilized women's groups, useful tactics for activism, and ideological foundations. Feminism has been instrumental in developing a critique of the lack of women's involvement in science and the lack of women's experience in knowledge production. Although most movement activists do not articulate a direct link between the EBCM and the broader women's movement, such a connection is often very clear through EBCM groups' philosophical stance and mission statements. For instance, one founder of a movement group, also involved in a network of environmental organizations, called breast cancer a "wedge issue" for larger gender equality and environmental issues: "We want to share leadership, and bringing ourselves [women] into balance in this issue will help to bring the whole planet into balance. I do think that environmental health, using breast cancer as a wedge issue towards that larger issue, is the issue of the millennium." An activist in another locale extended this idea: "We're just a breast cancer organization; you'd think we could stay focused. But breast cancer touches on every aspect of health, the economy, politics, so we get to do it all."

These quotes, coming from different sides of the country, exemplify the overarching approach that unites these "cultures of action" into one movement. This approach specifies breast cancer as an issue with broad ramifications for gender equality and larger sociopolitical issues. It is quite similar to the way that environmental justice groups have dealt with asthma.

AIDS MOVEMENT Although some EBCM activists generally did not overtly connect their activities to feminism, they did note how the AIDS movement influenced their methods: "Originally, the founders of the organization were following the lead . . . of the NBCC and the AIDS activists in drawing attention to the [breast cancer] issue. Such methods included public protest, as well as promoting the inclusion of women with breast cancer in advising research."

The AIDS movement has been an exemplary model for activists in the movement in terms of citizen-science alliances, public education, and social protest. The methods utilized by AIDS activists, such as working with researchers to change scientific study, provided an example for future breast cancer activism. In addition, EBCM activists saw how AIDS activists utilized public-protest tactics to subvert the social perception of AIDS victims and to generate more funds for research and activist involvement in research.[113] There were some cross-memberships between AIDS Coalition to Unleash Power and Breast Cancer Action, and several wealthy AIDS activists were key donors to Bay Area breast cancer groups. Involvement in public education and social protest varies across locales, but popular epidemiology and activist involvement in research permeates the entire EBCM. This hybridization of movements exemplifies the activities and methods of actors in a boundary movement. For example, some activists are based mainly in the breast cancer or environmental movement but also join in EBCM activities.

ENVIRONMENTAL JUSTICE MOVEMENT We are witnessing the early stages of linkages between environmental breast cancer activism and environmental justice activism. Both share a focus on toxics and are critical of chemical companies, oil refineries, and other polluters; they push the larger environmental movement in a more inclusive and radical direction; they view racism, sexism, and class inequality as central to environmental concerns; they focus on health inequalities in disease, death, and access to care; and they use the precautionary principle as a central theme.

For environmental justice activists, breast cancer is a disease with disproportionate burden. For a long time, incidence among whites was somewhat higher than incidence among blacks; that gap is narrowing. At the same time, however, mortality among blacks has always been higher, and that gap is getting larger. Moreover, environmental justice groups have had important experience in listening to their communities' expressed needs for health-related work, as demonstrated in the rapid way they took on asthma (see chapter 3). Women of color also have less access to mammograms and to breast cancer treatment, and environmental justice activists have a long-standing concern with such inequalities in the health industry.

EBCM groups understand that many of the contaminants that concern them also concern environmental justice activists. These EBCM groups have already extended their outlook to a wide range of political and economic areas, such as organizing to ban or restrict chemicals, to affect corporate production processes, to criticize health voluntaries and mainstream breast cancer organizations for failing to examine environmental factors, and to seek local, state, and federal regulatory policies to prevent disease. In the process, they have allied with many other progressive political organizations, and an alliance with the environmental justice movement makes sense.

Some of these linkages are expressly political, centered around organizing activities. Breast Cancer Action expanded its outreach, especially on the

Toxic Tour—a neighborhood tour of polluting industries coordinated by a coalition of environmental and cancer organizations since 1994. Breast Cancer Action worked with important area environmental and social justice organizations, Urban Habit and Bay View/Hunters Point Community Advocates, as part of the Bay Area Working Group on the Precautionary Principle, which successfully established a city ordinance on implementing the precautionary principle.

Other linkages are based more on providing health services and health education, but through community-led, environmental justice–sponsored clinics. One of the nation's best-known environmental justice groups, WE ACT, held the first-ever conference on breast cancer, the environment, and environmental justice on October 23, 2004, in New York in which they linked some traditional environmental justice issues with the concerns of many of their constituents facing breast cancer. Prominent members from the breast cancer movement attended, providing a strong perspective on environmental factors. Sistahs United, an African American–led breast cancer group in Tillery, North Carolina, is affiliated with the environmental justice group Concerned Citizens of Tillery, which has pioneered the struggle to protect people from hog farm contamination.

In 2004, Silent Spring Institute, along with Communities for a Better Environment, one of the nation's oldest environmental justice groups, and researchers at Brown University (myself and Rachel Morello-Frosch), received an NIEHS environmental justice grant, the first for the institute that included breast cancer. In 2005, a complementary grant from the National Science Foundation contributed more support for this project. Prompted by the emerging collaborations mentioned earlier, we embarked on a research agenda to extend the innovative work Silent Spring had been doing on environmental factors in breast cancer to a community populated by women of color and already the site of environmental justice organizing around the many oil refineries in the surrounding area. In the process of doing this work, we and our community partners assist residents of these neighborhoods to build their own capacity to research and challenge environmental contaminants.

We have conducted home environmental exposure assessments, collecting air and dust samples to assess indoor levels of pollutants, especially endocrine disrupters, that are potentially linked to breast cancer, reproductive disorders, neurological development disorders, and other health outcomes. Data collection, analysis, community education, and organizational linkages are ongoing in two locations—Cape Cod, Massachusetts, and Richmond, California, which is composed largely of people of color and impacted by industrial facilities, especially oil refineries. Study results will be shared both as aggregate information presented through community meetings, news media, and Web sites, and as individual report-back information to study participants. Using report-back approaches, we seek to maximize understanding of exposure data and their limitations, and to address the ethical issues of ensuring community and individual autonomy, right to know, and ultimately the right to act on scientific information by reducing exposures. We will publicize these ethical methods for reporting back study results to communities and individual study participants and will produce a handbook for general use in reporting results of community exposure studies. We will also engage in community-based outreach, including public meetings, Web sites, and publications. Last, we will test an intervention to reduce household pollutant levels.

The emerging alliance between environmental justice and EBCM activism offers a good example of the power of a growing environmental health and justice movement, to which I return in chapter 7.

CITIZEN-SCIENCE ALLIANCES

The citizen-science alliance serves a key role, especially in New York and Massachusetts, by supporting activism, changing attitudes and practices of scientists and activists, and providing a new value structure to some research. I have already described throughout this chapter many elements of the citizen-science alliance that Silent Spring Institute so well exemplifies. Although this approach is less widely utilized in the Bay Area, activists there support the idea of such alliances, and the involvement of Marin County Breast Cancer Watch in the new NIEHS Center of Excellence shows that they are moving in this direction. In this case, being asked to work with scientists on boards advising such research has not only provided the opportunity to influence research construction and potentially improve the knowledge base for environmental causation, but also helped initiate independent research projects directed by activist involvement.

Involvement in science was implemented differently in each area, largely depending on the structure mandated by the funding source. On Long Island, activist involvement was not a core element of NCI's support. In Massachusetts, the Silent Spring Institute explicitly seeks citizen-science collaboration. The nationally focused San Francisco organizations are not involved in research projects, though Marin activists are, and a few individual activists have been involved in research in a variety of ways. Overall, the Bay Area groups can be more radical in their approach because their freedom from state or federal research funding gives them more leeway.

In all locales, activism connects science to public concerns to promote public awareness. For example, activists on Long Island motivated the government to hold public hearings to inform communities about research studies it was funding. Massachusetts Breast Cancer Coalition's participation in the Precautionary Principle Project (subsequently the Alliance for a Healthy Tomorrow) is another way in which Massachusetts activists have joined science and the public in pursuing research and policy.

In addition to ramifications for research, these collaborations have altered scientists' and activists' perceptions. Activists described drastic changes in their expectations about what science could prove in terms of environmental causation and their perception of the length of time necessary to conduct research and of the processes involved. Activists without prior experience with researchers found that their feelings toward scientists changed from fear and anxiety to mutual respect and comfort. One woman who had worked on high-level government advisory panels recounted, "The thing that I came away with that was most surprising was how much the scientists and the MDs have come to value the activist perspective on these panels . . . not only just putting a face on the statistics, but also . . . they appreciate that . . . you ask the questions: Why is this relevant, who cares?" This quote exemplifies a theme running throughout interviews regarding the alliances: joining together was a transformative experience for both parties in which each learned to appreciate the other's very different perspective, despite initial reservations.

Although this process changed perspectives about science and researchers, activists raised concerns about becoming too accustomed to the process of research and potentially losing their activist perspective. However, such co-optation was not common. One older activist who had played a critical role in gaining access to the research table and who was one of the first to be involved expressed this trepidation: "I remember sitting at an NBCC board meeting years ago, and we were talking about, I don't remember the issue, and we were voting against it. It sounded like individuals really wanted it, but we decided it wasn't a good idea, and I said go back to our beginnings even a couple of years ago. I said that had this question been raised then, we would have been amongst the loudest voices saying we want this, and now we know it's not a good idea. [I] said, 'Oh we've become educated. But we never want to lose our fire.'"

We might expect scientists to worry that their legitimacy would be threatened if they worked with laypeople. Scientists involved in alliances in the EBCM described some initial fear, but general mutual respect between themselves and activists. They often greatly appreciated the input activists supplied, in addition to their efforts to bring research projects into existence. Of course, the scientists interviewed may be a biased sample of individuals who were previously open to activist involvement in research, even though they came from a wide diversity of backgrounds and experience.

Apprehension and prejudice on the part of both activists and scientists were the most serious obstacles to alliances. Still, most laywomen felt that they were respected and that their work was worthwhile and transformative. These initial fears and prejudices were verbalized by one activist whose experience demonstrated how this transformation took place: "I think there is a respect for bright people and I think the assumption often is that activists are going to be hysterical women. And I think once most scientists realize that we're not hysterical women, they find themselves, you know, intrigued. And they might come to the table with a lot of prejudices and worries, but I have rarely seen it continue to be a problem." Activists were savvy about how traditional science separated knowledge from lay people and were convinced this obstacle could be overcome: "I know that professionals like to build up their own vocabulary and their own aura and arena so that it can't be pierced by anyone because, after all, they paid their dues. But it's nothing more than communication and relationships. Everything is understandable and pierceable. The more people are willing to share their expertise, realizing that the others are not trying to replace their expertise or their judgment, I think the more effective we can be."

Another activist expressed the same sentiment of collaboration while still emphasizing each party's distinct role: "I just think that it's important that we all work together. Activists have a seat at the table not because we want to be scientists, but because we need to push along some of the work that should have been done long ago." In the words of one scientist involved in the Long Island Breast Cancer Study Project, "And so there get to be fads in research . . . dogmas . . . of what an acceptable area of research is, and so one thing I find very helpful about having a diverse group of advocates is that it can sometimes help to . . . loosen up whatever the current dogma is and get people . . . out of whatever dogma trench they're in."

These extracts also show how activist involvement contributes to a new value structure for science. In the case of breast cancer research, the value system of science has traditionally not questioned the biomedical model. Consequently, research has typically been focused on individual factors and responsibility in treatment and prevention. The EBCM subverts the status quo by taking a broader view of responsibility. In doing so, it has been able to create a new value system for science by pushing scientists to examine why they ask certain questions and not others, why they use certain methodologies, and, more important, how their research affects women with breast cancer. An activist provided an emblematic example of such questioning. A lay member of a review panel listened to a scientist give an extremely high score to a proposal simply on the basis of excellent methodology. When the laywoman pressed the scientist on the actual relevance of the project, the scientist realized the error and revised the score downward.

Another activist reported her experience in affecting a scientist:

> Yeah, I remember on one of the first panels I did for the California breast cancer research program, there was this guy who was very, very bright, and he was very invested in treatment research. And we had this whole long argument where he said to me, "Look, I can see that science is going to make breast cancer a chronic disease. Women could get diagnosed at thirty with it and live till they're sixty. And wouldn't that be great." And I said no, absolutely, that I didn't think it would be a success if breast cancer became like diabetes. That the toll on families and children and marriages and women would be overwhelming and that breast cancer was becoming a very common disease. And that if we just threw money at treatment, it would be like diabetes, and that would be a disaster. And that we had as physicians an obligation and responsibility to look at issues of cause and prevention and not just treatment, the way we were all trained and brainwashed in medical school. And so the next year he and I were on the panel together, and I forget what the topic was, a paper, a research proposal that we were arguing about, and he was arguing for it because he thought it would be good research into cause and eventually prevention. And for some reason I was arguing against it, and he said in absolute exasperation to me, "For God sakes, you won last year! You won, you convinced me breast cancer shouldn't be a chronic disease. Why are you fighting me this year?" [chuckle] And I think it was just that I didn't particularly like the research or something. But it was really very funny, and I realized that I had this enormous impact on him.

Alliances between scientists and laypeople are valuable not only for the EBCM, but also for many other arenas of environmental health. They represent one of the most significant legacies of this movement.

THE EBCM AS A BEACON FOR A BROADER ENVIRONMENTAL HEALTH MOVEMENT

The many connections mentioned in the section "Links with Other Movements" demonstrate the EBCM's powerful position. The points in the "10-Point Plan for Reducing the Risk of Breast Cancer and Ultimately Ending the Epidemic," as put forth in the fourth edition of the influential document

State of the Evidence: What Is the Connection Between the Environment and Breast Cancer? a joint effort of the Breast Cancer Fund and Breast Cancer Action,[114] demonstrate this position and actually lead to a very broad platform for environmental health and justice:

1. Establish environmental health-tracking programs at state and federal levels
2. Practice healthy purchasing through precautionary laws at all governmental levels
3. Protect workers from hazardous exposures, including toxics use reduction
4. Educate the public about radiation and reduce exposures
5. Hold corporations accountable for hazardous practices
6. Offer local, state, and federal incentives for green [environmentally safe and sustainable] production
7. Strengthen right-to-know legislation and public participation in decisions about toxic exposures
8. Enforce existing environmental protection laws
9. Require greater transparency in funding of scientific and medical training, research, and publications
10. Create a comprehensive chemicals policy based on the precautionary principle

This program is very far reaching, and the fact that it stems from two leading groups in the EBCM says a great deal about that movement. As the report concludes, "Waiting for absolute proof only means more needless suffering and loss of life. It is in our power to change the course we are on. It is time to act on the evidence."

SUMMARY

I have examined the disputes over environmental factors in breast cancer research by identifying three paradigm challenges posed by activists and some scientists to the broader scientific community: (1) to move debates about causation "upstream" in order to address the cause; (2) to shift emphasis from individual factors and responsibility to modifiable societal factors beyond an individual's control; and (3) to allow direct lay involvement in research, which may raise new questions as well as change how questions are approached, the methods used, and even the standards of proof. This three-dimensional model (figure 2.2) has enabled us to examine controversies in different arenas of the scientific process: doing scientific research, interpreting science, and acting on science.

According to the activists and researchers I interviewed, individual risk factors do not satisfactorily explain breast cancer etiology; therefore, the search for alternative variables, including environmental factors, and alternative methods remains an important agenda point for many breast cancer researchers and activists. Key studies of breast cancer and the environment have yielded equivocal results, showing that this area of research is a work in progress rather than a fait accompli. The lack of generally acceptable answers to the question "What causes breast cancer?" has prompted people to look at the question differently. Environmental breast cancer research has opened up pathways for more primary prevention research, such as endocrine disrupter research; has created a resultant focus on precautionary principle policy; and has identified novel methodologies with which to explore breast cancer etiology further. This alternative pathway orients research upstream to seek out preventable causes, includes lay input, and considers community-level and population-level factors. In addition, proponents of environmental causation theories have brought attention to interaction effects, both those involving two or more chemicals and those involving chemicals plus other factors (e.g., tobacco, alcohol, viruses).

When environmental factors are the subject of inquiry, discussions of science are very contested because they lead to potential challenges to the underlying status quo production, distribution, disposal, and regulatory practices of our society. For example, for axis 1 in figure 2.2—upstream prevention versus downstream action—corporations and government regulators will be unlikely to accept removal of many chemicals from circulation or to bear the exorbitant costs of extracting these chemicals from the environment. For axis 2—individual-level versus community-level research—corporate interests will likely oppose a research agenda that targets their products rather than individual susceptibilities. For axis 3—degree of lay participation—corporate and governmental actors will likely oppose strong lay participation because such involvement often stems from an approach that values community over corporate rights and that supports greater democratic participation in science, which in turn invites challenges on the other axes.

Innovative methods of knowledge construction that account for lay perspectives are pivotal in creating new means to study and understand environmental causes of disease. Although scientists involved with breast cancer activists will continue to uphold scientific rigor, the activist agenda and its reliance on the precautionary principle have also facilitated new methodologies and proposed a new set of norms for standards of proof. These struggles over approaches to breast cancer research demonstrate the many impediments to our comprehension of potential environmental causes of illness. My discussion of "impediments" does not imply that there can be an obstacle-free science; rather, I argue that the DEP obstructs science from considering environmental factors, which are a component of the broader science.

The paradigm challenges to traditional breast cancer research also occur around environmental health more broadly. By understanding what impedes construction of a new body of knowledge about environmental factors in breast cancer, we can be helpful as science and the public turn attention to environmental factors in other diseases, including asthma, Parkinson's, and autism. The three-axes conceptual framework can be applied to other areas of research on environmental health, and the doing-interpreting-acting approach to the scientific process can be applied more generally to other scientific issues where scientists and citizens advocate for "a lab of their own."

The EBCM has influenced the production of science and policy by beginning to develop a new public paradigm that informs the scientific study of women's health issues and the creation of environmental policy. This introduction of values into a supposedly value-neutral world has the potential to create a paradigm shift away from the disconnection between environment and health, as well as away from the methods through which health is generally researched. The potential shift in focus from individual to corporate and government responsibility faces much resistance, but may create a more broad-based political shift in accountability. Despite existing challenges to movement perspectives, the EBCM has become a unique example of a successful movement that combines health, environment, and women's movement ideologies. Its multilevel successes include the significant amount of public awareness generated, the amount of research performed, the dialogue created in the scientific community, and the development of citizen-science alliances.

This social movement developed as a boundary movement that crossed the lines of a number of other social movements. Although culture variations make each locale of the EBCM somewhat different, there are sufficient similarities in all locales to justify viewing the EBCM as a coherent national movement. It exemplifies social movement activity at the intersection of health and the environment, which is quite possibly the largest new arena of social movements, encompassing activism around lead poisoning, asthma, toxic wastes, nuclear power, food additives, biotechnology and genetically modified organisms, toxics reduction, and the precautionary principle. Future social movements, especially those involving health and environment issues, will likely arise in similar boundary-crossing fashion. This means we must be more flexible in defining a social movement and its development. Social movements have an enormous range of facets: political challenges to governmental authority, scientific challenges to medicine and science, organizational challenges to health voluntaries and related organizations, contention for power and authority among various organizations, cultural manifestations, and activities to increase public awareness.

The various issues of fluidity I have addressed force us to see the flexible nature of the actors within social movements. These movements are not discrete entities, but rather changeable social phenomena. By understanding these issues, we discover more complex approaches to discussing not just social movements, but social change more broadly. A better understanding of boundary movements such as the EBCM has the potential to reshape the priorities of science and medicine, while altering the delivery of medical care.

Boundary movements such as the EBCM are centrally concerned with democratic participation in science and in social policy involving that science. The EBCM shows ordinary citizens' ability to learn science, to demand a seat at the table for reviewing research proposals, to collaborate in research enterprises, and to press for broad extensions of citizens' right to participate in all aspects of society. Thus, it continues a long line of social movements that have sought to expand democracy, but it does so in a qualitatively new fashion that holds great promise for empowering citizens, while at the same time helping to improve scientific practice and the health of the public as well as to reshape the priorities of science and medicine.

|3|

ASTHMA, ENVIRONMENTAL FACTORS, AND ENVIRONMENTAL JUSTICE

IN THIS chapter, I show how laypeople, scientists, government, and community activist groups have shaped the discovery and treatment of the asthma epidemic. I discuss the larger debate among scientists, government regulators, and corporate interests concerning the link between air pollution and the exacerbation of asthma, and I examine the ways in which asthma activists, in particular those involved in the environmental justice movement, have used the framework of the environmental causation of asthma to address the underlying structural inequalities. Environmental justice groups have played an especially prominent role in their framing of the asthma epidemic as an issue of injustice within low-income and minority communities. Looking at the experience of two of these groups, ACE and WE ACT, this chapter traces how affected communities have used the broad base of scientific research on air particulates in combination with the lived experience of asthma in order to shape a collective illness identity and to mobilize this collective power to address social and environmental hazards in their communities. I use transportation issues as an example of an underlying cause that a range of community actors, including teachers, families, and friends, have sought to address. I focus specifically on how groups have used educational programs to support community empowerment and to allow students to develop leadership skills, give names to the problems they face in their everyday lives, and address these problems from a structural standpoint. I then discuss how these groups and scientists have worked together in the form of academic-community collaborations to gain community-level data and political legitimacy, and I use asthma as example of how communities and scientists can employ these mutually beneficial collaborations to address contested environmental diseases.

A NEW EPIDEMIC

Neighborhood residents, joined by environmental justice activists, have been prominent in calling attention to rising asthma rates, and health researchers, clinicians, and others have widely noted this lay identification. In addition, because asthma is primarily a childhood disease, members of affected communities, including teachers and clinic workers, have been crucial in recognizing high rates of asthma in children. Teachers have noticed a large number of their students using inhalers, coupled with high absence rates for children with asthma. Numerous asthma advocates and clinicians note that in some urban areas it is possible to find almost half a classroom suffering from asthma.[1]

Asthma rates have risen so much in the United States that medical and public-health professionals invariably speak of asthma as a new epidemic. This epidemic is one of today's most important public-health challenges because of the rapid increase in the number of people involved. It is also one of the most controversial because the environmental factors that cause or trigger asthma are connected to powerful financial interests and to highly significant and contested regulatory politics.

The number of individuals with asthma in the United States grew 75 percent between 1980 and 1994, making asthma one of the few diseases whose incidence and death rates continue to increase, despite broad-based medical advances in control and treatment.[2] In 2004, approximately 8.2 percent of adults reported having asthma.[3] In 2005, the CDC reported a 12.5 percent lifetime asthma prevalence among adults.[4] In 2003, it was estimated that 29.8 million Americans, including 6.4 million children, had been diagnosed with asthma.[5] From 1980 to 1998, visits to doctors' offices and hospital outpatient clinics because of asthma more than doubled. There are 1.9 million emergency room visits each year due to asthma.[6] Hospitalizations because of asthma increased from 408,000 in 1980 to 511,000 in 1995, before starting to drop to 451,000 admissions in 2001.[7] The estimated combined cost of medical expenditures and indirect costs to society from asthma is greater than $16 billion a year.[8]

Mortality rates from asthma, which have risen across populations in the past twenty-five years, are higher for blacks than for whites. Although the later 1990s brought a slight reduction in mortality in the overall population, this reduction did not hold for blacks.[9] Minority populations with the highest asthma prevalence in 2001 reported even higher prevalence in 2002.[10] Death rates are highest among older age groups, although the mortality rates for young adults experienced the greatest increase.[11] In 2002, the number of people who died of asthma was 4,261, and if current trends continue, deaths from asthma are expected to double to more than 10,000 a year by 2020, out of a projected 29 million people with asthma. What makes these figures even more notable is the contrast with the marked declines seen in asthma rates from 1960 to 1977.[12]

The incidence and prevalence of asthma have increased dramatically in all segments of the U.S. population over the past twenty years, but the effects on children are especially worrisome. Asthma is the number one chronic childhood disease, affecting approximately 6.3 million children annually and being the main cause of pediatric hospital admissions.[13] In 2003, the National Survey of Children's Health found that 8.1 percent of children younger than seventeen had been affected by asthma-related health issues within the past twelve months of their lives.[14] The most rapid increase in asthma from 1980 to 1994 was among children younger than the age of four, a rise of 160 percent. Among children five to fourteen, asthma sufferers increased by 74 percent.[15]

In many low-income urban areas, especially minority communities, rates are significantly higher than the national average. National prevalence of childhood asthma in 1997 overall was 7.8 percent for one- to six-year-olds and 13.6 percent for six- to sixteen-year-olds, but black children and poor children had a higher relative risk, being up to 20 percent more likely than average to have asthma—that is, with rates for black and poor children of 9.36 and 16.32 percent for those two age ranges.[16] Although the prevalence disparity is not so huge by race and class, the severity disparity is, as measured by hospitalization, with rates in New York City five times higher for minorities than for nonminorities.[17] In 2003, asthma was responsible for 12.8 million missed school days, making it the number one cause of school absenteeism nationally.[18] Asthma hospitalization and mortality rates correlate negatively with family income; the highest rates are reported in areas with the lowest incomes.[19] A Mt. Sinai School of Medicine study in New York City found that hospitalization rates were twenty-one times higher in poorer and minority neighborhoods than in wealthier neighborhoods with a whiter population. In neighborhoods in East Harlem, the hospitalization rate was twenty-two per thousand per year, whereas in high-income areas of Manhattan the rate was zero.[20] In 2003, the prevalence rate of asthma in blacks was 28 percent higher than the rate in whites.[21] Asthma now ranks in the top ten diseases and conditions that cause major limitations.[22]

As mentioned in *Healthy People 2000* and *Healthy People 2010*, the NIH's reports outlining national health objectives, asthma, as a result of its increased prevalence and social costs, has become a public-health priority for medical and public-health professionals and institutions, as well as for people affected by it.[23] Medical and public-health providers have expanded treatment and prevention efforts. Some environmental groups and community activists have also made asthma a key focus. In several locales, coalitions have formed that include activist groups, academic research centers, health providers, public-health professionals, and even local and state public-health agencies.

Among the public-health, scientific, governmental, and activist communities involved, there is widespread agreement that we do not know the causes of asthma, but that a variety of environmental factors trigger asthma attacks. The DEP focuses mainly on factors that are part of the indoor environment, including animal dander, cockroach infestation, tobacco smoke, mold, and other allergens. The primary outdoor factor is air pollution in the form of particulate matter, especially $PM_{2.5}$, particulate matter under 2.5 microns in diameter, that penetrate deep into the lungs and are linked to asthma and other chronic respiratory symptoms, especially among children and the elderly.

Health voluntary organizations (sometimes called health charities: private organizations that raise funds, seek policy changes, and in other ways advocate for health issues, especially surrounding a particular disease) play a key role as well, especially through medical and epidemiological research from the American Lung Association (ALA) and the ACS, including studies of air pollution and morbidity and mortality. Although the ACS is not interested in asthma, its interest in lung cancer enables relevant research support on air pollution. The ALA specifically takes a strong stance on environmental factors in disease, especially air pollution.[24] Ordinary citizens have increasingly come to identify pollution as a prime factor in the occurrence of asthma; as one parent of an asthmatic child noted, "We have the port [West Long Beach, California]. We have the refinery. We have the trucks. I think that's why this area is so bad [for asthma rates]."[25] Most significantly for this book, environmental justice activists have found asthma to be a core concern for their communities, and, using lay observation in combination with activist organizations, they have consequently turned asthma into a major organizing issue.

Scientists, activists, health voluntary organizations, and some government agencies and officials have worked through a variety of collaborations and alliances to tackle these issues using a variety of strategies and tactics, including demonstrations and other direct action, lobbying and advocacy, lay-professional collaborative research, support for increased scientific research, and community empowerment through community organization. An astonishing level of community organizing has occurred in the form of school-based programs, community clinic programs, and novel public-health initiatives in urban areas, especially among environmental justice groups. One example of a successful asthma collaborative, the Boston Urban Asthma Coalition, has united dozens of health, government, school, and environmental organizations and agencies. This collaboration enables them to support a variety of existing asthma programs and to develop new approaches. The coalition provides testimony on public housing, school nurse programs, access to asthma medications through Medicaid, and the switch to lower-emissions buses. It trains parents in grassroots organizing and leadership development to help treat their children's asthma. It works to create an affordable healthy home model and advocates for the use of healthy homes standards in all publicly funded buildings. It also works to improve indoor air quality in schools by advocating for healthy schools standards and by pushing the "right to know" about hazards and other potential health effects from the activities relating to construction and renovation of schools. This effort includes protecting custodians from the hazards of toxic cleaning materials, which are also linked to childhood and adult asthma.[26]

Federal funds support programs that cross typical boundaries of public health. For instance, in 2000 the Boston Public Health Commission received $1.9 million from HUD for asthma prevention. The program was designed to recruit five hundred households for home inspections dealing with dust, mold, and mildew; provide vacuum cleaners and air conditioners; replace ceiling tiles; and do other necessary home improvements.[27]

The asthma epidemic is also a source of much contention among scientists as well as between government regulators and corporate interests. This debate around the link between air pollution and the causation and exacerbation of asthma exists within a larger controversy regarding the science and politics of regulating air pollution in general. The scientists who study air quality are influenced both by regulatory disputes and the growing activism around environmental causes of asthma. Many progressive air quality researchers have increasingly become more accessible to environmental justice activists involved with asthma issues, in part because of the growing public debates around the potential environmental links between air pollution and asthma. Their accessibility is a logical step because environmental justice activists are among their most appreciative public audiences and because the activists, more than others, understand the need to combat governmental inaction and corporate practices.

I focus here on two of these activist groups, ACE and WE ACT. As shown in their approach, they, as well as similar organizations, are involved in a multipartner, multiorganizational, and intersectoral policy format. The multipartner and multiorganizational nature is important in that a diverse range of actors are involved, from community-based organizations to government health agencies to foundations. Further, a large number of organizations cross those boundaries, creating a large critical mass. The intersectoral approach means that people and groups address health problems as being caused by or exacerbated by or remediable by multiple spheres: health, transportation, sanitation, environmental policy and enforcement, housing, planning, zoning, and economic development. In that complex mixture, it is clear that regulatory disputes, scientific controversy, and environmental justice activism concerning asthma are inseparable.

SCIENCE AND CONTROVERSY OVER ENVIRONMENTAL FACTORS IN ASTHMA

A STRONG AND GROWING SCIENCE BASE

Activist groups' framing of environmental factors with regard to asthma is situated in the medical and scientific controversy over the broader link between air pollution and health. Through their work on addressing environmental causes of asthma, these groups have access to a long history of air particulate research. Despite the high quality of this work, the research on environmental causes and triggers of asthma within the larger literature on air quality is contested. Activist groups often understand that the scientific process is conservative and time-consuming, and that there will always be political opposition to doing good science in a contested area. They understand that air pollution research has had a long history of controversy. Mary Amdur, a leading researcher in the 1950s, was fired from her university position for pursuing health effects research. Lester Lave and Eugene Seskin faced corporate attempts to discredit their work in the 1970s; and Douglas Dockery, Jack Spengler, and Richard Wilson faced similar challenges in the 1990s.[28]

Traditional scientific research on air pollution tends to focus on health endpoints other than on asthma because of several difficulties in measuring an increase in the incidence of asthma.[29] First, medical and public-health science linking air pollution to asthma is hindered by a lack of good surveillance data. This lack of information feeds the debate around whether air pollution causes new cases of asthma or simply exacerbates current cases, which is the second major difficulty in measuring the increase in asthma cases. Fortunately, recent research does provide more specific data on asthma and air pollution, as discussed later.

To do a scientifically robust study of disease development often requires epidemiologists to conduct a cohort study rather than a cross-sectional study that examines data at one time. A cohort study would involve following children, preferably from birth, and looking at risk factors for asthma. This research process is very expensive, time-consuming, and complicated. In contrast to studies of *etiology*, studies of *exacerbation* can be done using time-series methods or other shorter-term approaches. Overall, if you look at the literature on chronic illnesses related to air pollution, such as chronic bronchitis, you will see that it is rather limited.

The difficulty in conducting environmental health research on asthma also leads to debate over the reality of the asthma epidemic; some skeptics argue that asthma has become a more common diagnosis, but one without adequate substantiation. In a period when many mysterious diseases are increasingly in the public light, asthma can appear to some skeptics as part of a large assortment of diseases and conditions that people are attributing to environmental causation.

Finally, there are clear links between sources of indoor air pollution (e.g., mold, animal dander, and dust mites) and asthma exacerbation, but less evidence linking sources of outdoor air pollution and asthma; again, such research has grown recently, though it lags far behind research on air pollution and other health outcomes. Public-health interventions are more likely to address indoor sources of pollution that are relatively easy to identify and remediate by people and by public-health practitioners. Because these indoor sources fall under the individual's responsibility, they can be addressed quickly and directly, in contrast to longer-range and broader-based projects to reduce pollution. The combination of these difficulties has lead to a contested body of research on outdoor sources of air pollution and asthma. There is, however, a much more developed body of research on the relationship between particulate matter and other health outcomes.

The research on particulate matter and health has been at the center of much environmental health research, but it is also the focus of scientific controversy and regulatory debate. Despite excellent research, corporate and governmental actors continue to fight against well-established findings. Current scientific thinking on particulates stems from the history of major air contamination crises. The association between air pollution and asthma was established as early as 1948, when 88 percent of asthmatics in Donora, Pennsylvania, had asthmatic exacerbations during a severe pollution episode.[30] Thousands of deaths during the "London fog" of 1952, when particles reached 2,800 grams per cubic meter (g/m^3), provided additional evidence.[31] This figure appears to be a 1,400-fold increase over the EPA's current recommended limit, though in fact it is less because of differences in averaging time and measurement. The exceptional nature of these episodes led scientists to believe there was a high threshold value of 500 g/m^3 for particulate-induced health effects. More recent research has shown that a 10 g/m³ increase in particulate matter of less than 2.5 micrometers (μm), or $PM_{2.5}$, from mobile sources accounts for a 3.4 percent increase in daily mortality.[32] Douglas Dockery's Six Cities Study provided powerful longitudinal evidence of particulates' responsibility for pulmonary morbidity and mortality, and in 1997 the EPA established a new $PM_{2.5}$ standard, but did not implement it because of a court order and other forms of opposition to stricter standards.[33]

Particulate researchers presented strong evidence in support of the new $PM_{2.5}$ standard, in part to counter attempts to halt enforcement through criticisms of scientific methods and negative findings by corporate scientists. Responding to many attacks, the particulate researchers showed that hospitalization from cardiovascular problems did not change when controlling for pollutants other than small particles (sulfur dioxide, carbon monoxide, and ozone). Those scientists established that the dose-response curve is linear (i.e., the higher the exposure, the greater the health effect), without a threshold.[34] In response to objections that it would be impossible to know which types of particulates to regulate, researchers disaggregated sources of twenty-five substances and determined that all were associated with mortality.[35]

Critics of the proposed new standard also claimed that there was no plausible biological mechanism by which particulates could kill people. Joel Schwartz, one of the leading air quality researchers, responded by pointing out that arrhythmia and myocardial infarction were the major causes of sudden death and, further, that low heart rate variability was the major risk factor in arrhythmia. Using heart monitors, researchers found that people's heart rate variability was reduced as their personal intake of PM_{10} increased, and in a later study they found the same with $PM_{2.5}$.[36] Critics then claimed the data were a spurious result of "harvesting" people who were about to die. If this were so, the excess of death in very sick people would be followed by a period of fewer deaths because the sickest would already have died. However, studies found the association between suspended particulates and increased mortality to be inconsistent with the harvesting hypothesis.[37] Other cohort studies demonstrated that long-term exposure to fine particulates is associated with increased mortality from cardiopulmonary diseases.[38] More refined research on mechanisms of injury by particulates has shown additional health effects: lung inflammation, increased neutrophils in blood, vascular injury, and direct toxicity to heart and lung tissue.[39] In total, particulates are estimated to account for more than 100,000 deaths annually—more than the deaths from breast cancer, prostate cancer, and AIDS combined.[40]

The most recent research has continued to support the Dockery Six Cities Study, the ACS study, and other key work. Pope and colleagues extended the well-respected ACS study from its initial seven years to sixteen years, using the most recent particulate matter–monitoring data for 1999–2000.[41] In correlating the health status of 500,000 people with the particular matter rate, they found that for every 10 micrograms per cubic meter ($\mu g/m^3$) there was a 6 percent increase in cardiopulmonary disease and an 8 percent increase in lung cancer. For the same 10 $\mu g/m^3$ increase, mortality for various cardiovascular disease increased from 8 percent to 18 percent.[42] One study using ACS data in the Los Angeles area found that the chronic health effects associated with exposure to $PM_{2.5}$ were even greater than previously reported.[43] A prospective cohort study on 5,000 people in the Netherlands from 1986 to 1994 found deaths increased among people living near roads with high pollution levels.[44]

The major studies involving particulates look at morbidity and mortality from severe lung and cardiovascular disease.[45] There is less research on asthma, but the literature on air particulates as a trigger of asthma attacks and hospitalizations has grown. Pope reported on a natural experiment in which when a Utah steel mill closed due to a strike, a dramatic drop in PM_{10} and in asthma exacerbation occurred.[46] Another natural experiment showed that a 23 percent reduction in auto use due to traffic control at the 1996 Atlanta Summer Olympics led to a 42 percent reduction on asthma claims reported to the state's Medicaid program.[47] Schwartz and colleagues found that increases in PM_{10} led to more emergency room visits in Seattle.[48] Averaging the health effects chronicled in studies in various countries, Dockery and Pope found that there was an average 3 percent increase in asthma attacks per 10 $\mu g/m^3$ of $PM_{2.5}$.[49] Several studies with samples as high as tens of thousand of children show increases in asthma and other respiratory diseases based on proximity to trucking routes.[50] Children living within two hundred meters of roads with heavy traffic are more likely to be hospitalized for asthma.[51]

A recent study, using new data from the Southern California Children's Health Study, followed 3,535 children for up to five years. The scientists found that children playing three or more sports in areas with high ozone concentrations were more than 3.3 times more likely to develop asthma than other children; this result did not hold in areas with lower ozone concentrations. Simply spending more than the median time outdoors made children 1.4 times more likely to develop asthma; again, this result did not occur in low ozone areas. Even though particulate matter was not implicated, these data provide evidence showing that some form of air pollution can actually cause asthma rather than just trigger attacks.[52]

Landrigan and colleagues constituted an expert Delphi panel to estimate the *environmentally attributable fraction* of acute asthma exacerbations.[53] Their definition of *environmental* excluded pets, insects, mold, and indoor tobacco smoke. They provided the panel of national experts with the latest research findings, then led them through a day-long meeting to refine their estimates, and came out with the conclusion that the environmentally attributable fraction of asthma was 30 percent of the total numbers of cases.

Using the Regulatory Modeling System for Aerosols and Acid Deposition, the research methodology employed by the EPA, the Clean Air Task Force published in 2000 a very widely reported study on the health effects of power plants in which it was able to show that the emissions—primarily sulfur dioxide and nitrous oxide—from these plants killed 30,000 people yearly. If emissions were reduced in accordance with legislative proposals current at that time, 18,000 of those deaths would be avoided. Those emissions also cause 630,000 asthma attacks a year, of which 336,000 would be avoided through those improved standards.[54] The task force also published in 2005 the summary results of a commissioned study conducted by Abt Associates indicating the effects of diesel vehicles' emissions on health, which found that fine particulate pollution from diesel shortened the lives of nearly 21,000 people per year and caused nearly 3,000 deaths. Furthermore, as shown in figure 3.1, results indicated that if diesel emissions were reduced by 50 percent by the

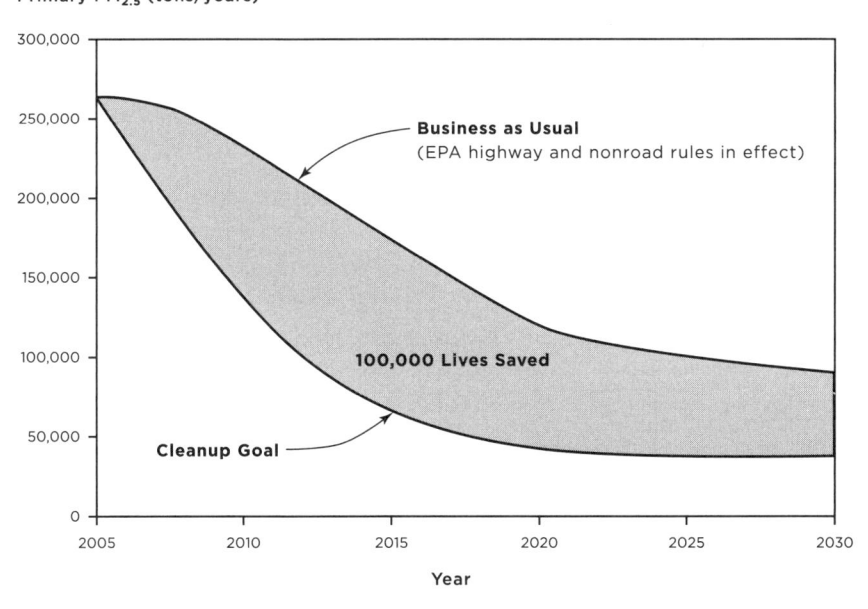

FIGURE 3.1 An aggressive program to reduce diesel emissions could save about 100,000 lives between now and the year 2030.

Source: Clean Air Task Force, *Diesel and Health in America: The Lingering Threat.*

year 2010, 75 percent by the year 2015, and 85 percent by the year 2020, the lives of nearly 100,000 individuals would be saved by the year 2030.[55]

Given the strength of the scientific literature linking air pollution to various health outcomes, many scientists and activists extrapolate the relationship between particulate matter and health to the relationship between air pollution and asthma. The search for a correlation between particulates and asthma rates, however, is plagued by what one researcher terms the "confounding curves": air quality has improved over time, whereas asthma rates have risen. There are various potential explanations for this seeming lack of association. First, indoor air quality may have declined as a result of tightening many buildings with insulation and weatherproofing in the 1970s, following the energy crisis. Second, particular matter is the only criteria pollutant (the EPA uses six "criteria pollutants" to measure air quality) without a chemical identity; that is, it is a combination of difference substances characterized by particle size rather than by chemical makeup. These fine particles contain varying mixtures of substances, and that variability may relate to variability in health effects.[56] Indeed, although overall air quality has improved, the amount of some pollutants has increased, and these pollutants may be responsible for the increase in asthma rates. Third, better air quality is measured in terms of $PM_{2.5}$, and, as I mention later, smaller ultrafine particles are not taken into account in regulatory procedures, yet there is increasing attention to their potential health effects.

For these reasons, environmental justice groups representing urban communities have begun to enter the scientific arena to support and conduct research on outdoor sources of air pollution and their link to the increasing rates of asthma, in terms of both new cases and the exacerbation of existing cases. For example, Community Action Against Asthma—a group representing a collaboration among scientists, health agencies, and community activist organizations—investigated the effects of environmental exposures on children in the Detroit area using a particulate matter exposure assessment in order to understand components of indoor and outdoor air quality associated with asthma.[57] To determine the range of environments and factors contributing to childhood asthma, this assessment included data collected on the community level, inside "microenvironments" such as schools and homes, as well as personal monitoring data collected from twenty children with asthma. In using community-based participatory research, members of the activist groups and community were involved in all stages of the research and thus were able to continue their activist roles in community organizations while producing results useful in the scientific arena. Similarly, El Puente, an environmental justice organization based in Brooklyn, New York, engaged in participatory research regarding asthma triggers in the communities, while continuing to pursue its environmental justice goals.[58] One of the key elements of this research was a series of health surveys designed and administrated by residents in order to use local knowledge to understand better the potential triggers within the community. The surveys were incorporated into the organization's environmental health and justice program and were thought to be successful in making residents more aware of actions they could take to address the poor environmental health conditions they had noticed.

In order to generate new science and regulations to reduce air pollution, activists are faced with a political controversy similar to the debate around whether global warming actually exists, a situation in which a few professional "skeptics" join corporate interests and antiregulatory elements of the government to fight against the broadly accepted science.[59] The scientific controversy surrounding the research fuels the political debate and is often used as a mechanism by corporate interests—such as coal mining, power production, and the trucking industry—to hinder regulations that would address the increasing rates of asthma and other diseases related to air pollution.[60]

CORPORATE RESISTANCE AND REGULATORY GRIDLOCK

The EPA is the primary government agency involved in the social discovery process, based largely on its regulatory mandate to analyze the National Ambient Air Quality Standards (NAAQS) every five years and to act on the basis of appropriate scientific knowledge and without considering costs. In the 1970s, the EPA regulated large particles, known as *total suspended particulates*, ranging in size from PM_{25} to PM_{45}, with a 260 g/m³ twenty-four-hour average. In 1987, the EPA moved to the smaller PM_{10} particulates, with a standard of 50 g/m³ annual average or 150 g/m³ twenty-four-hour average. In 1994, the ALA sued the EPA for failing to reset standards every five years, as required. In 1995–1996, EPA scientists estimated that 15,000 deaths were attributable to particulates each year, and the agency sought to add $PM_{2.5}$ standards of 15 g/m³ annual average or 50 g/m³ twenty-four-hour average. In November 1996, the EPA presented its proposals to tighten the NAAQS based on a review of 185 health studies of ozone and 86 studies of PM_{10}. The EPA's actions frightened many industries because they expected the higher standards to necessitate the purchase of expensive new equipment.[61]

Corporate resistance is very prominent in the contestation around the regulation of air pollution, particularly against the EPA's efforts to reduce the amount of particulate matter in the air by restricting allowable emissions releases and increasing enforcement. Opponents of the more stringent measures have argued both against the costs and against the credibility of what they call "junk science" supported by "hidden data" unavailable to challengers.[62] The EPA's response to the "junk science" charge was that there had been many peer reviews and that these data were very strong. In terms of costs, the EPA stuck to its belief that the law called for action based on its congressional mandate "to protect the public health, with an adequate margin of safety." According to the EPA, even with an accounting, the cost of 15,000 deaths was higher than the critics' estimate of the costs of compliance. Finally, responding to the charge of "hidden data," the EPA asked for a reanalysis of the Six Cities data, as well as of the ACS study by the Health Effects Institute, an organization jointly funded by the EPA and industry.[63] A team of investigators assembled by the Health Effects Institute were able to reconstruct the findings originally produced by the two studies and to confirm the findings that daily deaths increased with growth in measured levels of particulate air pollution.[64] Other reanalyses have been produced with similar results.[65] The Electric Power Research Institute sponsored a reevaluation of Schwartz's work linking particulate matter to daily mortality, which also confirmed the original findings.[66] As a result of the Health Effect Institute's reevaluation of the Harvard Six Cities Study, the EPA decided on July 16, 1997, to retain the PM_{10} standards and to adopt the $PM_{2.5}$ standards at 15 g/m^3 annual average or 65 g/m^3 twenty-four-hour average. But in a compromise, it agreed to monitor $PM_{2.5}$ particulates, do more research, and not implement the actual $PM_{2.5}$ standards until the next five-year review in 2002. The outcome of this conflict was that thousands of monitors are now in place around the country, and more research is strengthening the science.[67] The EPA subsequently kept the annual standard, but lowered the twenty-four-hour standard from 65 to 35 g/m^3.

When the rules were signed into law in 1997, a series of lawsuits followed. On May 14, 1999, the U.S. Court of Appeals for the District of Columbia, in its decision on one of these suits (*American Trucking Association, Inc., et al., v. United States Environmental Protection Agency*), agreed with the arguments that the EPA had construed sections 108 and 109 of the Clean Air Act so loosely as to render them unconstitutional delegations of legislative power. The court went on to note in another case that "although the factors EPA uses in determining the degree of public health concern associated with different levels of ozone and PM are reasonable, EPA appears to have articulated no 'intelligible principle' to channel its application of these factors; nor is one apparent from the statute."[68]

On May 20, 1999, EPA administrator Carol Browner commented on the court's decision in her testimony before the Senate Environment and Public Works Committee:

The court did not say that the air standards were based on bad science. Nor did the court find that the process that produced the standards were insufficient. In fact, the court explicitly recognized the strong scientific and public health rationale for tougher air quality protections. These proposals were based on a total of more than 250 of the best and most current scientific studies on ozone and PM—comprising thousands of pages—all of them published, peer-reviewed, fully-debated and thoroughly analyzed by the independent scientific committee. We stand by these standards. We stand by the science. We stand by the process. By finding this section of the Clean Air Act unconstitutional, the court has struck at the heart and soul of this legislation that is so critical to the health of our families. . . . To lose the ability to implement the new health standards for smog and soot would mean 125 million Americans—including 35 million children—are going to be placed at risk.[69]

This exchange between the EPA and the Court of Appeals reflects what has emerged as the most contested area of air pollution and public-health impacts. Industries argue that meeting the new standards require expenditures that far exceed potential savings in health care expenditures. But eventually the corporate argument on cost-benefit analysis lost out, and the Supreme Court rejected the Trucking Association's position in early 2001.[70]

Although cost-benefit arguments are clearly at stake here, there is also a dispute on the difficulty of setting a standard for a pollutant that does not have a well-defined threshold. Industry has been concerned that the EPA was being arbitrary and capricious because it is difficult to pick a standard in light of there being a continuous dose-response relationship.

Another major political outcome of the research on particulates stemmed directly from the Harvard Six Cities Study. Industry representatives asked the EPA to provide the primary data, but Dockery refused to share them for fear that research subjects' confidentiality would be compromised. Senator Richard Shelby (R-Ala.) added a provision to a 1999 appropriations bill (P.L. 105–277) that instructed the Office of Management and Budget to revise Circular A-110 to ensure that all data produced with a federal grant that is cited in federal policies or regulations would be made available to the public through a Freedom of Information Act request (Circular A-110, which has been in effect since 1976, governs the management of federal grants by institutions of higher education, hospitals, and other nonprofit agencies). The provision was attached to this appropriations bill without any public hearings, scrutiny, or comments. As one leading air quality scientist indicated, the Shelby Amendment on Data Access would have significant impact on future research: "I can no longer say to you 'I want you to participate in my study and keep what you tell me confidential.' I can say to you 'I can keep your data confidential unless it's used by the EPA in rulemaking. Then anybody can request access to your data.' . . . [It] changes the way we're going to have to do business in the future. I cannot give people the same assurances of confidentiality that I could in the past."

On January 6, 2001, Representative George Brown (D-Calif.), with support of science and social science organizations that feared impingement on independent research, introduced H.R. 88 to repeal this provision. At the present time, however, the Shelby Amendment on Data Access is in force. In fact, industry is asking that it be applied to past studies as well.

The George W. Bush administration is also working to hinder research on environmental causes of disease. In 2002, the scientific advisory committees within the Department of Health and Human Services (HHS) underwent major reorganization. Volunteer members of a committee assessing the effects of environmental chemicals on human health were almost entirely replaced. Some of the replacement experts have links to industries that produce the chemicals linked to health problems.[71]

As soon as Bush was elected in 2000, Vice President Richard Cheney presided over a series of secret gatherings of energy companies and federal officials with the goal to ease regulation and give energy companies more political power.[72] The new regime was adamantly opposed to research showing the problems of global warming, including the contribution from fossil fuels.[73] Cheney consistently refused to make public the records of the energy task force, which became a major issue in national politics and the 2004 election campaign. Given the influence of the energy industry over the Bush administration, it is logical that this administration would fight against the legacy of air pollution controls. Perhaps the most significant phenomenon was the alteration to New Source Review (NSR), whereby power plants had to add emission controls when upgrading their facilities. The new policy instead allowed plant owners to consider major renovations as merely routine maintenance, thus exempting them from pollution controls even if the renovations resulted in a significant increase in emissions.[74] Critics have argued that this alteration encourages industry to avoid regulation by continuing to use older, heavily polluting sources rather than build new, cleaner sources. One effect of the gutting of the NSR program was that in November 2003 the EPA dropped more than fifty investigations of power plant violations of the Clean Air Act for failure to upgrade pollution controls.[75] Because the new regulations on NSR were to take effect in December 2003, the EPA decided that past violations were moot.

State governments found themselves allied with national environmental organizations such as the Sierra Club and the Natural Resources Defense

Council in planning lawsuits to force more regulation, and industry advocates brought suits arguing that the NSR regulations applied to even narrower ranges of sources.[76] In December 2003, just days before the new rule was to be implemented, the D.C. Circuit Court issued a stay of the EPA's new rule based on the fact that petitioners against the rule had shown irreparable harm.[77] Due to the controversy surrounding the changes made to the NSR programs, Congress called for an independent assessment of the EPA's revisions by the National Research Council, which was to issue its final report in 2006. The EPA held a reconsideration of the rule in 2004; however, to date, it has released only preliminary rule-making proposals. Industry proponents and environmental organizations have continued to bring forth court cases concerning both the scope of the NSR program and the EPA's enforcement of the NSR regulations. In one such case in 2005, *U.S. v. Duke Energy,* the Fourth Circuit rejected the EPA's argument that Duke should be subject to Clean Air Act permits following the twenty-nine modifications made at its eight power plants. However, the Supreme Court announced that it would hear the lawsuit in early 2006, leading some to argue that the EPA should halt rule making while the Supreme Court hears the case.[78] The EPA has said that it will issue a final revised rule by January 2007, but the continued controversy surrounding NSR reforms have made it difficult for enforcement cases to go forward. Added to these difficulties, in 2004 the Bush administration forced the EPA to remove from its air quality report any mention of greenhouse gases.[79]

These challenges to the science behind federal regulations have made linking air pollution to asthma a difficult problem for researchers to overcome. Because air pollution politics is so central to the Bush administration's deep connections with oil and energy, the potential to reduce environmental asthma triggers is sharply limited.

A growing body of research shows that the much smaller ultrafine particulates ($PM_{0.1}$) trigger asthma and cause cardiac disease, respiratory disease, cell death, and brain damage. Ultrafines, a form of nanoparticles, are $PM_{0.1}$, far smaller than the fine particles of $PM_{2.5}$, which are not even regulated. Considerable recent research shows the dangers of these substances, some of which are produced in combustion processes along with fine and other particulates, and others of which are deliberately produced as part of a growing nanotechnology development. Their small volume means they have a larger surface-to-volume ratio and therefore can carry many more substances on their surface. This is precisely why pharmaceutical companies are interested in them as vehicles for delivering drugs. They are also being used widely in suntan lotion, auto parts, stain-repellant fabrics, and self-cleaning windows, and they represent a vast new area of technological development. But these particles are potentially dangerous because their increased surface-to-volume ratio allows more toxins to adhere to them. Also, they can pass deeper into the lungs than $PM_{2.5}$; enter mitochondria, where they can destroy them and hence their cells; and directly enter the brain. Ultrafines with metals and hydrocarbons increase the protein endothelin in the blood, which increases blood pressure. Size alone is a problem: even carbon nanoparticles with no pollutants attached cause lung damage in animals. Government support for nanotechnology is increasing the amount of nanoparticles being produced, and these nanoparticles will contribute to the ambient number of particulates available for ultrafine pollution.[80] As a result of the lack of necessary regulation as well as the increase in asthma prevalence, activist groups must continue to reframe the nature of the asthma epidemic in order to address its underlying causes.

ENVIRONMENTAL JUSTICE AND ASTHMA ACTIVISM

Many asthma activists have adopted an environmental justice framework that links a discourse of rights and social justice, drawn from the legacy of the civil rights movement, together with mainstream environmental values.[81] By drawing on an environmental justice approach, asthma activists frame the unequal burden of asthma in their communities in terms of disproportionate environmental exposures, which they place within in an overarching social justice framework.

Because of that social justice framework, I do not examine how asthma shapes *individual* illness experience, but rather how activist groups create a *collective* identity around the experience of asthma. It is collective in that people share a common bond of illness and a group sense of injustice about the causes of that illness. The organizations I have observed collectively frame asthma as an environmental justice issue and therefore transform the personal experience of illness into a collective identity aimed at discovering and eliminating the social and environmental causes of asthma—and, by extension, of other diseases of poverty. When people view asthma as related to both air pollution and the living conditions of poor neighborhoods, they reconstruct asthma narratives differently than the *narrative reconstruction* that occurs with other chronic illnesses. Narrative reconstruction describes how people produce a coherent explanation for their illness, thus providing a way to repair the rupture that chronic disease has caused in their relationship with the world.[82] Because asthma is increasingly framed in the language of air pollution and environmental justice, the disparities between those who suffer asthma and those who do not are translated into the rhetoric of illness experience. In the case of asthma, the illness narrative is broader than the typical narrative. The latter type of narrative typically incorporates perceived causes and effects of the disease with personal perception, work, family, relationships, and schooling. But asthma activists also include the political economic framework surrounding the causation and exacerbation of asthma and the political perspectives that situate asthma in terms of housing, transportation, neighborhood development, the general economy, and government regulations. This broader focus on the social and economic factors shaping the illness experience of asthma is reflected in the goals of one ACE organizer: "I think we have to look at how is it that our society has created such disparate environments for people to live in—from the kind of housing you have, to the kind of school you go to, to the kind of vehicle you ride in, to the kind of air that is outside your door. . . . I think that there's huge changes that are way beyond individual lifestyle changes that we need to look at about production of synthetic chemicals that may play a role, or about the way we're designing and building our cities, towns, and whatnot." Similarly, an ACE staff person noted:

> It feels like [asthma] has been taken out slightly from the context of everything else that is happening to people. . . . And I don't think that that is the way that community groups really approach asthma. They see it in the way environment justice sees it, defining the environment where people live, work, play, and breathe. And so it's the underlying conditions of poverty and social injustice that are contributing to all these things. And no matter whose fault it is, you can't just get rid of cockroaches and expect asthma to go away. For that matter, you can't just put in better buses and expect asthma to go away. It's all got to be approached in a social justice framework.

This kind of perspective enables people with asthma to place responsibility in part on social structural forces. This use of an environmental justice lens has led to the development of a public paradigm for asthma, focusing on the role of outdoor air pollution, in which there has been an extraordinary amount of public participation in defining the extent of the problem, its causes, and approaches for prevention and treatment. And this paradigm, like that for breast cancer, is accompanied by a "personal is political" framework because the collective illness experience so readily translates to a political challenge.

Asthma is a complex, multifactorial disease for which it is difficult to point the finger at a single proximal cause. There are distal causes, such as poverty, which are associated with indoor allergens, diesel exposures, inadequate health care, and poor diet. These causes link asthma more strongly with social justice campaigns than with source-specific control policies. Because some communities are disproportionately burdened with asthma and also with sources of air pollution, residents and activist make an on-the-ground linkage, even if there is not a strong causal chain in the scientific research base.

By viewing present conditions through a historical political-economic approach, groups such as ACE and WE ACT link the full range of social structural inequalities—including housing, transportation, employment, municipal services, land use, and education—to asthma prevalence. Environmental justice organizations have found asthma to be an excellent vehicle for addressing social and environmental hazards in their communities. Sources of air pollution such as automobile traffic, industrial emissions, bus depots, and waste storage facilities are predominately located in poor and minority communities. Environmental justice organizations, with their focus on eliminating the disproportionate burden of environmental and social hazards in such communities, are capable of identifying the increased rates of asthma in these communities and of linking those rates to the environmental hazards. Through teachers, families, and friends, community activists have been made aware of the sharp increase in childhood asthma rates, and they have rallied behind childhood asthma as a key example of the disproportionate burden of environmental health risks borne by their communities. They are able to use asthma as a framework around which they can rally and begin to implement programs that emphasize popular education and community empowerment practices. This community realization has been most pronounced in poor urban areas with large minority populations. Environmental justice ideas circulating in many of these communities have prompted people to consider asthma as another component of unequal environmental burden. For example, a strikingly large proportion of NIEHS's environmental justice grants for community collaborations focus on asthma. These grants have supported programs emphasizing community outreach and education campaigns such as the Community Asthma Prevention Program in Philadelphia, as well as projects such as the San Diego–based Clean Air for Barrio Children's Health program, aimed at collaborating with scientists to collect and analyze data on regional air quality in order to profile better the environmental hazard at hand.[83]

The activist groups ACE and WE ACT are two of the leading organizations in environmental organizing around asthma. ACE began in 1993 as a community-based nonprofit organization with the intent of providing both support for community leaders focusing on environmental justice issues and legal resources for other community groups in the area. Based in the Roxbury-Dorchester area of Boston, ACE has since become nationally recognized for its work. The organization is also active on issues involving brownfield cleanups (brownfields are contaminated sites where reuse or redevelopment is complicated by the presence of those contaminants), solid waste facilities, incinerators, and parking garages. It provides help from lawyers, public-health professionals, and environmental consultants to other groups in Boston and, through the Massachusetts Environmental Justice Network, to groups around the state.

WE ACT was founded by community members in 1988 in response to environmental threats to the community created by the mismanagement of the North River Sewage Treatment Plant and the construction of a sixth bus depot in northern Manhattan. Community residents suffering from respiratory ailments were able to rally around these issues, and WE ACT quickly evolved into an environmental justice organization with the goal of working to improve environmental protection and public health in the predominately African American and Latino communities of northern Manhattan.

As environmental justice organizations, ACE and WE ACT are influenced by the legacy of the civil rights movement and community organizing. As a WE ACT staffer noted, "I think one of the things that the environmental justice movement has learned from the civil rights movement is that . . . environmental degradation in communities like a place like northern Manhattan is linked in many ways to people's construct of race and quality of life and racism. You've got to fight that on many different fronts." This model of organizing illustrates how ACE, primarily an environmental and social justice activist organization, began to focus its efforts on asthma, a mainstream health issue. This decision came about through more than a year of talking with community residents. As one organizer recalled, ACE had expected at first to focus on issues such as vacant lots, but residents quickly established asthma as the number one priority: "And the one thing, the first thing that ever came out of their mouth was asthma. I mean, their mother was suffering from it, or their sister can't breathe or, you know, they got their inhaler, but they're still suffering from asthma." ACE realized that to tackle asthma requires addressing housing, transportation, community investment patterns, access to health care, pollution sources, sanitation, and health education. "Everything we do is about asthma," the same organizer continued. "Transportation is about asthma, development's about asthma."

As seen with asthma, environmental justice activism is often framed through the lens of health problems within the community. By linking the high incidence of health issues in low-income and minority communities with environmental conditions and access to resources, as opposed to linking it with individuals' behavioral choices, environmental justice activists are able both to empower community action and to improve the health of individuals within the community. The prevalence of a disease within a community serves as the focal point of activism, but the underlying disparities in living conditions are addressed as well. Many NIEHS environmental justice grant recipients target such health problems. For example, Robert Gottlieb's Healthy Food, Healthy Schools, and Healthy Communities project puts the high incidence of childhood obesity in low-income schools into an environmental justice framework. Obesity is often attributed to individual choice, but Gottlieb's project targets the environment and policies that support poor nutrition in these communities, which have resulted in a disproportionate distribution of obesity.[84] Such community-based projects educate lay members of the community both in how to overcome the health issue at hand and in how to seek out resources in the future. The Roxbury Lupus Project, also supported by an NIEHS environmental justice grant, has brought together activist groups, lupus patients, and health care professionals to address the environmental hazards linked with risk for lupus and connective tissue diseases. As with the projects focusing on obesity and asthma, the Roxbury Lupus Project aims at both addressing the health issue, through promotion of an early screening program and implementation of community outreach, and tracing the source of a noticed higher incidence of these diseases in the community, especially among African American women.[85] In all of these cases, the health issues being addressed serve as a way of looking at the larger environmental injustices in the community. In focusing on the diseases at hand, community-members and activist groups begin to alleviate the sources of the disparities facing the community.

TRANSPORTATION ISSUES

Transportation issues became very central to environmental justice organizing on asthma because so much particulate matter came from diesel buses. Activists argue that their neighborhoods have been chosen for the bulk of bus depots and garages in urban areas and are in the midst of traffic routes for highways and for industrial and commercial facilities. ACE encourages communities to take ownership of the asthma issue and to push for proactive, empowered solutions. Central to this community empowerment is the role of direct action and education, such as a campaign in which residents identified idling trucks and buses as a major source of particulate irritants. Significantly, the bulk of the work in this campaign came from children and teenagers, who organized an anti-idling march and began giving informational "parking tickets" to idling buses and trucks that explained the health effects of diesel exhaust. This approach to organizing illustrates the strong attempts made by groups such as ACE to develop new leadership and to engage residents of the neighborhoods in their campaigns.

Environmental justice groups' focus on trucks and buses is important because those vehicles run mostly on diesel gas, and diesel pollution is a very dangerous air pollutant that accounts for at least 70 percent of the total toxic risk for cancer from air pollution.[86] It is almost ironic that buses were icons of the civil rights movement because they typified segregation and Jim Crow laws, and now once again are part of a civil rights approach to the environment.

Because ACE identifies diesel buses as a problem, it also takes up transportation issues more broadly than just air pollution. It ran a major campaign targeting local and state government policies on the allocation of transit resources. Charging "transit racism," ACE argued that the estimated 366,000 daily bus riders in Boston were being discriminated against by the more than $14 billion of federal and state money being spent on the "Big Dig" highway project (the largest single public works project since the Depression), and that the Massachusetts Bay Transit Authority refused to spend $105 million to purchase newer, cleaner buses and bus shelters. ACE eventually did succeed in getting the transit authority to purchase numerous new compressed-natural-gas buses as well as some emission-controlled diesel buses and to retrofit many of its older buses with particle traps, significantly reducing emission rates over the past few years. An oversight committee (Massachusetts Bay Transportation Authority Bus Emissions Monitoring and Control Advisory Committee) set up to evaluate progress on these fronts included an ACE representative.

In tying dirty buses to higher asthma rates, ACE successfully framed an issue of transit spending priorities into one of health, justice, and racism. In 2000, the Transit Riders' Union, largely created by ACE, got the Massachusetts Bay Transit Authority to allow free transfers between buses because the many inner-city residents who relied on two buses for transportation had to pay more than others who had free transfers on subways.

Similarly, WE ACT has identified diesel exhaust as a major factor behind the disparate burden of asthma experienced in the community in which it organizes. In its "If You Live Uptown, Breathe at Your Own Risk" campaign, it featured a very effective graphic of people wearing gas masks as a way to prompt citizens to write letters to the governor and transit authority demanding that they put a moratorium on the number of diesel buses. Using publicity campaigns such as informative advertisements placed in bus shelters, public-service announcements on cable television, and direct mailing, WE ACT has reached a vast number of community residents and public officials and let them know that diesel buses may trigger asthma attacks. Though WE ACT's efforts increased public awareness of the group itself and its work to reduce asthma, the media campaign did not lead to a shift in the policy of the New York Metropolitan Transit Authority policy toward diesel buses. Thus, in November 2000, WE ACT filed a lawsuit against the transit authority with the federal Department of Transportation claiming that it advances a racist and discriminatory policy by disproportionately siting diesel bus depots and parking lots in neighborhoods with people of color. As a result, the New York City Economic Development Corporation developed a new master plan for the Harlem waterfront that was in line with WE ACT's demands and the community's vision.[87]

COMMUNITY EMPOWERMENT THROUGH ASTHMA EDUCATION

A key component of ACE's education and empowerment efforts is reflected in its Roxbury Environmental Empowerment Project (REEP). The project teaches classes in local schools and after-school programs, hosts environmental justice conferences, and through its intern program trains high school students to teach environmental health in schools. Classes are designed to educate students about environmental justice, and they use asthma as a focal issue. For example, REEP teachers discuss the potential process for siting a hazardous facility in people's neighborhoods, ask the students why this facility is being sited there, and what they would do about the siting decision. Through their "know your neighborhood" strategy, they teach students how to locate on local maps the potentially dangerous locations in their area. REEP provides teachers with an opportunity to teach environmental topics they might not otherwise get support for. One student project even identified mold spores in the school ventilation system, thus making progress in a location of frequent and direct exposure. REEP students were also involved in surveys about asthma in Dudley Square and other neighborhoods of Roxbury, which influenced subsequent studies (such as a monitoring study for diesel-related air pollution around Dudley Square and a variety of asthma initiatives). One of the ACE staff members pointed out that the REEP work helps combat stereotypes that people of color in poor and moderate-income communities are not apathetic about the environment or community issues.

ACE has helped some of its high school interns get into college as a result of the education they received in the REEP program, even though this outcome was not an intended one. ACE also participates in job fairs to help students find good employment prospects. On some occasions, it has brought Harvard School of Public Health air quality researchers along with it to present findings to school audiences. By having important scientists share relevant work with students, ACE demonstrates to children in underfunded and understaffed schools how important they are. REEP also shows children their importance by holding a graduation ceremony for them and their relatives, where they present their work, hear presentations by previous students, and get recognized in public. Local businesses that support REEP also attend, as do some state legislators and community leaders.

WE ACT's Healthy Home Healthy Child campaign reflects a similar community empowerment approach. This campaign, developed in partnership with the Columbia Center for Children's Environmental Health, works to educate the community on a variety of risk factors, including cigarettes, lead poisoning, drugs and alcohol, air pollution, garbage, pesticides, and poor nutrition. Educational materials, translated both from English into Spanish and from medical terminology into lay language, inform residents about the effects of risk factors and the actions they can take to alleviate or minimize those effects. In the case of air pollution, one of the actions that residents can take is to contact WE ACT and become involved in its clean air campaign. WE ACT believes that focusing solely on air pollution can be a disservice to the community, so it addresses all of the issues raised in the Healthy Home campaign. As with ACE's experience in identifying community issues, WE ACT's Healthy Home Healthy Child campaign began by focusing on specific asthma triggers, but soon expanded to include residents' key concerns, such as drugs, alcohol, and garbage.

REFRAMING ASTHMA AND CREATING A COLLECTIVE POLITICIZED ILLNESS EXPERIENCE

This environmental justice approach relies on community-level organizing and empowerment to respond to structural factors. But the same approach also has ramifications for the internalized self-perceptions that people with asthma have of themselves and hence for their illness experience. Part of the rationale for ACE and WE ACT's efforts to change social perceptions of asthma causation is to transform at the same time the self-perception of people with asthma. One of the REEP interns wrote an essay in which he characterized the kind of transformation that ACE engenders in people:

> There are things in my environment that truly outrage me. The fact that people have to wait hours for dirty diesel [Massachusetts Bay Transit Authority] buses

on extremely cold or hot days, the fact that someone I know is being evicted from their home because they can't pay their rent, and the fact that a small child I see everyday has died of asthma in a community where asthma rates are 6 times the state average. These things should not be happening where I live or where anyone lives. Everyone, no matter what community they reside in, should have the right to a safe and healthy neighborhood. So what is environmental justice is a hard question, but I know what it is to me. It is allowing everyone the right to have the best life has to offer from affordable housing to safe neighborhoods and clean air.

Thus, ACE's environmental justice framework allows students to see their individual lived experience in the context of a larger set of structural factors and injustices, and, rather than marginalizing and isolating their experiences, it thereby empowers them to participate in community changes.

The experience of illness plays a major role in the educational programs conducted by ACE and WE ACT. Because ACE, WE ACT, and various academic-community partnerships hold so many educational sessions to make asthma a very public concern, the stigma and denial normally associated with asthma are lessened. The power of destigmatization is central. We are only a couple of decades removed from the harmful psychoanalytic notion that asthma stems from overprotective mothers and that it is a psychogenic response to family dynamics.[88] It took better medical knowledge to overcome this psychogenic definition and to situate asthma's causes in environmental triggers such as household mold, animal dander, roaches, and exposure to second-hand tobacco smoke. Though medical research rejected the psychogenic model for causation, the stigma of asthma as a weakness rather than a chronic disease remains today. In an op-ed article in the *New York Times,* Olympic track medalist Jackie Joyner-Kersee addressed her difficulty in overcoming such stigma in order to be open about her asthma.[89] Draw-A-Breath (now the Providence School Asthma Partnership), a Providence, Rhode Island, asthma outreach program largely focused on school-based interventions and parent support groups, runs a summer camp for children with asthma. It points to the powerful message children can learn—that they can do athletics despite having asthma. Through community education and direct action campaigns, ACE and WE ACT help people with asthma to overcome the stigma that frames asthma both as a weakness and as a result of living in an unclean household. Asthma activists use the destigmatization process to politicize fellow sufferers and to gain allies in their effort to produce a healthy environment for their community.

A common theme to emerge from qualitative studies of people with asthma is the feeling of powerlessness. Both children and adults with asthma are without a potential cure for the disease and can rely only on management to prevent attacks. For children, managing asthma requires reliance on their doctors, parents, and teachers, which reduces their sense of individuality and exploration.[90] Children learn to associate various places with asthma exacerbation, leading them and their families to associate local environmental hazards with their asthma. When observing their child's asthma attacks, parents themselves may feel powerless to help him or her breathe normally. Limited access to quality health care also leads parents to feel that they cannot reduce their child's asthma suffering. Frequent trips to the emergency room are the norm for impoverished families seeking asthma treatment, resulting in poor management of the disease, greater financial burdens, the loss of control, and acceleration of the disease.[91] Inequalities in health care polarize the experience of asthma in terms of agency. For children and their families who cannot afford quality management of the disease, asthma becomes another problem beyond their control, exacerbating the feeling of powerlessness. Not only do they not have adequate access to health care, but they have little control of either indoor or outdoor sources of asthma triggers.

Community groups such as ACE and WE ACT work to reframe the illness in terms of the larger illness experience. They feel that the medical establishment has a limited ability to address many of the important factors in the experience of asthma, and they see their role as a bridge. One WE ACT organizer recounted the experience many people with asthma go through in a medical setting: "I think that doctors think that there is very little that they can do about [the factors of asthma]. They go through this checklist of risk factors at the beginning of a physical, which now includes, 'Do you wear seatbelts?' Like different questions assessing individual behavior and risk taking behavior. They focus on things that they feel they can change somehow. So they ask about the indoor environment." Groups such as ACE and WE ACT realize that doctors are often unable to address issues larger than individual behavior, as an ACE activist pointed out: "Even if a kid has really terrible asthma, they're in the hospital, you know, once every two weeks. Sometimes doctors aren't trained to say, 'Do you have mold in your home—where do you live?'" Organizers also recognize that even if the medical community incorporated questions about the home environment, many other important factors shaping the illness experience would still be neglected.

As this perspective indicates, public-health and transportation interventions can be helpful, but there are nevertheless overarching social inequalities that will continue to yield disproportionate exposure and occurrence of the disease. In painting this broad picture of the experience of asthma, ACE and WE ACT create the foundation for an environmental justice–based approach to reducing the burden of asthma.

Through the educational programs held by ACE and WE ACT, people with asthma learn to manage their disease while simultaneously beginning to see themselves as part of a collective of people with asthma who understand the importance of external factors beyond their individual homes. By learning that even indoor exposures through poor housing are a social phenomenon, they see themselves less as individual sick people and more as part of a group that has unfair disadvantages. The environmental justice approach informs these people that they can act to change their social circumstances, and in that sense asthma becomes a stepping point to a politicized view of the world. The ACE interns and many of the children they teach cannot separate out their experience of wheezing from their knowledge of the harmful effects of diesel exhaust from nearby buses. They cannot think about their inhalers without thinking about the excess of bus depots and trash incinerators located in their neighborhoods.

For a growing number of people and organizations, the experience of illness has transformed asthma from an individual disease into a social movement focused on health inequalities. Their role in building and maintaining this social movement is a growing concern for organizers, as noted by an ACE organizer: "The other part of ACE that's really emerged probably in the past couple of years is our role as movement builders, building an environmental justice movement both locally and nationally. And the leadership development fits under that as well. But it has changed the way we look at our programs. Now we're trying to figure out how we not only take our interns and train them as educators, but train them as organizers." People with asthma in Boston and New York are incorporating rhetoric from the environmental justice movement to address the social and political forces responsible for the disparate rates of the disease among urban communities of color. In approaching the overburden of asthma in their communities, ACE and WE ACT look beyond medical solutions to the social forces shaping the urban environment. As one ACE organizer noted, "I think our approach has been that if people are suffering from asthma, that's something we need to deal with now, today. And, yes, part of that answer is a medical solution, figuring out how to get people the right treatment. But in the meantime, if there are things we can do to reduce the level of triggers, if we can figure out how to take more a pollution-prevention approach to figure out how to keep these things from getting into the environment, then we ought to do that."

National networking encourages replication of the contested illness approach as it applies to asthma. WE ACT members in particular have been active in regional and national environmental justice coalitions as participants in the EPA's National Environmental Justice Advisory Council, the Environmental Justice Caucus, the Children's Environmental Health Network, and the Environmental Justice Fund. They have played a leadership role in regional environmental justice organizations, including the Northeast Environmental Justice Network and the Community-University Consortium for Regional Environmental Justice. This high-profile role in the environmental justice movement has allowed WE ACT to exchange campaign ideas in order to help spread some of the more innovative campaigns and tactics to other organizations and to pick up new tactics itself. Such collaboration and sharing of tactics represents a pattern among community-based asthma groups. For example, ACE borrowed its "transit racism" campaign from the Bus Rider's Union in Los Angeles and from WE ACT's current challenges to the Metropolitan Transit Authority's bus depot. Thus, even if there is no single national organization, strategies can be shared because there is an extensive network of organizations that work together both formally and informally.

Based on the illness experience described earlier, many people with asthma have developed what I term a *collective politicized illness experience,* in which their personal experience of illness, symptoms, coping, and adaptation has become linked with a broad social critique. This critique involves assessing both responsibility for the causes or triggers of the disease and responsibility for treating and preventing the disease. The use of an environmental justice discourse frames ACE and WE ACT's activism in terms of a social justice that has both local and national components. On the local level, ACE and WE ACT are directly involved in improving the living and working conditions of community members who suffer from asthma. By addressing transit issues and other socially structured forms of environmental inequalities, they are able to address local environmental problems. They have also been able to establish collaborative scientific endeavors to improve our understanding of the relationship between asthma and air pollution, and this work has had a national impact because many of the collaborating scientists rank among the most eminent air pollution researchers, including John Spengler, Jon Levy, and Frederica Perera. For many environmental health scientists, these kinds of collaborations are satisfying because they allow for a meaningful application of their findings and because the environmental justice groups are so committed to this effort. Environmental organizations' dedication encourages in scientists the fundamental social reform principles regarding public health that many of them were imbued with in their training or that they came to on their own as they found the traditional public-health sector less receptive to such efforts. For example, these scientists want stronger federal regulations based on the excellent science base they create, but they discover obstacles to effective regulation; activist pressure can help them achieve regulatory reform and hence improve people's health.

Some of these collaborations between activists and scientists also advocate for social justice through their focus on outdoor source of air pollution rather than on indoor causes, the latter of which places much of the burden of responsibility on the individual. Interestingly, however, even the indoor focus does not mean people are in charge of prevention. For example, even if people keep a very clean home, they very likely will have cockroach problems because of structural deficiencies (holes in the wall) and neighbors' behaviors. There are also indoor air pollution sources, such as gas stoves, which cannot be replaced by the tenant and which may be used inappropriately because of structural issues (e.g., if the furnace is out and the landlord is not responsive, renters may use their stove to heat the apartment in winter). Overall, the politicized illness experience links the local burden of disease together with broader social factors, which, in the words of an asthma activist, means working toward "building an environmental justice movement both locally and nationally."

APPROACHES TO SCIENTISTS AND ACADEMIC-COMMUNITY COLLABORATION

ACE's use of an environmental justice frame means that the organization is not wedded to "traditional" science. It sees that other groups have gotten entangled organizationally in complicated scientific debates over epidemiological methods and statistical significance that can last many years. At the same time, though, it understands the need for scientific evidence and scientific legitimacy, recognizing the long-term importance of establishing links between air pollution and asthma. ACE's decision to work selectively with science and to insist on the role of science in empowering community residents is central to its asthma work. Hence, it uses science as a tool in the larger arsenal of political and social movement tactics.

Support from some researchers at the Harvard School of Public Health and the Boston University School of Public Health provides an opportunity for ACE to work with science in its own way. ACE's AirBeat project monitors local air quality and then analyzes the relationship between air quality and the frequency of medical visits. ACE mobilized researchers and government agencies to install a monitor at ACE's Roxbury office. Community members are also directly involved in the planning and implementation of these studies, as evidenced by REEP students' involvement in identifying data types to be collected from community clinics. On one level, ACE collaborates with scientists to produce quantifiable outcomes that its members hope will lead to greater understanding of air pollution and asthma. This collaboration has resulted in jointly authored articles on air particulate concentrations, published in major environmental health journals. However, AirBeat is useful on other levels as well. ACE derives legitimacy from the involvement of government agencies and scientists in the campaign process—for example, in the presence of Harvard scientists and the then head of EPA Region 1, John Devillars, at the press conference when the air monitor was unveiled.

These environmental justice groups remain cautious about what they can gain from scientific research, however. As one scientific collaborator with ACE noted:

> [Community groups] are often very disappointed when . . . we cannot show in a scientific study that there is a direct linkage there. And usually it's an issue of having not enough numbers, weak associations, and you're never going to be able to say definitively that the bus station or the toxic waste site or whatever, is the cause of the illness that is being observed in the community. So that can be very frustrating. ACE recognizes the weakness of epidemiology [to explain] and so hasn't gone a long way down that path. And [ACE] has rather focused on what the environmental exposures are and trying to deal with just [that] on the basis of controlling exposures.

A researcher in partnership with WE ACT concurred:

> Their agenda is not a scientific one as to discover the exact causes—it's more to identify causes that are known and that may be harming a community and actively doing something about it. So . . . it's a mixture of politics of discrimination and environmental kinds of things. But they do a very good job. . . . It's very hard in some sense to also answer the questions that WE ACT asks. You know, I sort of share their goal of thinking that diesel sources of point pollution are really a bad thing for people and that it's a good idea to have, either have them disappear all together or at least have them more equitably distributed.

One WE ACT staffer expressed his frustration with the scientific process:

> In the scientific arena, I think that the debate is always there about what's a definitive study. How is it proved beyond a reasonable doubt? Yet nothing in science is ever proved beyond a reasonable doubt, but we're always looking for that definitive study in the scientific realm. You know, science, unlike any other social endeavor, places itself on this pedestal that says that, in essence, if it could

speak as human being, "I am science, and this is the end all and where with all." And if I don't have a definitive answer, then there is no definitive answer. So there is that struggle within the scientific community. Is this the definitive study? Have we taken a large enough sample? Did we control for all the uncontrollable variables? Was it perfectly done? Of course it can't be perfectly done. And why are we waiting for—quote, unquote—"the definitive answer"?

For scientists, ACE and WE ACT present opportunities to collaborate in the collection of community-level data and to gain local political legitimacy and connections. Although many of the scientists we interviewed and observed noted that this collaboration made the scientific inquiry process easier, they also expressed a sense of obligation to the localities in which their research occurred and to individuals with asthma. As one researcher noted, "Connecting to people that are having real problems is very important. You just feel a lot of empathy for people that are suffering. And it's very important for us to get out of the academic world and have those connections to really ground our work."

Researchers working with both ACE and WE ACT emphasized the importance of joint decision making and goal setting, in part to avoid the perceptions of exploitation that have damaged other citizen-science alliances in other locations. They noted that other alliances between citizens and scientists often do not achieve these goals, as one scientist pointed out: "The ideal way that everybody describes the way community partnerships should develop is you should become partners. . . . You should really share agendas about what kind of work you want to do and then jointly reach decisions about what you are going to go after and then apply for money or support to do that. And that probably happens in about 5 percent of the cases [laughs]." The same researcher continued: "One of the biggest difficulties for me, I mean I'm used to projects where I get the money. I hire the staff. They have to do everything that I say basically. And it's not like that working with community groups. And in fact they know stuff that I don't know. I'm hopelessly naive in many ways related to how things will play out." In addition to providing each other with legitimacy, WE ACT's work with Columbia researchers has been both a learning and a teaching process, as an activist noted:

> [In] reaching out, we haven't always come upon the most community-conscious people. And in the process of that relationship, it has taken us a lot of work and time to educate those researchers, to give them a broader understanding of the complexities and the diversity of this community that they actually work in. And I think that overall I can say that has its rewards, but it hasn't been all smooth,

and it hasn't been all easy.... I think that the relationship has also yielded for us an increase in our own sort of scientific knowledge, and for them I think it's increased their knowledge of this community, its history, its complexity, its diversity. It's taught them how to be better communicators, how to be better neighbors in some ways.

Researchers and public-health practitioners rely on groups such as ACE and WE ACT to provide what one researcher called a "direct finger on the pulse of what's going on in the community from the health and environmental perspective." ACE and WE ACT are able to bring real-world community problems to the attention of academics and government and to present them through an interface accessible to researchers and public officials. In turn, because of their trusted community role, they are valuable in translating for and disseminating to the community all public-health information about indoor and outdoor environmental asthma exposures.

Nonetheless, scientists, officials, and policymakers require a very different discovery process than do community activists and community organizations such as ACE and WE ACT. For example, ACE does not wait for definitive scientific support regarding the impact of air pollution on asthma, but rather acts immediately. Because ACE chooses to work with science in a limited fashion and often draws conclusions and benefits from the research beyond the actual findings, scientists and activists perceive their work with each other as problematic at times and occasionally become frustrated with each other's approaches. According to one ACE member, "We had the launch of our air monitor and had an article in the [*Boston*] *Globe,* 'New Tool in the Fight Against Asthma.' And some of our [science] partners were like, 'This is not necessarily the case.' That's how we look at it, that's how a lot of residents look at it, but the folks like the DEP [Department of Environmental Protection] and others [felt] that's not necessarily why this air monitor is here, but that's the message that the media got." That ACE person followed up on this point:

> I know that we've been challenged even by our partners, like at the Harvard School of Public Health and the folks we worked with on the air monitor, who always are saying, you know, if you really want to deal with the asthma problem in Roxbury, it's more than diesel buses.... And we know that the problem is more than traffic, and we know the problem is more than diesel buses. You know, it's housing, it's access to health care. There's a whole rubric of things. But, at the same time, clearly, you've got to start doing whatever you can, and I think that folks know that even if science hasn't proven [it], people know that there's a link.

A researcher working with WE ACT noted a similar tension:

> Being affiliated with advocacy work . . . runs counter to the classic scientific process where the scientist approaches data in an impartial way and is, takes whatever comes out of the data in very balanced unbiased way. Whereas, from the community perspective, often they're going into a study with a particular bias about what the outcome will be, and they're going to take the data and may . . . be inclined to twist it or portray it in a way that supports their perspective, and that's something that scientists, you know, obviously are not that comfortable with. That hasn't been a real big problem for us . . . When I publish results, I just say what I think is justified. WE ACT can take those data and say whatever they want to say, and that's okay with me. But it is an issue. I think it's a little bit of a friction that . . . just has to be worked out.

Such a response suggests that only researchers with some degree of commitment to environmental justice goals would be willing to work with these groups. The presence of friction and the willingness to work with this friction require an initial sympathy to the activists' goals.

ACE and WE ACT's intent is to win public debates, not scientific ones, and they see no reason to place high stakes on controversial scientific opinions because science is only one component of their overall approach. They see themselves as being trendsetters in the amalgamation of science, policy, politics, and community organizing, as formulated by one WE ACT staff member:

> I think that the environmental justice movement has actually been a hybrid of . . . the idea of working from the grassroots, but also recognizing that there is a power structure that you have to be sophisticated enough to affect and change. And I think that as WE ACT, we have sort of been right on the cusp of that hybrid or that synthesis, if you will, of working both within sort of the grassroots arena or the broader social arena, yet at the same time recognizing that you've got to sort of be a strategic operator, political operator, in order to move public policy. And that you pick from the best of both worlds in order to advance the goals.

This quote, especially in the speaker's use of the term *hybrid,* points to an interesting facet of environmental justice groups. By being hybrid groups that cross the boundaries of activism and science, they have an impact on both science and activism. As multifaceted boundary movements, they also build the capacities of their staff and constituencies.

Groups such as ACE and WE ACT choose to keep the focus of their work on their communities. Like other environmental justice organizations, they believe that if they become too nationally focused or involved in too many governmental and academic meetings, they will forsake their long-term local base. They are aware that even if there is national implementation of $PM_{2.5}$ standards, local injustices will remain, and hence local action will always be necessary. Nonetheless, ACE and WE ACT's local work does have national impact. Community-initiated citizen-science alliances with national-level research institutions such as Harvard, Boston, and Columbia universities have a cumulative national impact. The organizations involved can influence research methods and techniques, how findings are presented, and, in some cases, the findings themselves. For example, WE ACT's institutionalized partnership with Columbia's Division of Environmental Health Sciences has led not only to useful studies on the health effects of diesel exhaust in northern Manhattan communities, but to cooperation in the development of new methodologies for traffic counts, ambient air monitoring, and the use of biomarkers as a measure of an individual's exposure to diesel exhaust.

WE ACT has been slightly more visible than ACE in its collaborations with university-based scientists. It is partners with Columbia University School of Public Health in a federally funded project on collaborative academic-community research and advocacy. Together, these community and academic entities have published numerous papers in scientific journals, and in 2002 they edited a supplement of the prestigious *Environmental Health Perspectives* titled *Community, Research, and Environmental Justice* (which also contained an article by ACE members). WE ACT coordinated a conference in 2002 called "Human Genetics, Environment, and Communities of Color: Ethical and Social Implications," cosponsored by the NIEHS, the NIEHS Center for Environmental Health at the Columbia School of Public Health, and the Harlem Health Promotion Center. The conference brought together community advocates, policymakers, and scientists from across the United States to educate one another and deal with critical ethical, legal, and social implications of research on human genetics for communities and people of color. ACE has had collaborations with Harvard, as mentioned earlier, and has had ongoing involvement with Boston-wide meetings that bring academics, policymakers, and researchers together. The group plans to develop a database of academic researchers who can be linked with community organizations, and it has submitted some grant proposals jointly with various university groups.

THE FUTURE OF SCIENCE AND ACTIVISM ON ASTHMA

Without agreement on the scientific evidence linking air pollution to the increasing rates of asthma, the prospect of stricter regulations at the federal level is low. Environmental justice groups choose to address air pollution at the local level rather than wait for the political and scientific controversies to be settled. Through innovative collaborations in community-based participatory research, local groups are conducting their own science in conjunction with some of the foremost scientists in air pollution research. For example, staff members of WE ACT and researchers from the Mailman School of Public Health at Columbia University published preliminary results of a cohort study of the health effects of urban indoor and outdoor air pollution in minority neighborhoods of New York.[92] Similarly, using GIS technology to monitor particulate matter emissions around Boston's Roxbury neighborhood, ACE members and Harvard School of Public Health researchers were able to publish their findings in a leading environmental health journal. This study in particular demonstrates the innovative methodologies made available through community-based participatory research: community members carried portable air monitors, allowing the community to participate in the data-collection process and to determine which streets to investigate.[93]

In the previously mentioned 2002 special issue of *Environmental Health Perspectives*, edited by WE ACT staff, both ACE and WE ACT presented studies coauthored with public-health researchers and designed to assess the link between asthma and air pollution.[94] Prior to the advent of academic and community scientific collaborations, community groups could only make requests for increased research on the link between air pollution and asthma. Through community-based participatory research, these groups play a larger role in the political debate as well as in the scientific controversy.

But activist groups such as ACE and WE ACT are not entirely committed to relying on science to provide absolute proof that air pollution causes new cases of asthma. Instead, they push for regulatory action based simply on the fact that their constituents are directly exposed to unequal levels of air pollution. Thus, the environmental justice groups involved in asthma activism push for more science at the national regulatory level, but they rely on political action at the local level to reduce the disproportionate burden of asthma in their communities. The accumulation of many such local efforts lends more substance to national campaigns; so many major cities have become host to such activism that it is difficult to ignore. At the same time, there is room for activists and advocates to work at whatever level is most comfortable for them. Even though committed to partnerships, the scientists often demand a higher level of scientific rigor than community groups might feel is necessary. Both ACE and WE ACT believe they are pushing their scientific allies to be continually more community oriented in defining problems and in designing research and interventions.

We have seen here the multifaceted approach toward science by environmental justice groups. They know they need a strong science base, but they are often reluctant to put a major part of their effort into that base because it can take away from their community capacity building, political organizing, regulatory pressure, and other efforts. The different lay and professional perspectives I have mentioned make for contradictory beliefs and actions, which is typical of citizen-science alliances. Such alliances do, in fact, put much effort into talking about these differences and in what ways they may alter the nature of the collaboration. The time and effort required for collaboration is, for some scientists, a barrier to working in such community-based participatory research, but for others it is an interesting challenge and an ethically valuable approach.

The strength of the science, the spread of public-health action, and the growth of public advocacy through citizen-science alliances have engendered powerful corporate opposition. Major companies find it important to counter the policy implications of air particulate research. For instance, ExxonMobil took out a large advertisement on the *New York Times* op-ed page that argued that air pollution was not a factor in asthma and that medications were the best we had to offer in dealing with this condition.[95] In the antiregulatory regime under George W. Bush, it will be less likely that the EPA will press forward with more stringent regulations. Nevertheless, the very strong science base for the connection between asthma and air pollution, relative to other environmentally induced diseases, provides strong legitimacy to professionals and activists. The linkage of health with other social sectors creates an opening for a broad-based effort at improving health and democratizing society. Federal agencies' interest in multisectoral efforts that bring together activists, researchers, and providers offers a strong advance in collaborative, participatory models of health action.

Most participatory public-health models of health intervention fail to address environmental causes of asthma. For example, they do not lobby for power plant closure or regulation or for litigation to force federal compliance with Clean Air Act statutes, and they do not demand alternative energy sources to reduce fossil fuel usage. Given the extent of the asthma epidemic, it is understandable that many clinicians, social workers, and community activists want to do front-line work on personal behaviors to get rid of mold, mites, and roaches and to use protective covers for bedding. They know that such measures are effective in reducing asthma suffering. But even if these programs reach a significant portion of inner-city residents, they cannot offer any protection against the outdoor air pollution that according to activist groups continues to trigger asthma attacks. Some activists believe that the focus on indoor environmental factors burdens people with individual responsibility for dealing with the triggers in their homes, when instead the target of change should be broader-scale environmental factors that affect whole neighborhoods or cities. They note that there are still important players in the asthma world who do not address air pollution. For example, the Asthma and Allergy Foundation of America, New England Chapter, puts out a sixty-two-item checklist for parents and providers, in which only one item concerns outdoor pollution: "outdoor fumes (such as from car exhaust, idling vans or buses, or nearby businesses) are prevented from entering the building through open windows or doors."[96] Checklists in research literature are similar and often do not even mention vehicle exhaust.

SUMMARY

The social discovery of asthma and its environmental correlates by lay, scientific, and political actors has been a unique example of action on environmentally induced disease in that there are so many shared points between the actors. As I have shown, a considerable amount of attention to the new asthma epidemic comes from empowered laypeople who are concerned about environmental triggers of the disease. Their approach to asthma includes action in a number of social sectors, such as health services, public health, education, housing, transportation, and economic development, and it is framed in environmental justice terms. Even among more mainstream progressive public-health approaches to asthma, there is considerable attention to various social sectors and the environmental justice perspective. But even though this progressive public-health approach shares some beliefs with the environmental asthma activists, the programmatic work of the more mainstream organizations is often based solely on indoor hazards. Environmental asthma activists, in contrast, focus their attention also on political and economic action, addressing outdoor hazards, although they understand the need for the household level of attention.

ACE and WE ACT, as representative of this environmental asthma approach, define themselves as environmental justice organizations, with asthma activism serving as only one part of their work. These environmental justice groups have found creative ways to work alongside scientists, while not placing primary emphasis on research. The environmental asthma activism that we have addressed here offers lessons for future activism regarding contested illnesses in its use of a combination of community organizing, social support for sufferers, creative political actions, and citizen-science alliances for treatment, prevention, and research. Because so many environmental justice groups have worked on asthma, they and the larger environmental justice movement have been strengthened in the process.

Given my point that ACE and WE ACT emphasize community action rather than scientific research, we must ask why they began to work with scientists and why they continue to do so. For both groups, working with scientists was not an initial focus. They began their organizing work without emphasizing collaborations with medical and scientific professionals. But this type of collaboration developed in part because a sufficient number of researchers were clearly sympathetic to the groups' environmental justice framework. These scientists may have become sympathetic because their research showed the centrality of major industrial pollution sources, combined with federal laxity on regulations, to the health issues at hand. Perhaps they resented corporate attacks on their research. Or they just witnessed the vast inequality of disease manifestations and saw that the sufferers had few allies. It became clear that medical and scientific allies would provide legitimacy for these environmental justice groups' aims, allowing for the development of citizen-science alliances. In addition, activists understand that government typically wants a strong science base in order to justify action. They also understand that corporate opposition will continue and that even current government support for environmental asthma connections may not be permanent. By allying with scientists, these groups can help the continued production of science to strengthen their case. They are also able to push their science collaborators to greater support for citizen participation. ACE and WE ACT have developed into hybrid organizations that are doing public-health work and research, as well as engaging in community organizing. The community groups are attuned to the possibility that their collaborations with scientists might bring them away from their original goals, leading to mission drift. However, through partnerships with university-based researchers, environmental justice groups maintain the option of being more science oriented at some times and more activist oriented at others.

The hybrid nature of academic-community partnerships also benefits professionals, who receive support from the people who are affected by the problems on which the researchers are working. By becoming attuned to the value of lay involvement, researchers become more well rounded, expand their repertoire of knowledge and research questions, and are able to compete for federal grants that require such partnerships.

The social discovery of asthma reveals a complex picture of the types of activism, a diverse array of actors, and collaborative scientific endeavors. Some activist groups choose to address indoor sources of air pollution and asthma triggers and are involved in public-health science and intervention. Others, such as ACE and WE ACT, choose to address outdoor sources of air pollution, arguing that focusing solely on indoor sources of pollution fails to address the larger social and environmental causes of asthma. They do not deny that asthma may be reduced by domestic regimens such as cleaning and using hypoallergenic products, but they also seek to expand the focus to outdoor air pollution. At the same time, some activists move beyond the scientific debate, arguing that the unequal exposure to sources of air pollution should be reason alone for stronger environmental regulations. Yet it is the combination of various kinds of activist and advocacy approaches with various scientific approaches that has yielded a formidable advance in air pollution science. Because of the shared perspectives among actors, the hybrid nature of the environmental justice groups, and the significant successes in asthma research and particulate regulation, this case offers lessons in cooperation for action on other contested environmental diseases. Although other diseases may have less solid science bases, the willingness of scientists and government to work collaboratively with lay organizations is crucial.

Such collaboration through citizen-science alliances marks the actions and accomplishments of activists and professionals concerned with asthma and breast cancer. These alliances are some of the strongest examples of the interaction between health social movements and scientists, and they offer excellent models for people concerned with other diseases. In breast cancer groups, laypeople have shown their ability to educate themselves very well on the medical issues as well as on causation and prevention. But illness contestation around asthma has some interesting differences from illness contestation around breast cancer. Asthma sufferers, activists, and professional allies have had to argue for the importance of the rise in asthma and have found incomplete acceptance of the diagnosis of asthma. For breast cancer, in contrast, there is no dispute over the disease. It is clear, however, that for both asthma and breast cancer, challengers to the DEP share a new view on etiology. This new view has been easier to advance for asthma because there is considerable scientific evidence on environmental causation through air pollution, whereas endocrine disrupters and other environmental causes are weakly implicated in breast cancer. The similarities and differences between these two diseases, along with GWRIs, are the subject of chapter 5.

|4|

GULF WAR–RELATED ILLNESSES AND THE HUNT FOR CAUSATION

THE "STRESS OF WAR" VERSUS THE "DIRTY BATTLEFIELD"

IN THIS chapter, I show how GWRIs are a different type of contested illness from breast cancer and asthma. I begin by examining the lay discovery of GWRIs by veterans returning from the Persian Gulf and the formation of a collective illness experience by veterans seeking explanations for their conditions. I discuss the scientific controversy and scientific complexity surrounding the existence of a unique Gulf War Syndrome by tracing the government and scientific research. This chapter covers the many obstacles faced by veterans who were forced to challenge a DEP that did not adequately address their needs: (1) missing information, misinformation, and secrecy; (2) distrust, disputes, and fragmentation; and (3) issues of legitimacy for both illness sufferers and scientists working on GWRIs. I show how citizen-science alliances were more difficult to form than for asthma and breast cancer due to military and VA control of data and domination of research, veterans' patriotism, and a lack of educated, politically experienced, and well-connected members. The controversies and scientific challenges surrounding GWRIs can serve to inform future dealings with illnesses associated with military involvements, as well as situations in which people's medical complaints escape routine medical understanding.

AN OUTBREAK OF MYSTERIOUS SYMPTOMS

Many veterans have maintained that 1991 Persian Gulf War exposure to a variety of toxic substances caused numerous types of ailments, especially fatigue, joint pain, dermatitis, headaches, and memory loss. More than 110,000 have sought treatment for service-related illnesses, according to a leading clinical researcher who deals primarily with these Gulf War veterans. This number is large, given that a total of approximately 697,000 troops were deployed. Research has been plagued by poor data and by disputes over how to study health effects.[1] The VA and the DOD have significant differences on their lists of who served, who received immunizations and vaccinations, and what dates constitute possible exposure. Military personnel faced multiple potential sources of disease, including organophosphate pesticides, nerve gas, pyridostigmine bromide (nerve gas antidote), oil fire smoke, depleted uranium used in ordnance, toxic paints, and diesel exhaust.[2] Federally funded research on GWRIs was extensive: $7.1 million in 1994, $17.3 million in 1995, $18.8 million in 1996, $34.2 million in 1997, and $37.9 million in 1998. Among the VA, DOD, and HHS, $247 million was spent on GWRI research between 1994 and 2003.[3] Yet since 2000 funding has steadily declined.

GWRIs represent a different kind of contested illness than asthma and breast cancer, as evidenced early on by conflicts over the widespread denial that such illnesses existed and subsequently by arguments concerning various stress-related explanations. I first study identification—the basic social discovery of GWRIs—then the ongoing controversies about these illnesses, and, finally, the effects of these disputes on health and public policy. I then develop a new approach that demonstrates how sufferers' self-reported symptoms and exposures offer valuable data and more humane sensibilities to the medical community, policymakers, and health professionals.

Unlike the more restrictive definition of environmental factors I used in looking at asthma and breast cancer, environmental factors for GWRIs often include treatments such as inoculations and antidotes because it is hard to disentangle the iatrogenic effects of those treatments from other environmental effects, because sick veterans have lumped them together, and because some GWRI research focuses on such combinations.

In order to study disputes over GWRIs, I focused on the work of the BEHC, a collaborative project of the VA Medical Center and the Boston University School of Public Health in Boston. It was one of only four such centers in the country and was also committed to strong involvement of Gulf War veteran activists. I attended monthly meetings of the center staff, as well as periodic meetings of the science advisory committee, the veterans' advisory committee, and other VA Medical Center meetings that involved the group. Further data came from interviews with various veterans activists groups, other researchers and officials, and two annual national conferences of federally funded Gulf War researchers.

With GWRIs, the DEP originally attributed veterans' symptoms to war-related stress. Veterans and some of their scientific allies strongly opposed this version of the DEP because they felt it blames the victim and discourages research on environmental causes. This opposition, in conjunction with the more sophisticated thinking of some researchers and the Institute of Medicine's recommendation for new research on stress-environmental interactions, has initiated a shift to viewing veterans' symptoms as similar to other multisymptom diseases and conditions and believing more firmly in the reality of those symptoms.[4]

A U.S. Army doctor who has treated military personnel with Gulf War illnesses referred to another source[5] when he explained in an interview that he approaches treatment of GWRIs with the belief that "if you have to prove you are ill, you are never going to get well." In that situation, sick GIs and veterans face a very unique problem in comparison to the sufferers of asthma and breast cancer.

IDENTIFICATION: THE SOCIAL DISCOVERY OF GWRIS

LAY DISCOVERY As in most cases of occupational and environmental illness, GWRI sufferers were the first to identify their health problems. Soldiers were aware of the potential for exposures in the Persian Gulf just because of the military vaccinations they received, the training drills they underwent with protective equipment and gas masks, the history of Agent Orange's health effects from the Vietnam War, media coverage of Iraq's biological and chemical capabilities, and knowledge of Iraq's widespread chemical warfare in the 1980–1988 Iran-Iraq War and of its chemical attacks on the Kurds.

Veterans returning from the Persian Gulf reported ailments they did not experience prior to service. Media stories highlighted hardy people who found themselves debilitated to the point where they could not function at work or home and who still faced problems getting the government to recognize and treat their illnesses. Veterans did not initially attribute their ailments to service in the Gulf. However, after learning of other veterans' illnesses from media reports, they began to speculate that service in the Gulf might be a cause. All the veterans I interviewed spoke of informal social networks that shared strategies for getting treatment after VA physicians refused to treat them. One remembered: "Back in 1994, a lot of veterans were yelling in the newspapers. And every time something happened, the VA would get a workload. You know, they'd get this rush of complaints. If the U.S. government put out a list of possible complaints they could have had during duty in the Gulf War theater, I think that would have been a nicer way of doing it, not with all the veterans blurping up all over the United States."

As several veteran interviewees explained to us, a growing collective identity helped them see the potential to pressure the government to take responsibility for the illnesses they were experiencing. In one case, an organization that was initially formed to provide emotional support to families whose loved ones were in the Gulf shifted its focus after the war to providing support for families whose loved ones had returned from the Gulf with unexplained illnesses: "We got a call in regards to a sick soldier in August 1991 . . . the Department of Veteran's Affairs did not recognize him as a veteran because he was still active duty. . . . So the man couldn't get any help. . . . He had a wife, he had a kid, and he was looking for help." The organization developed into a clearinghouse of resources for veterans and family members trying to get treatment and began working with families all over the country who were trying to help their loved ones who had been deployed and then returned sick. As the organization's founder, the father of a Gulf War veteran, explained, "What happened after August of 1991 was that . . . we started to get [calls saying], 'This is not the guy that I sent to the war. This is not the same person.' . . . 'I sent a teddy bear to the Persian Gulf, and I got a grizzly bear back.' . . . 'My husband is sick, my son is sick, what do we do?' Well, the point was that we turned around and we called the Department of Defense, and we called the VA, and we asked them what did these troops need to do?"

A tip from a VA employee helped the organization begin taking the necessary steps to get the government attention it wanted, demonstrating how disease groups often rely on the support of other actors within the DEP process—certainly other social movement organizations, but even some government agencies or components of them. According to the founder of the clearinghouse organization, "We got to know a young man up there who worked in the VA, in Washington, D.C., who was also a Vietnam vet. . . . And his recommendation to us was that this is beginning to get significant . . . and what we needed to do was to start documenting this. Because if it really became a major issue, then we would need all that documentation."

Veterans also relied on existing institutions in seeking explanations for their conditions—especially, for example, existing veteran service organizations (VSOs), such as Veterans of Foreign Wars, Disabled American Veterans, the American Legion, and AMVETS, all of whom had dealt with Vietnam veterans sick from Agent Orange. As one interviewee explained, however, not all VSOs were initially sympathetic to ailing Gulf War veterans: "When we went to the VSOs . . . they weren't really cooperative on the whole. They weren't there for us. But a lot of things made sense, and they slowly got on the ball. A lot of us said, 'The hell with them, we gotta lead this parade.'" There still exists, however, tension between and among veterans and the VSOs who represent them over the proper role of science in identifying the causes of GWRIs.

Specific Gulf War groups, such as the National Gulf War Resource Center and Operation Desert Shield/Storm Association, eventually formed, though without large national memberships. The Internet was a major resource for collaboration. Veteran activists regarded their action as crucial to the research enterprise, as evidenced in this quotation from an interview with one of the researchers: "That's really what got this thing up and running in the first place was the veterans groups and individual veterans with incredible stories to tell that became public news stories. I would say that without that you wouldn't be getting any of this . . . Gulf War–related illness work. Wouldn't have done any of this and wouldn't still be doing it."

Gulf War veterans were fighting not only to get medical recognition, but also to get treatment, and their anger was often focused on the difficulties of getting that care. They initially wanted simple acknowledgment of their symptoms and adequate care and compensation. According to one interviewee, this was the least they should expect, given the contract all military personnel sign with the government: "When the recruiter sat on my couch, and I signed for my son to go into the service, the recruiter stated, as an authorized representative of the United States government, that if my son got hurt, sick, wounded, or injured in the service of his country, they would take care of him. The big thing, even today . . . the thing that the Gulf War veterans and their families want is for the United States government to honor its contract."

Another interviewee expressed a similar concern when she described her understanding of the risks inherent in military service: "Healthwise you can't do anything about what the soldier was exposed to. I mean, that's what we got the hazardous duty pay for. But you can do something about how we respond to it." During a debate between veteran activists and federal officials at a conference my research colleagues and I observed, a representative of a major VSO summarized his assessment of the situation: "There are three fundamental beliefs: one, doctors make people well; two, if I serve in the military and I am injured, a nation grateful for my service will come to my assistance; three, Congress will fulfill that promise. Sick vets lose all three beliefs, and Congress just doesn't care."

Even when Congress required the VA to begin providing care to veterans suffering from undiagnosable symptoms, veterans found themselves being denied care due to other preexisting policies, such as means testing (requiring a certain low-income level to qualify). For example, as one VA researcher explained, "The perception out there is that Gulf War veterans are being taken care of. . . . [But] the VA continues to create policies that deny some vets care. Means testing means no care for vets if their disability level and income are at certain levels."

GOVERNMENT AND SCIENTIFIC DISCOVERY Once disease sufferers identify an illness phenomenon, they often pursue legitimization of their beliefs through government and scientific channels. Although those channels have their own discovery processes, governmental and scientific organizations would not likely be engaging in discovery without veteran awareness of problems. Government discovery proceeded slowly, beginning with the DOD's denial that there were toxic exposures. Motivated by veterans' dissatisfaction over DOD resistance and VA slowness to treat veterans, some federal legislators began investigations of chemical exposures and pressed federal agencies to probe deeper into existing research on potential toxic health effects. In addition to DOD, VA, and CDC studies, several interagency task forces were set up to coordinate research. The Institute of Medicine was mandated to provide an analysis of the existing body of research, resulting in the creation of the Committee on Measuring the Health of Gulf War Veterans, and later the RAND Corporation received a contract to do the same. In 1998, Congress established the Research Advisory Committee on Gulf War Veterans' Illnesses, which was appointed by the VA secretary in 2002 to advise the secretary on research relating to GWRIs. The twelve-member committee, headed by James H. Binns Jr., a Vietnam veteran and former principal deputy assistant secretary of defense, and assisted by a panel of scientists, was not intended to conduct research, but rather to submit annual reports evaluating the status of ongoing research and to review federal plans regarding GWRIs.[6] The work of Robert Haley, often criticized as a maverick, was in part responsible for this new push by VA head Anthony J. Principi. Although this new committee appeared to lend more legitimacy to Haley, who was placed on it, it also may have been largely a way to assert fairness in representation.

Disease recognition also involves scientists and physicians realizing that there is a new phenomenon among illnesses. Though some VA physicians and researchers dismissed veteran complaints, others began to see patterns in the complaints and explored them further. Scientists researching GWRIs came from a variety of backgrounds and motivations. Some were on the VA, DOD, or other federal staffs assigned to do such research. Others were medical researchers with backgrounds in other medically unexplained physical symptoms (MUPS). Some were occupational and environmental illness researchers who saw this issue as a logical extension of their work. Among the latter group, some were very sympathetic to the veterans' complaints and believed they were not being taken seriously.

COMPLEXITY OF THE ISSUE

Research on GWRIs has been hampered by much complexity. GWRIs, more than many illnesses believed to have environmental causes, are ambiguous and lack a clear case definition. Some epidemiologists and physicians believe that the VA demand for a specific diagnosis was too strict, and thus veterans were unable to legitimate their illness in the early period.[7] Although most researchers have abandoned the concept of a single Gulf War Syndrome, there is no consensus on how to delimit the range of symptoms and illnesses, primarily because of the multiplicity of exposures and diffuse nature of complaints. Epidemiological studies have identified different clusters of symptoms, but no conclusive connection to specific exposures. Some symptoms resemble those for chronic fatigue syndrome, fibromyalgia, and multiple chemical sensitivity, all of which are poorly understood conditions and subject to much dispute.[8] In a 2005 report, the Institute of Medicine found a connection between pollution from oil well fires and combustion products, on the one hand, and increased risk of lung cancer, on the other, but was not able to conclude based on this correlation whether veterans' health problems are associated with these exposures.[9] Other diseases, such as amyotrophic lateral sclerosis (ALS), have been found among veterans, but do not necessarily suggest a Gulf War Syndrome. The uncertainty further confounds the process of research because there is no consensus on which of the many potential approaches, methods, and hypotheses to pursue. An epidemiologist pointed out: "I think it's very difficult for researchers. Even though we say we're totally objective, we state hypotheses, we develop a questionnaire where we, you know, do you permit five questions or fifty questions about something? An infectious disease scientist is going to look for infectious mechanisms. A neurotoxicologist is going to [look for other things]. So the scientists kind of design their studies in a way that they hope to find things that are in their field, what they'd like to find."

THE CURRENT RESEARCH BASE FOR GWRIS

Despite the complexity of the issue and early resistance to disease recognition, the identification process has moved forward from initial lay discovery and then government and scientific discovery to the production of a body of literature addressing the prevalence and possible causes of veterans' illnesses. Officials and researchers initially developed a stress-based DEP, and veterans engaged in political struggles to counter it.

Though I emphasize lay discovery, in the case of GWRIs there is anecdotal evidence that DOD officials anticipated exposures because most troops were administered vaccinations against anthrax and received pyridostigmine bromide to protect against nerve gases such as sarin. The military received special permission to bypass FDA regulations for these drugs.[10] Pentagon officials were also concerned about the health effects of oil well fires and sent a research team from the U.S. Army Environmental Hygiene Agency to the Middle East immediately after the ground conflict to begin monitoring the pollution, but the team's report concluded negative health effects were unlikely.[11] In addition, the military expected many of the stressors present during the war to cause some troops to experience posttraumatic stress disorder, as had been the case with Vietnam veterans. Thus, the military set up centers to detect psychological stress at several bases through which deployed troops were processed upon return. The fact that many of the DOD's medical personnel were primed to treat psychological symptoms may have caused them to jump too quickly to attribute physical symptoms to psychological stress, even though very few returning soldiers were exhibiting classic posttraumatic stress disorder symptoms.[12]

When DOD medical professionals were sent to study reserve units in Indiana and Louisiana in 1994, their reports concluded that psychological stress was the cause of self-reported physical symptoms. By 1994, when the NIH Technology Assessment Workshop took place, most experts who testified before the panel had been involved in the early psychological stress diagnoses in Indiana and Louisiana. This connection may explain the panel's conclusion: "It is possible that the expression of post-traumatic distress may be distinct in the Persian Gulf experience and may take the form of somatic and multisystem symptoms rather than classic post-traumatic stress disorder numbness and flashbacks." Although the panel members' report went on to emphasize that they did not mean to suggest "that there is no physical basis for the reported symptoms," it nevertheless added that "expression of the reported post-traumatic stress disorder symptoms represents a psychophysiological response that needs to be evaluated."[13] The culmination of this stress-based paradigm was the Presidential Advisory Committee's 1996 report that attributed veterans' illnesses to stress and ultimately recommended that the "entire federal research portfolio should place greater emphasis on basic and applied research on the physiologic effects of stress and stress-related disorders."[14]

During this mid-1990s period, however, James J. Tuite III, a staff member of the Senate Committee on Banking, Housing, and Urban Affairs, was assigned by the committee chairman Senator Donald Riegle to investigate possible U.S. exports to Iraq of biological materials used to manufacture warfare agents. In response to Tuite's confirmation of Iraq's biological weapons capability and the testimony of veterans who believed they had been exposed to chemical warfare agents, Congress allocated money for VA investigations. The VA's research agenda was weighted heavily toward projects examining toxic exposures and their physiological and neurological effects.[15]

At the same time, VA research examined whether there was a higher prevalence of symptoms in veterans and explored links to particular exposures that might be the cause. This work was hampered by unresolved debates over the case definition for GWRIs and whether GWRIs represent a unique syndrome or a series of unrelated symptoms. Even though most researchers no longer believe that GWRIs represent a unique syndrome, some veterans groups do, so the debate continues. The 2001 conference on federally funded GWRI research featured a plenary session entitled "Is There a Gulf War Syndrome?" It included Dr. Robert Haley on the panel, one of the only researchers currently arguing for the syndrome definition.[16]

SIGNIFICANT FINDINGS ABOUT GWRIS

Relying on DOD and VA registries, researchers found excess mortality among Gulf War veterans only for motor vehicle accidents.[17] Other scientists concluded that there were no excess hospitalizations and no excess birth defects.[18] Veterans countered these findings, noting that although many symptoms did not require hospitalization, they affected the quality of daily life and impeded the ability to work. Eula Bingham, former Occupational Safety and Health Administration director and head of the VA's Persian Gulf Expert Scientific Committee, criticized the Presidential Advisory Committee for its failure to understand this point.[19] Critics also mention the fact that only inpatient admissions to military hospitals were included in the study and that the study period extended only twenty-five months after the end of the war.[20] It is possible that excess hospitalizations might have occurred in nonmilitary hospitals because veterans had already experienced the military's reluctance to treat them. In addition, veterans covered by their spouse's health insurance might have chosen private care due to the military's reluctance to deal with Gulf War complaints and the perception that regular hospital care would be superior.

Veterans' claims are supported by studies that show an excess of self-reported symptoms among deployed versus nondeployed troops, including chronic diarrhea, other gastrointestinal symptoms, memory loss, concentration difficulty, trouble finding words, fatigue, depression, posttraumatic stress disorder, bronchitis, asthma, alcohol abuse, sexual discomfort, and anxiety.[21] The validity of self-reported symptoms has been repeatedly questioned because physical examinations in major studies failed to find medical conditions. Proctor and colleagues found increased symptoms in many organ systems in two Persian Gulf cohorts, as compared to soldiers in Germany.[22] When veterans diagnosed with posttraumatic stress disorder were removed from the analysis, virtually no changes were noted. This finding supports the idea that people were not reporting symptoms simply because they had a stress-related problem. A further contribution of this study is that the researchers found positive correlations between several exposures and symptoms reported for the specific organ systems where the effects would be expected. Some epidemiologists consider this work on self-reported symptoms and exposures to be central to a new perspective on how to conduct research on environmental health.

Other research has investigated the possible effects of pyridostigmine bromide (a widely administered nerve gas antidote), depleted uranium, and oil well fires. Several DOD-sponsored RAND reports and the Institute of Medicine's summary of the knowledge base concerning these and other exposures concluded that there is no concrete evidence to support the notion that any of these exposures cause GWRIs. These reports noted, however, that dose-response effects at chronic low-exposure levels are not understood well enough to rule them out completely. They pointed to poor record keeping as a major obstacle. Because these reports determined that there was not sufficient evidence to decide whether there were toxic health effects, some concerned people may infer that the door is still open to conducting environmental health research. Other researchers examining the effects of pyridostigmine bromide on veterans did so by using tests of attention, visuospatial skills, visual memory, and mood to compare Gulf War veterans who reported taking this drug against veterans who did not do so, and they found that those who had used pyridostigmine bromide performed worse on executive system tasks.[23]

Indeed, research examining the neurological effects of nerve and chemical agents including sarin, DEET, permethrin, and other organophosphates has continued. Animal studies examining the effects of sarin have observed the behavioral, biochemical, and neurophysiological effects of exposures to be similar to those that may have occurred in Gulf War veterans, including decreased enzyme production, reduced brain activity, and immune system repression.[24] However, these researchers recognize that evidence is still too weak to determine fully the effect that sarin may have had on Gulf War troops.

Other researchers have found correlations between GWRI symptoms and a lowered ability to metabolize organophosphates that is rooted in genetics and that may have contributed to veterans' development of illnesses.[25] Further work has been done using magnetic resonance imaging to examine brain damage in veterans. This research led Haley to conclude that damage to brain regions may be one of the etiologies of a Gulf War Syndrome, but symptoms differ across individuals due to varying levels and locations of damage.[26] Haley, whose work the U.S. government and other scientists have regarded with much skepticism, has made a plea to researchers in the United Kingdom to continue the hunt for a disease in veterans that, Haley has argued, resembles early stages of diseases such as Parkinson's.[27]

Epidemiological studies have also countered initial conclusions regarding birth defects in children of Gulf War veterans. One study found a higher prevalence of tricuspid valve insufficiency, aortic valve stenosis, and renal agenesis or hypoplasia among infants conceived postwar by male Gulf War veterans, as well as a higher prevalence of hypospadias among infants conceived postwar by female Gulf War veterans. Owing to statistical limitations, researchers were not able to determine causation; however, these studies suggest that environmental or synergistic factors might contribute to defects in veterans' infants.[28]

Researchers continue to identify biological markers that might provide evidence of exposure. The researchers I interviewed at the BEHC believe that studies of neurophysiological and neuropsychological symptoms were a promising direction, and they found subtle changes in nerve conduction, cold sensation, finger dexterity, and executive functioning. There are also alterations in eyeball movement, weakness of lower extremities, indices of neuropsychological dysfunction, and alterations in some auditory functions. They, however, are uncertain about whether some of these changes are markers for diseases or merely clinically insignificant differences.

Using GIS techniques in combination with veterans' self-reported symptoms, spatial analysis studies have recently attempted to associate troops' locations of deployment with subsequent symptoms and diagnosis of "chronic multisymptom illness," determined by the presence of at least two of three symptoms: fatigue, musculoskeletal pain, mood changes, and cognition. Although no global spatial distribution was observed, researchers did find certain spatial clustering and a few significant hot spots, meaning that several people who were in a certain region at the same time experienced what was termed "severe postwar CMI [chronic multisymptom illness]."[29] This research suggests that despite some limitations, GIS might serve as a beneficial tool in future research.

The research base also includes studies from allied nations. A study in the United Kingdom compared a random sample of British veterans with two control groups: nondeployed personnel and troops returning from Bosnia. Deployed veterans had higher rates of unusual self-reported symptoms but no increase in mortality.[30] Likewise, a Canadian study found an increase of self-reported symptoms but no increase in mortality.[31] Other European studies report similar findings. France originally reported no increase in mysterious symptoms among their returning veterans, but they had never conducted a study of health consequences. Interestingly, the French also generally report little chronic fatigue syndrome or multiple chemical sensitivity, which may have kept French researchers from doing studies.[32] More recently, the French government responded to a veterans' suit in 2000 and instituted a study of all deployed military personnel in 2002, including both self-reports and medical examinations. The most common symptoms were headaches, neurological and psychological problems, and back pain, occurring at similar or slightly higher rates than for troops in the United States and the United Kingdom.[33]

Efforts have recently been made on the part of the U.S. government and scientists to synthesize the work of GWRI researchers. In focusing on veteran health care registry studies, hospitalization studies, studies of outpatient encounters, and health care utilization studies, researchers concluded that although the research done over the past few years has been successful in areas such as increasing veteran access to medical care, no constellation of symptoms identifiable as "Gulf War Syndrome" has been found.[34] The Institute of Medicine's fourth report for its *Gulf War and Health* series, issued in September 2006, found that despite the increased rates of certain ailments among veterans, there is no evidence of a unique syndrome.[35] In reviewing the existing literature, researchers continue to conclude that both neuropsychological and health symptoms have unclear etiologies and argue that psychological, historical, and environmental exposure factors must be considered in attempting to determine causation; it is becoming increasingly clear that a variety of variables have contributed to the reported symptoms.[36] Thus, many questions remain unanswered. However, the call for more research has coincided with a shift in federal research priorities that has resulted in a consistent decline in funding for GWRI research since the year 2000, as seen in figure 4.1. As of September 2003, 80 percent of federally funded medical research projects had been completed, and an additional 19 percent of projects were ongoing. The VA has failed to address many of the research questions laid out in 1995.[37] One leading researcher noted that some of these research questions were unrealistic from the start because of inadequate technology and insufficient time—for example, in trying to find a sensitive diagnostic screening test for leishmaniasis. The DOD, which had provided the bulk of federal funding for GWRI research, does not plan to fund new projects. This decision is not surprising; prevailing questions regarding postwar somatic symptom syndromes are typically pursued until the next war diverts attention from them, at which time they remain fundamentally unanswered.[38]

CONTESTATION: THE ONGOING CONTROVERSIES

As sufferers and their supporters challenged a DEP that did not adequately meet their needs, they faced many obstacles: (1) missing information, mis-

Dollars (in millions)

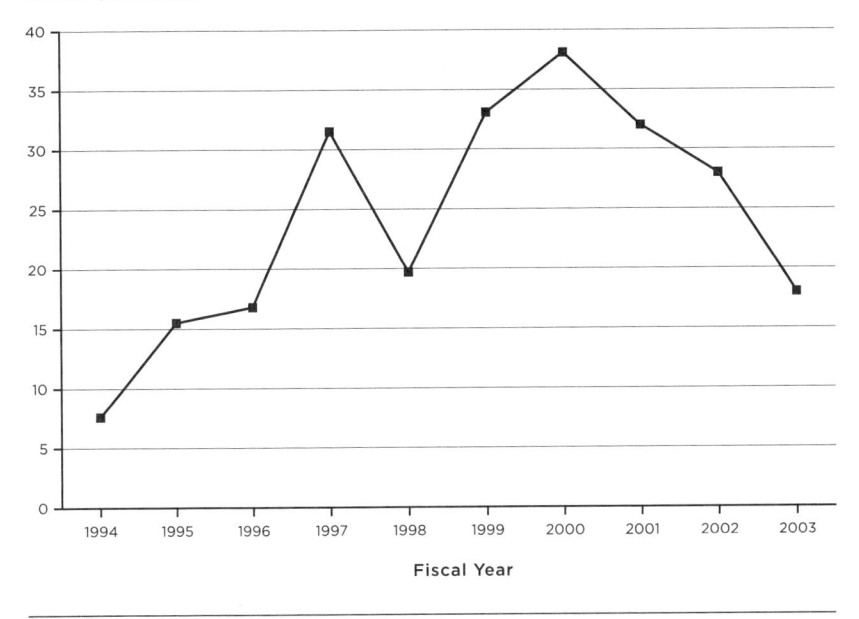

FIGURE 4.1 Funding for Gulf War research projects, fiscal years 1994–2003.

Source: U.S. GAO, *Department of Veterans Affairs: Federal Gulf War Illnesses Research Strategy Needs Reassessment,* GAO-04-767 (Washington, D.C.: U.S. GAO, June 1, 2004).

Note: GAO analysis of VA data. Only direct costs for each agency are included. Direct costs cover the actual research activities and materials and have not been adjusted for inflation.

information, and secrecy; (2) distrust, disputes, and fragmentation; and (3) issues of legitimacy.

MISSING INFORMATION, MISINFORMATION, AND SECRECY

Research on GWRIs is plagued with insufficient or incorrect information, especially the lack of predeployment health assessments and exposure records. Researchers must contend with the difficulties of retrospectively determining predeployment health status. In lieu of those data, scientists must rely on self-reported predeployment health, an approach that is typically criticized for its lack of reliability and validity.[39] Government realization of the lack of predeployment health data has led to numerous proposals for improvements in the recording of possible exposures through predeployment health questionnaires and the creation of a centralized health-tracking database to monitor the health of patients throughout the DOD medical system.[40] In addition, based on the recommendations regarding the systematic collection of health data from both the 1999 Institute of Medicine–sponsored report *Strategies to Protect the Health of Deployed U.S. Forces* and the DOD's report *Effectiveness of Medical Research Initiatives Regarding Gulf War Illnesses*, the Millennium Cohort Study was initiated in 2001 in order to monitor the health of military personnel during and after deployment.[41] This longitudinal study is the first of its size and nature (more than seventy-nine thousand individuals had enrolled as of 2005), and its planners hope it will provide a better understanding of the health risks associated with deployment. Despite that enrollment, the Millennium Cohort Study has only a 30 percent response rate, and key researchers doubt it will answer questions that have not yet been answered.

Interviewed veterans and researchers identified the denials and disputes over exposures as major barriers to progress. One veteran remarked, "From a political standpoint, the biggest job for us is getting the government to recognize the findings that they don't want to hear about." One VA researcher commented:

> The government hasn't been real forthcoming with releasing information, and that's causing problems with the research. Right now we have a grant [from] the Department of Defense, looking at some people who were actually around where the detonation of chemical warfare agents occurred. And getting the information from them is extremely difficult. You feel like you're trying to do research with your arm tied around your back. . . . So I have to go scrap around and beg and go through all the channels to see if I can get this released even though the Department of Defense has funded this study. Because I'm not in the Department of Defense, I constantly get the feeling that I just get the information they want me to have.

For many veterans, it seemed clear that they were exposed to toxic substances, as one explained: "When Gulf war veterans would say, 'You know, I was exposed to sarin in the Gulf,' the people at VA would very diligently follow up with the appropriate people in the Defense Department, and the Defense Department's line at that time was, 'There was no sarin in the Gulf.' So VA would come back to the veterans and say, 'Sorry, Joe, you weren't exposed to sarin.' And the veteran is looking at this person like they're crazy and saying, 'Wait a minute, I was in the Gulf, and you were never even a vet, no less in the Gulf. How could you possibly tell me what I was or wasn't exposed to?'"

Similar experiences disillusioned many veterans. Until public disclosure about the presence of nerve gas at Khamisiyah, a major depot bombed by the United States, the DOD maintained that no exposures occurred, thereby denying vital information to veterans and researchers. When the President's Advisory Committee asked the DOD to turn over the extensive logs of exposure data from the Gulf, the DOD was able to locate only thirty-six of an estimated two hundred pages.[42] Whether the DOD hid the missing logs, as some claim, or something else happened to them, the missing information is a serious impediment to research.[43] Furthermore, in 2004 the government issued a report claiming that the DOD and the Central Intelligence Agency's (CIA) 1996 and 1997 plume model studies of chemical exposures at Khamisiyah were flawed. The government report found several errors in the DOD research, including inaccurate models for gauging long-range dispersion of chemical agents, source data based on incomplete information, underestimated plume heights, field tests unrepresentative of the experienced conditions, and inconsistent results.[44] Using the plume model designed by the DOD, researchers found in a 2005 study that exposures at Khamisiyah were associated with an increased risk of brain cancer death, but that was the only association they found out of fifteen diagnoses.[45]

Despite the fact that chemical alarms sounded daily, and soldiers were trained to don gas masks upon hearing them, military commanders later told Congress that the Defense Intelligence Agency and CIA said that many were false alarms, yet they had not communicated this information down the chain of command. Thus, soldiers believed they were being exposed to chemicals when they were not. As the 1994 NIH panel observed, "Although warfare has always been stressful and fear-inducing, the Persian Gulf War was the first combat experience in which the real threat of chemical and biological warfare was known to troops before entering the combat area."[46]

According to many researchers, there is no serious dialogue between the researchers performing studies on GWRIs and the physicians who are treating the veterans. As one epidemiologist noted, "I know in the veterans' eyes it's a real issue. They have a lot of complaints about their medical treatment and their clinical care that they come and voice to us. But as a resource center we're not directly responsible for taking care of them clinically. I tried to help and direct them into proper channels within the clinical care side of it, people that I know here. But there is a division or separation, and so we could spend our whole life trying to clinically get them treated."

As a consequence of this lack of communication, researchers told me, uninformed clinicians are not asking their patients questions that might provide clues to etiology. Thus, patients may not receive the best possible care, and valuable research data are not gathered. In addition, although the VA funded several research centers to do dedicated work on GWRIs, it did not provide ongoing channels of collaboration between those centers.

DISTRUST, DISPUTES, AND FRAGMENTATION

DISTRUST OF THE GOVERNMENT Many veterans whom my colleagues and I interviewed distrust government research on toxics because of past practices. For decades, the government denied the existence of diseases among "atomic veterans" and civilian workers in nuclear weapons production.[47] For Vietnam veterans, the Agent Orange Act that specified compensation for Agent Orange–related illness was not passed until January 30, 1991, sixteen years after U.S. troops left Vietnam. The Institute of Medicine did not link Agent Orange exposure to Hodgkin's Disease, Non-Hodgkin's lymphoma, soft-tissue sarcoma, chloracne, and spina bifida until 1994, sixteen years after the first congressional hearings on the issue and twenty-four years after the spraying.[48] The government did not validate connections between Agent Orange and diabetes in veterans or birth defects among the Vietnamese until 2000.[49] Some researchers point out that it took this length of time to conduct studies that actually showed the data, though others believe there was inadequate attention to the subject. As the director of a veteran's group noted, "How long did it take those exposed to the atomic bomb . . . they actually had people, and they wanted to find out what effect this stuff had on them. They tested it out—radiation. . . . So there's a pattern, it looks like to me, that, 'Hey, if we wait long enough, those of you who are sickest are going to die off, and we don't have to spend all this money to give you the benefits.'"

After the Gulf War, sick veterans complained that officials offered only a psychiatric explanation and would not take seriously their reports of toxic exposures. One remarked:

> When . . . Gulf War veterans went to clinics and hospitals run by the Department of Defense and Veterans Affairs, they weren't met with open arms by many doctors. Many doctors thought that they understood readily what was occurring with these veterans. It was a short war; this must be mental illness or posttraumatic stress disorder. . . . And then, when veterans complained about being exposed to chemical weapons in the Gulf—seeing them, the alarms going off, reporting symptoms consistent with exposure to chemical weapons—these claims were not investigated seriously or with any sort of vigor at all by the Department of Defense. . . . When, after all that, these agencies now are tasked with investigating Gulf War Syndrome, do you think veterans are gonna believe the kinds of conclusions they come up with? Absolutely not.

Similarly, Dr. Michael Hodgson, a former American Legion medical advisor and later a VA researcher, remarked, "It is hard for me to believe that you can announce every other week that yet another 10 or 15 thousand may have been exposed to nerve gas and then convincingly tell veterans they don't have nerve gas disease."[50]

There are also disputes between government bodies: the DOD, the VA, the CIA, the Presidential Advisory Committee, congressional committees, the NIH panel, the Institute of Medicine committee, and the HHS. At the center of these disputes is the DOD, accused by veterans groups of concealing exposure information and strongly criticized by the Presidential Advisory Committee for mishandling the issue, often in surprisingly strong terms: "DOD's slow and erratic efforts to release information to the public have further served to erode the public's trust."[51] Even though the Presidential Advisory Committee criticized other agencies, it too tried to censor information. One of its researchers, Jonathan Tucker, claims he was fired for gathering information from veterans and whistleblowers as he sought to show that the Presidential Advisory Committee overemphasized the stress paradigm at the expense of research into chemical exposure.[52] Another government panel, the VA's Persian Gulf Expert Scientific Committee, headed by Eula Bingham, was very critical of the Presidential Advisory Committee's support of the stress perspective.[53] There were also significant disputes among committee members.[54]

Congress responded to the DOD and VA shortcomings. Congressional representative Christopher Shays (R-Conn.) requested that the General Accounting Office investigate why federal studies failed to confirm the existence of GWRIs or their potential environmental causes, and the resulting report was quite critical. Shays chaired a House committee that severely chastised the Pentagon and the VA for poor work and asked that they be removed from further oversight of GWRI research.[55] Dissatisfied with VA research and treatment, Representative Lane Evans (D-Ill.) introduced a bill in 1998 to have the Institute of Medicine compare Gulf War veterans with other veterans; any conditions found to be different would lead to a presumption of disease, treatment, and compensation without concern over cause.

Because veterans view the government as the cause of the problem, they question its ability to perform objective science. (Note that this skepticism is different from the distrust that asthma and breast cancer activists have toward what they consider appropriate science. For the latter activists, research is typically conducted by academic scientists who may be funded by federal grants, but so is a large part of all academic science. Gulf War activists face a situation in which a very large amount of research is conducted directly by the federal government, and even the federally funded work is largely directed by government officials' agendas.) The issue of government objectivity is clarified by determinations of what would be seen as a conflict of interest in other areas of health; if, for instance, corporations were responsible for researching occupational health issues in their industry or facilities, objectivity would immediately be a major concern. A 2001 report by the National Gulf War Resource Center, one of the more vocal veteran organizations, recommended that "Congress should reprogram funds allocated to GWS [Gulf War Syndrome] research away from . . . DOD-DVA medical research programs (including all examining stress) into private sector or state-run medical research initiatives (overseen by an independent body, including veterans service organizations)."[56] Even though veterans are invited to be part of panels and conferences, they continue to feel slighted. This was apparent at the 1999 conference of federally funded Gulf War researchers. The conveners dedicated a panel to veterans' concerns, with veterans as speakers and DOD officials present to answer questions, but poor attendance by researchers and officials suggests a lack of interest in veterans' perspectives. At the 2001 conference's "Public Availability" session, at which veterans' concerns might have been made known to a large audience, the conference organizers arranged the format as a series of small, informal roundtables. Veterans in attendance said they felt this approach was an attempt to fragment and silence them.

DISTRUST OF THE SCIENTIFIC PROCESS Because many veterans felt alienated by the government, they did not trust scientists whose research was funded by government sources, much less government scientists themselves. The following exchange among veterans, government officials, and researchers on a panel at a Gulf War research conference captures this strain in relations:

> VA OFFICIAL 1: My concern as a scientist has to do with how best we can communicate to vets some of the complexities of research results that many of us went to school for years to try to understand. These really are difficult issues in terms of the science and trying to interpret it for you as well as for ourselves. What better things can we do to do that?
>
> VETERAN ACTIVIST 1: I was only a political science major, but the following passage [reading from a military report on depleted uranium] doesn't need any further interpretation. It's clear that it is concluding there are health effects.
>
> MILITARY OFFICER: [reading from another passage in the same report] Based on this, it is clear there are no health effects from [depleted uranium].

VA OFFICIAL 2: It's confusing when study A gets this and Study B gets that. Science is complicated.... How many times have you watched TV, and it tells you eat margarine, then two months later, don't eat margarine? So this process is very complex.... The bad news is this takes time and our patients want answers now.

GULF WAR RESEARCHER 1: You all have to ask yourselves, "What does it take for you to believe something?" It takes more than one of us and one piece of research before we can agree on something. Please keep that in mind when you try to market findings instead of using good science.

VETERAN ACTIVIST 2: When have we had enough research? Vets are very confused by all the research. How important is a finding? There is a correlation between soda and polio, but we don't blame Coca Cola for the disease. Researchers should report when they have accepted a null hypothesis. It might be a pain because then you have to design totally new research to get your funding for the next year, but it could be important to vets because it answers a question for them—it rules out that possible explanation.

GULF WAR RESEARCHER 2: I'm not sure we're ever gonna know. There's a lot of data that weren't collected. It doesn't really matter what causes their illness. We need to know what we can do for treatment.

At this point, the focus of the discussion could have shifted to the lack of data and either how that obstacle could be overcome or how in its absence policy might be made on a value basis instead of on a scientific one. Instead, when two veterans tried to push the issue of shifting funds from research to compensation, a researcher attacked them for moving the discussion away from research.

VETERAN ACTIVIST 3: The VSOs are saying to you, "Treat our sick vets and compensate them." We don't want to be involved in DOD policymaking. Listen to the population of veterans. DAV [Disabled American Veterans] has one million members. We want answers to very simple questions. VSOs are not out screaming for additional research studies. We've created a cottage industry for research on Gulf War illnesses.

VETERAN ACTIVIST 4: One thing about communication is I would just encourage openness to the fact that all the results are not yet in—openness, communication, and listening to both sides. We've talked about how much research has been done, but not about the other side, how much compensation has been paid out. And I know this is a research conference, but still.

VA OFFICIAL 2: This is a research conference . . . and we have this habit of presenting research at these sorts of things. I spent twenty-three years directly caring for vets. They didn't want to know about uncertainty. They wanted certainty. And when there isn't any, they want to know what I would do in their situation. If anyone sitting here thinks we have definitive data, it belies their ignorance. . . . It's been instructive to me to see how we all see the same research and pick out the line in it that supports our position. The difficulty is that we can always keep doing that. What you have to deal with is at a policy level, with the degree of certainty we have, how are we going to treat human beings?

An example of how veterans were steered away from discussing political obstacles to the science took place at the 2001 conference as well. Following a panel titled "Approaches to Case Definitions: Is There a Gulf War Syndrome?" Patrick Eddington, a veteran and former executive director of the National Gulf War Resource Center, attempted to address the panel of researchers: "An issue that I want to return to very briefly is this whole concept of not having all the exposure data. I think one of the biggest problems that all of you who are receiving funding to look at this issue is that we continue to have a huge volume of information that is classified by the Department of Defense and Central Intelligence Agency."

But before he could continue, Tim Gerrity, the panel's moderator and former assistant chief research and development officer of the VA's Office of Research and Development, interrupted Eddington to inform him that he was going beyond the bounds of science. When Gerrity told Eddington, "That is not the topic of this meeting," Eddington replied, "Dr. Gerrity, I'm afraid it is the topic of conversation here because we're talking about ammunition . . . from Iraqi military units, ammunition very similar to the kind we destroyed at Khamisiyah." Before he could finish his point, Gerrity interrupted once again: "We're going to have to go to the next question." Eddington, speaking over the next questioner, commented, "I think that answers my question." Continuing to interrupt the next questioner—Peter Spencer, a toxicologist on the panel—Eddington explained how the DOD thwarted his attempts to do the necessary science: "To the extent that it would have assisted our study to have individuals who had positive evidence of prior exposure to sarin . . . would have been most helpful. We made repeated attempts to obtain such data, but they were not forthcoming."

Veterans' attempts to introduce the problem of inadequate data can be seen as an implicit claim that science lacks the ability to arrive at conclusions with any certainty. As the previous exchanges illustrate, some scientists would rather keep the debate focused on their limited but privileged realm of understanding. The alternative is to acknowledge that the limitations posed by inadequate data require value-based rather than science-based policies. Nevertheless, in the face of these struggles, many veterans have continued to call for more science, but in areas of their choosing and with their involvement. In the next section, I discuss the extent to which veteran participation in the scientific process has both helped and hindered their cause.

LAY INVOLVEMENT Veteran involvement in the scientific process has ranged from bringing the issue to the attention of the public, government, and researchers to shaping research agendas and protocols. Both veteran and researcher interviewees felt veteran activism was instrumental in getting the government to acknowledge that veterans were sick and that investigations into the causes needed to take place. One researcher reported, "I think they've had the critical role. . . . I still think back to when I went to this meeting . . . and there was a panel of Gulf War veterans, some of them with unusual illnesses, who provided very emotional, compelling, and scientific testimony about what happened to them. And I think that that type of involvement and intervention really helped push the agenda."

One example of successful involvement on the part of veterans is found in the VA's recent promise to offer compensation to Gulf War veterans diagnosed with ALS, also known as Lou Gehrig's disease.[57] Despite the VA's initial denial that Gulf War veterans were suffering disproportionately from ALS, a joint study by the VA and the DOD was launched after veterans continually voiced their concerns about a seemingly high number of cases of ALS among Gulf War vets. Not only did action on the part of veterans result in compensation, but in March 2003 the VA announced it would be developing a registry of living veterans suffering from ALS. Some researchers commented that this gesture was somewhat disingenuous because ALS affects less than fifty veterans and costs the system little to take on as a problem. Other medical conditions and complaints of symptoms are far more widespread—up to one-third of veterans by some estimates—and would cost the VA a large amount to cover.

When it comes to the actual inclusion of veterans in the design and implementation of research, however, both veterans and researchers describe both frustrations and positive experiences. One veteran who had served on a VA expert scientific committee explained that although "representatives from the veterans organizations played a positive role with the scientists in shaping study design," there were a number of veterans on the committee who wanted to be more involved. Merely providing a link to the community of veterans who serve as research subjects is not meaningful participation for many veterans.

One interviewee explained that including veterans on advisory boards is becoming a standard in research, so much so that in order to receive federal funds for research on Gulf War illnesses, researchers are required to have an advisory board that includes veterans. Yet, as he explained, scientists may not always accept such a requirement with open arms: "The government's making them do it because it's a really good thing to do, but . . . overall I find ambivalence. There are some [researchers] who want to do it, and there are some who say it's really going to be a pain in the neck. Many of them are wary to let not only nonexperts but in some cases angry nonexperts involved in the process. I don't see the scientific community embracing this."

Another veteran explained that his participation on advisory boards had left him feeling like he was giving more than he was receiving: "I'm still involved in the circle of veterans, and I don't mind giving advice [to researchers] on what's happening, where to go for help, and how to get it. What I found out was that they're not doing the same thing for us."

Researchers, on the other hand, spoke of having to balance positive veteran contributions with scientifically ill-informed requests and demands. Some researchers also reported that veterans misperceived them: "They come in and just think that we're here to get grant money and salary money, and we don't care about the veterans." Others explain that they cannot win no matter what their research finds because of pressure from both their employers (the DOD) and veterans: "So let's say this study I'm working on, I find a positive association between the environmental contaminants [and veteran health]. So on one hand you have DOD that basically fights that. On the other hand, I have, if I don't see an association, the veterans groups that are outraged that you're not showing anything."

Despite the tensions that emerge from lay participation in the scientific process and the failure of such participation to lead to answers regarding the health of veterans coping with GWRIs, both veterans and researchers I interviewed hoped that in the future the collaboration would continue to take place and that it would happen from the beginning of a project so as to avoid suspicions, resentment, and other feelings that hinder cooperation. According to a navy admiral and GWRI researcher who spoke at the 1999 conference, "The greatest fallout from Gulf War veterans' illnesses has been the loss of trust. If they don't trust me, I can't help them. When you lose trust, you are in a hole, and you have to dig your way out." For many activists, the obstacles they believe the government placed in the way of sound scientific research into the causes of GWRIs were precisely the reasons for widespread mistrust and suspicion, and no amount of veteran participation in the research process would be sufficient to erase these feelings.

Although the mistrust that resulted from these obstacles as well as the obstacles themselves have perhaps prevented science from ever identifying with any degree of certainty the causes of GWRIs, many veterans I interviewed continued to be optimistic regarding the future of government handling of veteran health and more generally lay participation in the scientific process. One veteran explained that a growing number of officials in key institutions have come to embrace the idea of lay participation in the research process: "They really believe that the collaboration was not only important, but that it's always got to happen that way. And so I think from that perspective, even though they have some scars from this, I think the next time around the lines of communication are going to be open. The lines of communication are open now. . . . I think they're in a better position to help folks next time."

Another veteran also emphasized improvements in communication between veterans, the government, and scientists: "Communication doors are open that weren't before. They're starting to realize that they've got to deal with us." In a general reaction to the process of lay participation, one veteran explained: "It's an ugly process, and I've been disappointed by it often. But I think in the end better than average policy is made. If it weren't more democratic, I think people's suspicions would have boiled over by now. And so even though it's ugly when you get everybody involved, and I know scientists sort of cringe and hide in the corner when nonexperts are in the room ranting and raving, I think everybody benefits ultimately."

These sentiments suggest that lay-professional collaboration, even if it does not result in a solution to the problem, may at least leave participants optimistic about their future ability to be involved in risk assessment decisions that affect their lives.

Yet it is very clear that there are no effective citizen-science alliances for GWRIs as there are for asthma and breast cancer. Asthma and breast cancer alliances have been more likely to be initiated by activists, including many asthma activists with extensive organizing experience and breast cancer activists with extensive political connections. Both asthma and breast cancer citizen-science alliances have contained many laypeople with considerable education and connections to academics, features less common for the Gulf War veterans. Thus, the Gulf War activists had a much weaker lay input and little influence on the conduct of research. Further, alliances were forged late in the controversy, after many veterans had lost trust in the government and, in some cases, in science's ability to find any answers. These alliances were hard to form because the military and VA maintained very tight control of data and research. In addition, Gulf War veterans adhered to a military culture that made criticism of government research and treatment seem like an unpatriotic attack. Though many of the researchers I interviewed provided examples of ways in which veterans possess privileged information about their own exposures and illnesses that can be used in health research, and veterans themselves firmly believe that their knowledge is crucial to understanding the causes of GWRIs, incorporating this knowledge and these experiences must happen earlier in the problem-solving process.[58]

A brief comparison of Gulf War veterans' organizing efforts with asthma and breast cancer activists' efforts provides a better understanding of the veterans' problems (see chapter 5 for a more detailed comparison). Gulf War veterans would have been satisfied to have the government say it was sorry they had become ill and provide the care and compensation they were seeking. Their anger developed when they were either turned away from health services or told that the problem was all psychological. In contrast, asthma and breast cancer activists coming from the environmental end were not seeking apologies, were less concerned with access to care, and were focused on prevention.

For the most part, the veterans did not want to get involved in scientific research. They did push for certain chemicals and other exposures to be studied. For example, many activists felt that exposure from the Khamisiyah chemical weapons depot and the oil well fires were not studied enough. But veterans were not clamoring to be partners in the research. A small number of them became well versed in the research literature, but they did not approach the level of education and participation attained by environmental justice activists engaged in community-based participatory research work on asthma or by breast cancer advocates, as seen in the NBCC's Project LEAD, which goes around the country training women to become eligible to serve on funding review panels. Asthma and breast cancer activists were more eager than veterans to ask for a seat at the research table, for several reasons. Environmentally focused asthma action was steeped in an environmental justice movement and its civil rights movement background. Many breast cancer activists had come out of the women's health movement, where women had been very active in scientific areas, especially around reproductive rights. In addition, many of the women activists were professionals, including scientific professionals, even if not in the field of breast cancer research, and they readily developed a desire to be involved in the science. Finally, they had the education, resources, and connections to pressure government agencies to allow them in.

Unlike Gulf War veterans, asthma and breast cancer activists benefited by having recognized disease entities that continue to occur, whereas GWRIs were an exceptional and mysterious occurrence and as a result solicited less attention than illnesses with ongoing causes. Breast cancer activists also benefited from the feminist, women's health, and AIDS movements. Gulf War veterans, in contrast, received support from Vietnam veterans, but relatively little support from veterans of other wars and even less from other social movements more generally. Furthermore, their experiences with disability went against traditional notions of masculinity pervasive in the military. Years of advocacy and activism also put asthma and breast cancer into the public light, casting off much prior stigma. Asthma camps, athletes' testimony on their histories with asthma, and extensive school-based education brought about a new public perception of asthma. Races and walks for the cure, elegant fund-raising dinners, and other public displays made breast cancer very publicly noted and discussed. In comparison, Gulf War veterans faced the stigma of being accused of malingering or having a psychiatric disease. As Gulf War researchers increasingly found Gulf War illnesses to be similar to chronic fatigue syndrome and other MUPS, there seemed to be no special illness or syndrome for which veterans could attract support and sympathy.

As research failed to turn up environmental causes for GWRIs, veterans became disillusioned, and their social movement lost strength. The asthma and breast cancer movements, however, continued to grow. Asthma researchers are succeeding in showing environmental sources of asthma attacks and even some causation. Even if breast cancer researchers are not finding environmental causation, increasing numbers of activists have taken up the position that environmental causes are worthy of much study. Asthma activists and the breast cancer movement focused on environmental justice have had strong local, state, and national organizations and a framework in which activists can carry out many specific tasks, thus giving them a tangible sense of doing something. For Gulf War veterans, no similar level of organizational connectiveness took place. Further, veterans were demobilized to disparate locations and lost the geographical cohesiveness that has been shared by many women with breast cancer.

FRAGMENTATION AMONG VETERANS As mentioned earlier, there were disputes among veterans, with more established VSOs being wary of more radical challenges from Gulf War activists. Veterans' collective strength was diminished by the division between those who had completely lost trust in the government and the science it was producing and those who wanted to work within the system to shape the research agendas in ways that might benefit veterans. As one veteran working for a VSO explained, "A lot of people who sort of lead the charge . . . are very upset and very angry, and vociferous and energetic, and they usually scare the hell out of policymakers when they show up at meetings and start screaming. And I've learned to empathize with these people because they're very frightened, . . . and they haven't gotten answers, and policy hasn't been always very helpful. . . . I think those people have been an impediment because they've skewed policies, and they've actually caused some policymakers and elected officials to try to avoid the issue because it just seems so poisonous."

The tension between mainstream VSOs trying to work within the system and sick veterans issuing passionate pleas for care and compensation was also observed at a 1999 conference on federally funded Gulf War research. At a plenary session during the conference, Jack Feussner, chief research and development officer of the VA's Office of Research and Development, explained that some people become too emotional about issues surrounding veterans' health, a claim commonly made by experts confronting toxic waste activists. He then acknowledged representatives of several VSOs for their efforts to remain calm and rational.

At the same conference, during the only session organized explicitly to give veterans a voice, veterans from different VSOs began debating with each other instead of directing their questions to government officials on the panel. As a veteran representing a mainstream VSO explained during a debate over the possibility of depleted uranium's role in veterans' illnesses, "The people I served with are not sick from [depleted uranium], though they are sick. I hope that the administrators and grant makers here will refrain from listening to who is screaming the loudest and put together a body of research that can help us help sick veterans. Our biggest concern is getting these people treatment and compensation for their illnesses."

This veteran then qualified his remarks by adding that decision makers should not wait for a scientific explanation to come together perfectly and that if they know something today, they should inform the VSOs so that these organizations can attempt to get laws passed to help the veterans. The debate among veterans and VSOs at its core is a debate about the role of science and values in policymaking: Do veterans deserve care and compensation because a country grateful for their service owes it to them or because scientific evidence links their illnesses to their service?

Another VSO representative's comments further illustrate the struggle to find the proper balance between relying on science and moving ahead with treatments and compensation in the absence of certainty:

I think that if we completely ignored what the science was saying we never would have gotten anywhere on Capitol Hill. If I was dismissive of the science, staff would have been dismissive of me. So we ... would always recognize what the science was finding.... We didn't know how they [postwar sick veterans] were doing before the war, so we ... just have to assume that they were in good health. Being in a veterans organization, I'm going to make all those assumptions in the vet's favor.... I'd go up to Capitol Hill and say, "Yeah, we don't know, but we're going to have make a number of assumptions. We're going to have to assume they were in perfect health before they went to the Gulf and that if the population itself is reporting poor health at a greater rate than matched controls, then it was the Gulf." And there's a method to the madness. If there's no nexus between Gulf War service and veterans' current health status, the government's not obligated to pay any compensation. The government's not obligated to treat them for their illnesses.... So we'd take sometimes shreds of evidence, the little drips and drops of evidence from the science, [and] we'd make more of it than there was there, a lot like lawyers do in court cases.

In striking a balance between engaging the science and demanding that veterans' needs be met regardless of the science, veterans and VSOs have taken two primary approaches: calling for a change in research methods and standards of proof; and shaping the direction of research by suggesting possible exposures and interactions the veterans might have experienced. One VSO representative's comment demonstrates the attempt to push the science toward treatment trials, despite the absence of any understanding of the causes of GWRIs: "It was recognized really around 1995, 1996 that we had to start doing clinical trials in addition to all the epi[demiology] that was going on. And the reason we figured that out so fast was all I do is talk to sick Gulf War veterans on the phone and visit with them when I travel. So unlike some of my colleagues over in the federal agencies who don't get out much, I spend a lot of time outside the Beltway." However, this same interviewee contrasted his organization's attempts to force the government to begin clinical trials with a mild dismissal of science altogether: "We were pushing really hard for disability compensation benefits to be made more generous. And again, we don't let the science get in the way, and that's sort of not our role. We don't pretend to be scientists. If the science is fuzzy, we always err on the side of the vet. We've been criticized for that by some policymakers, that we don't let the facts get in the way, as they say. But there aren't an awful lot of facts, other than that we have a lot of sick vets."

The fragmentation of the veterans movement has been more problematic than fragmentation in the breast cancer movement. Major struggles have occurred within the mainstream breast cancer movement and between parts of the mainstream movement and the environmental wing. Still, breast cancer groups have enormous organizational resources and skills, and they can still achieve many gains despite such struggles. Further, the more environmental and political wing has been able to win over many others who have more traditional approaches, thus creating more potential for unity.

STRUGGLES FOR LEGITIMACY AND THE EMPHASIS ON STRESS IN THE DEP

The struggle to legitimate both the study of GWRIs and the illnesses themselves creates tension and conflict for both veterans and scientists. Veterans often explained to me that they understood the need for a scientific understanding of their illnesses, but that they nevertheless needed treatment until that understanding emerged. As one veteran put it, "I understand the protocol for studies, but I need treatment." Another explained the need for a better procedure for handling situations of uncertainty with respect to war-related illness: "I talk to other veterans, and they're just confused. They're really adamant [sic] about going to the VA and getting tested, probed, pricked, and all that stuff. So they have to have a better way. . . . They have to come out with a better systematic approach to quality care. And that issue just isn't there."

The father quoted earlier who founded an organization to help veterans and their families also captured the struggle between rejecting and accepting science. On the one hand, he felt that the science is necessary to identify treatments, but, on the other hand, he felt the government is obligated to fulfill its contract with veterans: "I would love to be able to stand up and say, 'it is the [pyridostigmine bromide] pills.' Okay? Because all we're interested in, all I'm interested in as a father, and all that [my organization] is interested in is finding out what's wrong with our troops. Let's treat it, get it taken care of, and we can get them back to being viable contributing members of the community."

But even those veterans who pursued a scientific explanation that could support service-connection claims soon realized that various obstacles—both scientific and political—would make their case that much more difficult to state. Veterans' experiences of their illness were simply not consistent with the growing medical and scientific knowledge of their illnesses.

Veterans want the government to recognize their illnesses as legitimate and to ensure eligibility for treatment and disability benefits. They also want a sign of official belief that they are not mentally ill—which has real psychological value to ailing soldiers. For many veterans, the reliance on a model that ties illness only to stress, with the VA's accompanying cognitive behavioral treatment trial, is a form of delegitimation.

Though my data suggest a shift in the DEP from a *primary stress perspective* (stress as the primary causal factor) to what I term a *contextual stress perspective* (stress as one part of a complex equation of factors), many veterans feel their illnesses are not legitimate as long as researchers claim that stress plays any role. It is understandable that veterans would oppose the primary stress perspective, which is often framed around the idea that all wars are stressful and that soldiers have always returned with some stress-related illness.[59] That approach focuses on individual psychopathology and minimizes the effects of toxic substances and other environmental conditions. It also taps public stigma concerning stress as a cause of physical health problems, as noted by Senator John D. Rockefeller IV (D-W.V.): "When you say 'stress' to the American people, when it's diffused through the media, they think it's something psychological, it's something of the mind, when in fact these people—maybe 50,000 or more of them—who went over there completely healthy and came back who are now very, very sick, and it's not just a stress syndrome."[60]

This resistance to stress explanations is captured by Congressperson Bernard Sanders' (I-Vt.) reaction to the Presidential Advisory Committee's conclusion that GWRIs are primarily stress related. Sanders drafted a letter demanding that the committee chair rewrite the conclusions.

Nevertheless, some might think that a primary stress model would be beneficial. After the Vietnam War, veterans stricken by nightmares and intrusive memories and some sympathetic mental health professionals lobbied for the inclusion of a combat-related disorder in the third edition of the American Psychiatric Association's *Diagnostic and Statistical Manual*. They were successful, and the establishment of posttraumatic stress disorder was the result.[61] However, although the creation of posttraumatic stress disorder validated Vietnam veterans' suffering, many mental health professionals have concluded that Gulf War veterans do not meet the clinical criteria for posttraumatic stress disorder.[62]

Whether they meet the criteria or not, the fundamental difference between the experience of Vietnam veterans and that of Gulf War veterans is the government's attribution of responsibility. Vietnam veterans accepted posttraumatic stress disorder diagnoses because it meant they could get treatment and that the government was actually attributing their condition to their experiences during the war. The government has not taken responsibility for Gulf War veterans' illnesses in the same manner. In many respects, it was willing to take responsibility for the stress effects, but veterans held on to a toxic contamination model. In the words of one researcher, "the government wanted to call it 'stress of war,' while the veterans wanted to call it 'dirty battlefield.' " Gulf War veterans view a posttraumatic stress disorder diagnosis or other stress-related explanation as the government's attempt to place responsibility on them for their problems.

Because of the valence of the primary stress explanation, many veterans still oppose the contextual stress perspective, even though it acknowledges an interaction between stress and possible toxic exposures. One veteran formulated the problem this way:

> Many of the vets, in the early nineties after the war, were either diagnosed with major depression or posttraumatic stress disorder or anxiety disorders and things like that, and they reacted ferociously, many, to these diagnoses and suspected that physicians, especially physicians who worked for the government either in the military or VA, were trying to deny the legitimacy of their symptoms by suggesting that they were psychological in nature. So after a couple years, you couldn't say "stress," and you couldn't say "mental illness" in a room of Gulf War veterans without really risking your health, and so the debate was hampered by that, because, you know, being exposed to oil well fires I'm sure isn't good for you, being in a deep funk, and being exposed to oil well fires is probably worse for you.

The contextual stress perspective establishes the link between psychological stress and poor health, seeking to understand how stress impacts the body physically. If this framework is applied, the reality of veterans' illnesses is not denied, but because of the earlier use of the primary stress perspective veterans are equally resistant to the contextual stress perspective. As already discussed, the National Gulf War Resource Center has called for the transfer of funds away from all DOD- and VA-funded studies examining stress. If sick veterans were thought to be manipulating the system to get compensation and benefits, we might expect them to accept a stress explanation because it would justify benefits as happened with posttraumatic stress disorder in Vietnam veterans. That veterans continue to challenge even the emerging DEP, which addresses the interaction of stress with environmental exposures, illustrates their deeply held belief that their symptoms are attributable largely to environmental exposures.

Although veterans have struggled with legitimacy in terms of defining their illnesses, scientists have also dealt with threats to legitimacy, especially those who pursue alternative hypotheses or work with lay people as coparticipants in the research process. Many scientists assume that involving lay people means lowering their own standards of proof and compromising the practice of good science. Consequently, researchers who embrace community action models are subjected to greater scrutiny. One lay advocate even suggested that funding may have been withdrawn from some studies due to lay involvement. GWRIs are not included in medical and public-health curricula, thus leaving it unfamiliar to potential researchers. Other researchers may decide against investigating GWRIs after they note how long it took to research Agent Orange and thereafter to compensate Vietnam veterans. Researchers' institutional requirements, such as tenure and the need to obtain part of their salary through grants, may also influence their willingness to study such a controversial and difficult topic.

One key barrier is the determination of what counts as evidence. For veterans, both self-reported symptoms and self-reported exposures are important pieces of evidence. For the military and most researchers, self-reported symptoms have marginal acceptability, and self-reported exposures are virtually never acceptable. Gulf War veterans give much credence to personal narratives of healthy people who returned from the war with illness, whereas the military and researchers see these narratives as merely anecdotal material that is contradicted by large epidemiological studies. When veterans uncover examples of military secrecy, such as the reluctance to admit that the U.S. blew up the chemical weapons depot at Khamisiyah, they take it as evidence that their ill health is due to exposure to toxics. But the military relies on data that show no difference in health status regardless of proximity to the depot. In addition to these differences, which might be explained as expressions of lay explanatory models, veterans also focus on the few scientific studies that report positive findings on toxics and health effects. Despite widespread methodological criticism of those studies among scientists, veterans hold them as key evidence of damage from toxics.

The case of GWRIs illustrates how external pressure and internal efforts can lead to a change in the DEP. For example, a former VA employee reported that over a five-year period, he saw the VA "moving from a defensive posture and a posture of denial to a posture of addressing veterans problems and acknowledging that this isn't a matter of blame, that this is a matter of, 'we have a duty to help them get better.'" He explained this change:

> I think it was a combination of things. I think the average veteran was just saying, "Would you please acknowledge there is something wrong with me and don't tell me it's in my head?" The other is . . . despite the fact that we were being beaten up by certain corners of the community, particularly Congress, scientific panels were providing confirmation of what our findings had been up to that date, and I think that, even though there were criticisms by those groups too, it allowed us

to be a little more bold, and also those committees taught us really good lessons. The Presidential Advisory Committee . . . was one because . . . they took to task VA and DOD for not doing the best job to reach out and take care of veterans and also the failure to communicate effectively.

In challenging the initial stress-based DEP, veterans began by seeking assistance from their representatives and senators. Some felt that with only one Gulf War veteran in Congress (Rep. Stephen Buyer of Indiana), Congress was not as great an ally to them as it was to veterans of other wars. Nevertheless, they maintained that veteran pressure on Congress has led to many beneficial outcomes: the 1992 legislation requiring the VA and DOD to create health registries; 1994 legislation permitting the VA to provide benefits to veterans with unexplained symptoms; allocation of $150 million for research; the 1997 requirement for deployment health assessments; the 1998 creation of the Research Advisory Committee on Gulf War Veterans' Illnesses; and the 1998 legislation guaranteeing presumption of exposure, similar to Agent Orange legislation. However, veterans have encountered problems in gaining the intended benefits from legislation. Although the Persian Gulf War Veterans' Benefits Act of 1994 was intended to provide compensation to veterans suffering from disability from undiagnosed illnesses, the VA interpreted the legislation such that undiagnosed illnesses were excluded. Furthermore, the Persian Gulf War Illness Compensation Act of 2001 (HR 621)—a piece of legislation that redefined "undiagnosed illness" such that veterans could receive compensation regardless of etiology and that was passed by the house in July 2001—was later incorporated into other pieces of legislation. By the time this changed legislation was made into Public Law 107–94 as the Veterans' Compensation Rate Amendments of 2001, much of the essence of the original bill, including mention of undiagnosed illness, had been lost. Veteran activists also report much less success with other areas of the government, as evidenced by the ongoing tension with the VA and DOD. Similarly, veterans have had only minimal success within the scientific community. Though their interests in studying depleted uranium and other exposures have resulted in concerted efforts by the VA to fund research addressing their concerns, this research has failed to show health effects. Veterans have also had mixed success with the media. At first, sick veterans were media heroes. Even into 1996, coverage of GWRIs was prominent, with a shift from the human-interest story of sick veterans to the story of a potential Pentagon cover-up. More recently, however, according to our content analysis and the veterans we interviewed, the media has shown little interest in GWRIs. As noted earlier, Hyams, Wignall, and Roswell view this loss of attention as a typical phenomenon once a new war comes along, but indeed the drop-off came before the Iraq War.[63]

At present, there has been a shift in the DEP, from a primary stress approach to a contextual stress approach. The latter is still a paradigm in formation, without the firm consensus that would make it generally applicable. In place of the primary stress model, where complaints were seen as psychological symptoms, the new approach sees symptoms as real, even if unexplainable. In a contextual stress approach, researchers understand the stress of potentially real exposures—for instance, the stress of experiencing alarms in the face of known and threatened chemical exposures. Although there is not much research in this area yet, more scientists are calling for studies of the interaction between actual exposures and stress reactions. Of the completed studies examining these interactions, some suggest that chemicals and stress may be interacting synergistically, resulting in greater biological effects than those produced as a result of exposure to individual agents.[64]

Researchers also take seriously the stress experienced by veterans over the inability to find explanations for their mysterious symptoms. There is no major research program to confirm this paradigm in process. Indeed, the transition is more apparent in the clinical sphere. One clear implication is that clinicians should not immediately resort to psychiatric consultations as a chief form of care. Instead, a more multifaceted treatment program is necessary.[65] This developing paradigm also has a strong policy component. One part is the pressure to collect predeployment health data in future engagements. Another part is the realization of the political and clinical costs of denying health problems in the first place. Therefore, in response to both internal shifts and veteran pressure, this new DEP is moving forward, especially in terms of clinical and policy developments, without the scientific evidence to justify it.

Despite obstacles to obtaining adequate data, most researchers looking at GWRIs are seriously trying to understand this mystery. Many have tried to find relationships between symptoms and a variety of exposures, and, in the absence of support for those associations, they have attempted to find respectful explanations for veterans' problems. In this process of learning to understand the unexplained symptoms better, we witness an ongoing negotiation of the meaning of illness. Veterans are trying to transform their ambiguous suffering into an understandable phenomenon. This attempt is an example of what the medical sociological perspective speaks of as "organizing" the "unorganized" illness through clinical interaction and negotiation. In the initial formulation by Balint, organizing is typically done in doctor-patient interaction, but in this particular case it is done through a larger series of interactions that extend beyond the doctor-patient relationship.[66] The negotiation is on a large scale, wherein the meaning of mysterious symptom complexes are questioned in the absence of tangible etiology.[67] As the case of GWRIs illustrates, negotiation of meaning becomes more and more complex as the illnesses themselves grow in complexity and as the number of interests competing for acceptance of their definition grows.

MYSTERIOUS DISEASES AND MUPS

Unlike with breast cancer and asthma, GWRI researchers are not typically from environmental epidemiology backgrounds. Instead, they largely became interested in GWRIs as a result of their background in investigating conditions such as fibromyalgia, chronic fatigue syndrome, and multiple chemical sensitivity. Further, unlike for other diseases where there is potential environmental causation (e.g., asthma and breast cancer), researchers have found very limited evidence of environmental causation for GWRIs.

This discovery led me to examine GWRIs as an instance of MUPS, which includes chronic fatigue syndrome, multiple chemical sensitivity, and fibromyalgia.[68] These conditions share a problem of "contested causation," with that contestation being very public. Multiple chemical sensitivity is a controversial disorder characterized by recurring multisystem symptoms in response to low-level chemical exposures deemed safe for the general population.[69] Chronic fatigue syndrome is a clinically defined syndrome that is characterized by chronic fatigue and a constellation of other symptoms and physical findings. Fibromyalgia is a common form of nonarticular rheumatism associated with chronic generalized musculoskeletal pain, fatigue, and a long list of other complaints. People with these conditions suffer various combinations of fatigue, joint pain, dermatitis, headaches, memory loss, blurred vision, and diarrhea. There is considerable overlap among the various MUPS. The term *medically unexplained physical symptoms* is interchangeable with other terms, such as *functional somatic syndromes* and *chronic multisymptom illness*.[70] Because these illnesses have similar case definitions, the likelihood of a patient's being diagnosed with more than one of them is great.[71] Gulf War veterans received diagnoses of other MUPS.[72] A 2005 study in which physicians and nurses performed medical, gynecologic, and psychiatric histories and examinations on deployed and nondeployed veterans found increased incidence of fibromyalgia and chronic fatigue syndrome among deployed veterans, leading the researchers to suggest that chronic fatigue syndrome in Gulf War veterans may be a different disorder than that occurring generally in the United States.[73] However, Wessely suggests that all MUPS may represent a single entity rather than multiple different illnesses.[74] Not surprisingly, the treatment approaches for different MUPS are also quite similar. The fact that there are both overlapping diagnoses and a variety of umbrella terms for these constellations of symptoms demonstrates the confusion surrounding them. Locating GWRIs in this class of illnesses shifts the focus from how illness sufferers seek to establish environmental causes to how they negotiate diagnosis for mysterious diseases or symptom clusters. This shift is interesting to study because an increasing number of mysterious diseases, not traceable to known pathogens or toxins, are challenging medical science and creating disputes among sufferers, physicians, scientists, and government officials.

The intersection of two key trends—declining patient trust of the health care system and increasing incidence of chronic and often unexplained illness—suggests a need for an improved understanding of when and how patients mobilize to acquire or challenge a diagnosis. In fact, the two trends are inextricably interwoven. Medical professionals' frequent inability to treat patients with chronic illnesses, even if well defined, often leads to even greater mistrust.[75] Likewise, a growing public awareness of environmental contamination and physicians' inability to explain some illnesses with purported environmental causes set up mistrust and a greater need for information. I do not mean to imply that environmental causes are typically out there waiting to be discovered, but rather to note that it is easy for sufferers to turn toward an environmental explanation when modern medicine fails to find other causes. In light of widespread governmental and scientific resistance to environmental explanations of many diseases where there is evidence of such etiology, it is easy to see how people with unexplained symptoms would gravitate toward such explanations.

Keeping in mind the quote from an army doctor who believed that "if you have to prove you are ill, you are never going to get well," we can understand the veterans' plight. When people who are experiencing symptoms and feeling sick are compelled to spend their efforts proving they indeed are sick, they have less opportunity to engage in a clinical interaction that will yield treatment. The contested causation of MUPS can erode patient-provider trust and mutual respect. The uncertainty of MUPS, added to the disputed causation, makes it harder for providers to share decision-making power with patients. Further, searches for blame, responsibility, and reparation may distract from therapeutic activities.[76] By extension, the stress of struggling to achieve recognition of one's sickness can exacerbate symptoms. Hence, a fundamental motivation of patient activism is the challenge of legitimating symptoms and receiving a formal diagnosis. In lieu of a formal diagnosis, even the labeling of a condition and the creation of a collective narrative can provide a sense of structure and order that helps sufferers to understand their experiences.[77]

More than a decade of research on GWRIs has resulted in an emerging consensus among medical professionals that there is no unique Gulf War Syndrome. One medical researcher concluded, however, that "irrespective of the emerging professional consensus, Gulf War syndrome is established as a popular, media and social reality anyway. Investigating how and why this concept developed is important, but the answers will not come from statistics, but social sciences."[78] Indeed, social science research into GWRIs seems to be increasing. Thomas Shriver and his colleagues have directed much of this research, examining internal and external conflicts within the movement, looking at the cultural and political institutions that have shaped the movement, and analyzing the movement through interviews with veterans in order to understand how government and veteran perspectives have differed.[79] In a 2001 study in which fifty-five interviews were conducted with Gulf War veterans and their family members, Shriver examined respondents' experiences through a social constructionist framework. Researchers found that veterans who reported experiences emphasized a top-down resistance to legitimating any claims relating to an environmental illness frame, a resistance that included the withholding of medical information and inconsistent handling of compensation claims. Shriver argues that this refusal to validate claims of GWRIs stems from a desire to protect the government's legitimacy with regard to its ability to protect public health.

Researchers are also beginning to examine strategies used by veterans and government actors to promote their perspectives throughout the conflicts surrounding GWRIs, and they argue that similar conflicts over environmental exposures will continue to arise as we move forward in a society reliant on high-risk technologies.[80] Research has also focused on the role of women in the movement, examining how activism has empowered women and posing questions regarding the gender differences in the movement.[81] Some social science approaches to these illnesses have described these diseases as postmodern illnesses, arising out of the intersection of biology and a form of late-capitalist consumer culture in which popular media messages construct reality. This view of the cultural shaping of illness is shared by Wessely, who suggests a cultural shift has occurred in which perceived environmental hazards and threats shape popular models for illness and disease.[82] At the very least, it seems that technological advances bring with them increasing public concerns about technological risks. And there is even evidence that concern with environmental conditions makes a person more likely to attribute symptoms to environmental causes.[83]

Another consequence of modern technological developments is what Gray calls "postmodern medicine," in which laypeople lose trust in trained medical professionals and instead become medical experts themselves.[84] This perspective is very similar to Kroll-Smith and Floyd's account of multiple chemical sensitivity sufferers who often engage in their own "controlled experiments" to determine which chemicals and what amounts trigger which symptoms.[85] All of these views suggest that the landscape of medicine, doctor-patient relations, and even the experience of illness has changed dramatically.

In addition to the low level of medical legitimacy for MUPS, there is also variation in diagnostic legitimacy across specialties. In the United States, for example, individuals who suspect they have fibromyalgia are likely to find their way to a rheumatologist. Because most rheumatologists recognize the fibromyalgia diagnosis and understand the cognitive behavioral therapy and graded exercise therapy approaches to treating it, their patients are likely to have a more positive experience, even if their symptoms are not resolved. Sufferers of GWRIs and multiple chemical sensitivities may try any number of specialists, but because there are no clinically proven effective treatments, even a supportive specialist can provide little more than psychological support. Chronic fatigue syndrome patients, however, may get recognition from generalists, but have not been embraced by any of the most relevant specialists: infectious disease experts, neurologists, or psychiatrists.

Diagnostic legitimacy is further compromised by what we might term the *psychiatric card*. There is widespread belief among medical professionals that MUPS are psychological in origin and that these patients have personality problems or a psychiatric illness.[86] Physicians may be aware of the common knowledge that a significant fraction of patients seeking medical care have had primary mental illness diagnoses, even in the era before the current MUPS were known. In addition, the very low mortality for MUPS may make clinicians and researchers view these diseases and conditions as less worthy of attention than other more threatening diseases. It is important to reiterate that neither medicine as a whole nor particular specialties are monolithic in their approaches to MUPS. There are contradictory possibilities for medical recognition of MUPS. Physicians may get more clients by expanding the realm of problems they deal with, yet they may also be stigmatized for championing an unscientific approach. Clinicians and researchers are not very eager to research conditions that are so ambiguous and difficult to study, that are seen by their colleagues as less amenable to scientific research, and that have little financial support available for research. A link to environmental causation may actually hinder the search for diagnosis and the resultant treatment and benefits that sufferers need. When the putative causes of diseases are environmental factors, illness groups face an uphill battle.

The multiple ambiguities of these mysterious diseases or conditions make it hard for sufferers to receive a diagnosis and the support stemming from it. This result calls into question the ability of traditional medical practices to understand patients who suffer such problems. If we are in an era of postmodern illnesses shaped as much by culture as by organic causes, then the process of diagnosis carries greater significance.[87] These diseases also have greater salience because they bring forth a politicized form of illness experience. Even if that experience does not require social movement organizations, it does necessitate politicized support groups that engage in the struggle over the very labeling of each disease.

The added importance of diagnosis also points to a whole new approach to doctor-patient interaction, as captured by Gray's notion of "postmodern medicine."[88] If laypeople are increasingly losing trust in medical experts, efficient, accurate, and effective diagnoses become more and more important. As laypeople become medical experts themselves, medical professionals must also begin to understand how diagnoses and all their attendant social and medical consequences shape people's illness experiences. Ultimately, what may be needed is a model for understanding how a doctor and patient can work together to arrive at a diagnosis that is medically sound and consistent with the patient's experience of illness and expectations for recovery. This is the medical negotiation process that medical sociological analysis has long seen as central to illness experience, clinical interaction, and medical care. Such an approach would be more consistent with the shift Sharpe and Carson call for in which MUPS would be understood as nervous system dysfunctions rather than as "mental" dysfunctions.[89] Such a shift would result in the integration of psychiatric treatments into primary medical care, making them more acceptable to patients.[90]

Part of this transformation in doctor-patient relationships and in medical care more generally may require a willingness to admit to the uncertainties involved in medical practice and to abandon the dualistic view of the mind and body.[91] It may be time for medical experts to embrace medical uncertainty as precisely the reason why patient expertise and illness experience are essential to the diagnosis, treatment, and recovery process. Further research is needed to determine what embracing uncertainty might mean for diagnosis, treatment, and recovery, and whether this approach is better than the current practice of withholding diagnoses when uncertainty exists. For instance, will embracing uncertainty alter patients' expectations and taken-for-granted assumptions about medicine, perhaps in turn reducing the frequency with which dissatisfied individuals form illness groups that mobilize to challenge what they see as an unresponsive medical system? Conversely, is avoiding patient mobilization a desirable goal, or is mobilization necessary to question and challenge diagnostic practices in ways that otherwise would not occur? Furthermore, how will the answers to these questions vary across cultural and political boundaries?

Much as researchers had to come to grips with the transition from the primacy of infectious diseases to chronic diseases, they may now face a new transition toward a growing preponderance of (though not a replacement by) the mysterious MUPS. At the same time, they will need to be prepared for a possible step backward, as has occurred in modern medicine wherein AIDS and the resurgence of other infectious diseases has put a dent in our previously accepted notion of epidemiological transition.

SUMMARY

It is striking how the contestation over Gulf War ailments, like a growing number of other medical concerns, has become a public dispute, with citizens playing a central role in disease identification, though in this case not in research. Disputes concerning GWRIs have led researchers to utilize less established methods (e.g., the determination of self-reported symptoms and exposures) and have pressed clinicians and researchers to take seriously the symptoms experienced by veterans.

At present, there is little evidence that environmental exposures during the Gulf War affected soldiers' health. Meanwhile, researchers have had to face declining funding. For instance, the BEHC finally lost its special financial support in early 2000 when the VA changed the center's mandate from a unit dedicated to Gulf War research to one studying general environmental health. Several research projects were simply terminated. Following the shift in the VA's research priorities in 2000, federal funding for research concerning GWRIs has continued to decline, as shown in figure 4.1. Despite continued calls for research examining the synergistic effects of chemical agents, the government has largely given up on establishing causation and is shifting attention away from GWRI research to research examining the effects of veterans' deployment to areas other than the Persian Gulf, including Iraq, Afghanistan, Bosnia, and Kosovo.

Some researchers and many veteran activists want lessons from the Gulf War to inform future military involvements so that caution is followed regarding inoculations, antidotes, and toxic exposures. Indeed, there has already been post–Gulf War resistance to anthrax inoculations as well as large-scale departures from the military reserves due to compulsory vaccination.[92] Anthrax vaccinations, although unapproved by the FDA, were issued to military personnel beginning in 1998 and were reauthorized by Congress for mandatory use during military emergencies under the Project BioShield Act of 2004. This practice was deemed unlawful by a judge in October 2004 following a lawsuit filed by six military personnel who had been expelled from the military for refusing anthrax inoculations and following a widespread refusal from service members to comply based on other troops' complaints of fatigue, joint pain, and memory loss after they were given the vaccinations.[93] In 2005, a judge ruled that the Pentagon could resume giving anthrax vaccinations on a voluntary basis with the informed consent of troops.[94] Some attention is now being given to health issues associated with Lariam, the antimalaria medication being given to troops deployed in Iraq and possibly associated with mental health issues.[95] Media and government officials alike have already begun to pay attention to high rates of posttraumatic stress disorder in troops returning from the ongoing Iraq War. This attention may indicate a heightened awareness of the dangers to which service members are exposed in deployment and their potential health effects.

In addition to the concerns of military personnel, the government hopes to learn from the experiences of 1991 Gulf War veterans. In its 2004 report, the Research Advisory Committee on Gulf War Veterans' Illnesses made several recommendations for future military deployments based on the Gulf War experience, including improved record keeping, communication among government agencies, and the development of a health data system. Indeed, new provisions have been made with the mobilization of troops in the Persian Gulf again, including pre- and postdeployment health assessments.[96] In May 2001, Secretary of Veterans Affairs Anthony J. Principi established two War-Related Illness and Injury Study Centers, one in New Jersey and one in Washington, D.C., designed specifically to address deployment-related persistent fatigue, pain, cognitive and sleep problems, and otherwise difficult-to-diagnose illnesses in veterans.[97] Many sick veterans and some researchers believe there are environmental causes for Gulf War ailments, and they intend to keep alive their quest for confirmation.

Gulf War veterans have not successfully forged the strong citizen-science alliances we see in other contestations over environmental and occupational exposures. Compared to the other VA Gulf War centers, lay involvement at my research site, the BEHC, showed an atypical level of participation, with veterans helping to develop questionnaires, pressing for study of certain exposures, and working to boost response rates. The low level of citizen-science alliance is in part attributable to the military's and the VA's tight control of data and research, as well as to the veterans' patriotism and a military culture that discourages what are perceived as attacks against the government.

The contestation over GWRIs offers different lessons than the struggles over asthma and breast cancer. Because it is unlikely that we will find much evidence for environmental causation for GWRIs, the lessons are largely about how to accept people's complaints when they appear out of the ken of routine medical understanding. Although the episodic nature of war and its resultant conditions means that such mini-epidemics are not part of the everyday medical and public-health gazes, there are important things to learn about the interaction of stress and environmental factors, even if research has yet to tease them out. The similarities between GWRIs and other MUPS put pressure on medicine to seek better ways of understanding these new mysterious diseases, including the most basic level of accepting them as diseases or conditions, something that is not an issue for asthma and breast cancer. The need for better surveillance is something that all three conditions hold in common. For GWRIs, predeployment health monitoring is crucial, though it is different than the survey-type monitoring we need in diseases caused or exacerbated by toxics.

In chapter 5, I make many comparisons between asthma, breast cancer, and GWRIs to show different policy outcomes and to show the relative role of science and social movements in legitimating the search for environmental causation.

|5|

SIMILARITIES AND DIFFERENCES AMONG ASTHMA, BREAST CANCER, AND GULF WAR ILLNESSES

I CHOSE asthma, breast cancer, and Gulf War illnesses because I knew they differed in significant ways in terms of the science base, as evidenced in both biomedical acceptance of the disease or condition and its potential environmental causation. I knew that the strength of social movements across these diseases and conditions differed, and I was aware of the differences in government regulatory practices and other policies. Each of these diseases and conditions (hereafter it makes sense to speak of "diseases" for simplicity) is interesting in its own way, and I believe each of the prior chapters on asthma, breast cancer, and Gulf War illnesses stands alone as an examination of contested illnesses. Yet there is also much to be gained by comparing these diseases in order to see why there have been different outcomes and to think about future disease contestation. This chapter, of necessity, repeats some of the points made in the previous three chapters, but I offer "reminders" rather than full repetitions of the material.

RESISTANCE TO LOOKING FOR ENVIRONMENTAL FACTORS

Resistance to looking for environmental factors differs according to the disease being considered. For asthma and breast cancer, the resistance comes primarily from corporate actors whose production, distribution, and disposal practices depend on dangerous substances. Secondarily, the resistance is from governmental actors who support corporate interests or do not wish to be strongly regulatory on ideological grounds (which is, in effect, the same thing as straight-out support for corporate interests). Resistance also comes from scientists who find the evidence unconvincing or cannot reorient their perspectives to account for environmental causation, either because this view goes against their intellectual framework or because it will affect their careers. For GWRIs, resistance comes primarily from the government, which does not want to acknowledge potential diseases stemming from "routine" warfare. This perspective is both ideological in that it seeks to avoid conflict over the conduct of war and economic in that it seeks to avoid high medical and disability costs. Ultimately, all forms of resistance to environmental causation are linked: they stem from people, entities, and institutions that are integral parts of a social order based on profit over well-being, unfettered economic development that benefits the few over socially conscious planning that benefits the many, and consumer-oriented production over well-rounded personal development. Hence, any perspective that targets environmental factors in health is a perspective that calls into question the dominant social structure.

Thinking back to the core concepts I introduced in chapter 1, we can view this shift in perspective as a process by which popular epidemiology and critical epidemiology combine, typically in the form of health social movements and their allies, to construct a public paradigm that counters the DEP. This process is pushed forward by a shift in people's experience of illness, whereby they transform individual suffering into collective realization and collective, politicized illness experience.

Such a process is why these three diseases are very prominent in the public eye, with much debate over identification, environmental causation, treatment, and lay involvement. Research into each disease is under public scrutiny in terms of funding, hypothesis formulation, research design, interpretation, and resultant regulation. And because medical and governmental institutions do not take the lead in identifying environmental factors, social movements play a large role in identification, advocacy, and debates over environmental causation.

Environmental causation is difficult to demonstrate because exposures are usually not controlled or easily measurable in terms of quantity and location. Further, exposures to multiple agents are common, and there is usually not a signature disease for a specific agent. In some cases, such as GWRIs, the symptoms are diffuse and fail to fall into the categories of diagnosis defined by the biomedical model of disease. For those reasons, research capacity is weaker for environmentally induced diseases than for most other diseases, which subsequently results in much policy being shaped in the absence of scientific certainty. In addition, even in cases where a disease has a solid funding and research base—for example, breast cancer—little of that funding goes toward research for environmental factors.

POLICY OUTCOMES AND WHAT INFLUENCES THEM

I organize this comparative analysis by first looking at four types of policies. *Research policies* include allocation of funds for both intramural and extramural research. *Regulatory policies* include legislation, official regulations, and agency directives that put limits or restrictions on industry, health care providers, or others. When litigation results from regulation, I include it under regulation. *Compensation/treatment* includes the certification of diseases and conditions as legitimate for provision of health care services, alterations in health care surveillance, and determination of the types of disability and therefore of the compensation provided to sufferers. *Citizen participation policies* include measures that mandate a variety of types of lay involvement, including citizen participation on advisory boards and scientific review panels and collaboration in research. Although citizen participation is not a common outcome in policy studies, I include it because it is central to much of the activity around these three diseases, and I expect it will become important for a growing number of other diseases.

After examining the differences in policy outcomes, I focus on four factors that shape those policies: the science base supporting the environmental causation hypothesis; prevalence and perception of risk; the sources of support for the environmental causation hypothesis; and the strength of health social movements. All four factors contribute to policy outcomes, but I find the strength of health social movements to be particularly important for these three diseases. In some cases, social movement activity can be more important than the strength of the science base in terms of policy outcome success.

To help set the stage for looking at how policy is determined, I show in figure 5.1 a model of this process based on stages of action. In stage 1, prior to the mobilization of a health social movement, a disease receives greater or lesser attention from policymakers as a result of the strength of the science base supporting the environmental causation hypothesis, the prevalence and perception of risk, and the sources of support for the environmental causation hypothesis. In stage 2, those three factors, combined with existing policies, influence whether a health social movement will emerge and how it will operate. The movement then becomes a fourth factor, in tandem with the first three, to continue to shape a third stage of future policymaking. The significance of health social movements is that they are often able to promote policies even in the absence of support from one or more of the first three factors. For example, environmental breast cancer activists lack a strong science base supporting the environmental causation hypothesis, but they can nevertheless mobilize the support of

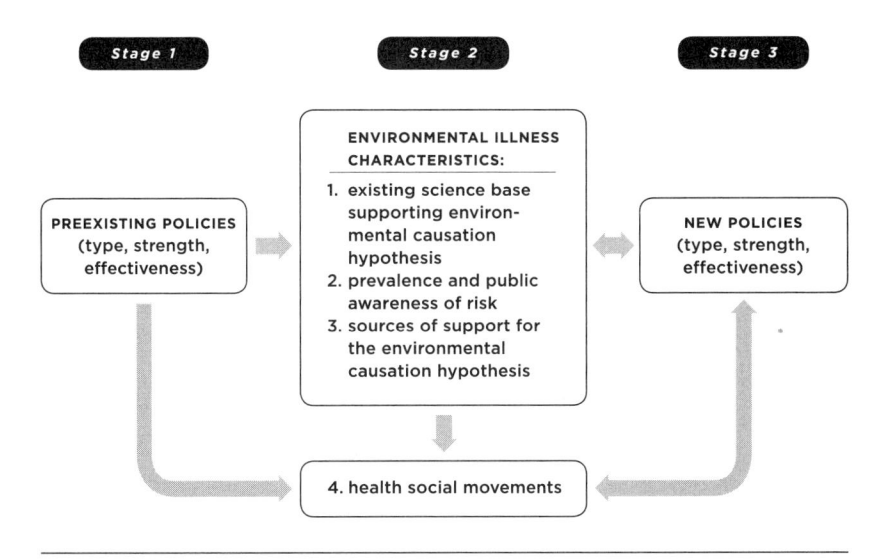

FIGURE 5.1 The dynamic relationship between health social movements and policy.

Diagram: Stage 1 — PREEXISTING POLICIES (type, strength, effectiveness); Stage 2 — ENVIRONMENTAL ILLNESS CHARACTERISTICS: 1. existing science base supporting environmental causation hypothesis; 2. prevalence and public awareness of risk; 3. sources of support for the environmental causation hypothesis; 4. health social movements; Stage 3 — NEW POLICIES (type, strength, effectiveness).

physicians, scientists, and policymakers who find such a hypothesis plausible and worth investigating.

The policies that exist in each of the four policy-type categories for the three diseases are summarized in table 5.1. The table lists some key *current* policies both at the general level and in terms of my research sites. It includes policies that existed prior to any mobilized effort to address the disease and policies that emerged after sufferers of an environmental illness organized and lobbied for new or stronger policies. In my discussion, I make clear when certain policies developed and changed.

ASTHMA

For asthma, existing legislation provides a substantial head start to illness sufferers and advocates in that the Clean Air Act creates a context within which illness groups can develop strategies of action. The activist groups I examined in chapter 3 focused primarily on local-level transit and housing policies that would result in a cleaner environment and cleaner air to breathe.

RESEARCH FUNDING POLICY Under the Clean Air Act, the EPA is required by law to provide scientific evidence when defining the NAAQS. Air pollution standards cannot be based on cost-benefit analyses, but rather only on the best science available at the time. With federal legislation driving a large

TABLE 5.1 Current State of Policies for Asthma, Breast Cancer, and Gulf War Illnesses

	ASTHMA	BREAST CANCER	GULF WAR ILLNESSES
Regulatory Policy	• Clean Air Act, NAAQS; diesel reduction from buses in Boston	• Endocrine Disrupter Screening and Testing Advisory Committee	• anthrax vaccination policy lifted • DOD predeployment health monitoring • DOD required to adhere to FDA human subjects regulations
Research Funding Policy	• EPA research on particulates • NIH, CDC, and HUD Healthy Homes project • Children's Environmental Health Centers • National Institute of Allergy and Infectious Diseases	• DOD Breast Cancer Research Program • NIH, NCI CARA funding • Long Island Breast Cancer Study Project (joint NIH/DOD) • Mass. state funding for Silent Spring Institute • allocation of tax money to research through voluntary check-off box	• Persian Gulf War Veterans' Benefits Act • VA and DOD funding of internal research • Environmental Hazards Research Centers • Institute of Medicine, CDC, President's Advisory Committee
Citizen Participation	• NIEHS community-based participatory research • Community Outreach Education Project (community involvement in educational programming, not research)	• Consumer advocates on DOD advisory panels • Long Island Breast Cancer Study participation requirements • NIEHS Breast Cancer Centers of Excellence participation requirements • Silent Spring Institute citizen participation policy	• Informal, none required
Detection, Treatment, and Compensation	• Treatments covered by most health insurers and Medicaid • costs of prevention (e.g., hypoallergenic products) not covered	• Breast and Cervical Cancer Prevention and Treatment Act of 2000 • Endocrine Disrupter Screening and Testing Advisory Committee	• Veterans Programs Enhancement Act • Persian Gulf War Veterans Act • Veterans Millennium Health Care and Benefits Act

amount of the research into air pollution's respiratory effects, organizations advocating for cleaner air need not devote extensive resources to passing legislation that would produce further research.

Despite the Clean Air Act's guarantee of ongoing science into the health effects of air pollution, much of this research is on respiratory conditions other than on asthma. In addition, in agencies other than the EPA that fund asthma research, much of the research is on indoor causes of asthma or on treatments of asthma rather than on environmental causes. For example, the NIH, the CDC, and HUD also fund asthma research. However, the funding provided by agencies such as the NIH typically goes to treatment rather than to etiological studies. Seventy-one percent of NIH asthma grants goes to pathophysiology and treatment studies, but only 17 percent goes to etiological studies. Of that 17 percent, roughly two-thirds are directed toward environmental factors.[1] In looking at environmental factors, the NIEHS's environmental justice and community-based participatory research programs have funded a very large number of community asthma groups and played a key role in advancing science and community empowerment.

COMPENSATION/TREATMENT One significant issue in asthma treatment is whether Medicaid and health insurance will cover anti-allergy products (e.g., pillow and mattress covers, air purifiers) and remediation of household factors contributing to asthma (e.g., mold, carpets). At present, such expenses are not covered, with the exception of a small number of intervention programs. As for treatment, many individuals with asthma have access to health care that allows them to manage their condition, although the problem of the uninsured population deprives many others of care.

REGULATORY POLICY The most prominent form of regulatory policy affecting asthma is the 1970 Clean Air Act and its 1990 amendments. The Clean Air Act authorizes the EPA to establish the NAAQS to protect public health and the environment. The goal of the NAAQS is to set maximum pollutant standards every five years based on current scientific knowledge. Though the Clean Air Act is a federal law, NAAQS are created on a state-by-state basis. I noted in chapter 3 the controversy over the EPA's actions on air particulate regulation, in which it retreated from $PM_{2.5}$ regulation even after the Supreme Court supported its right to regulate. The George W. Bush administration gutted the Clinton administration's NSR process to clean up old power plants, a major retreat in air regulation.

At the local level, regulatory bodies are working to meet demands from resident activists to improve urban pollution problems that contribute to asthma.

In Boston, regional authorities responded to ACE's efforts and closed one bus depot the activists felt was sited unjustly in poor or minority neighborhoods. Similarly, the Massachusetts Bay Transit Authority agreed to implement an alternative fuel prototype program, to participate in a public process for service improvements, and not to buy any new conventional diesel buses.

CITIZEN PARTICIPATION The NIEHS, CDC, and EPA have funded individual asthma projects that involve the community, but the NIEHS's Community-Based Participatory Research program and its education-based Environmental Justice: Partnership for Communications program are the only ones still operating. The NIEHS has increasingly strengthened its requirements for deep and fundamental community participation, going so far as to expect in most new cases that principal investigators will be part of community-based organizations rather than academic partner institutions that fund prevention and intervention hypothesis-driven research.

There are five Centers for Children's Environmental Health and Disease Prevention Research cofunded by the NIEHS, CDC, and EPA, which include or focus on asthma as a major area of study in conjunction with community-based participatory research projects. One example of these centers is the Columbia Center for Children's Environmental Health in New York, which works in collaboration with the activist group WE ACT. The latter group has published several articles for which community members were coinvestigators or actually involved in data collection or both.

BREAST CANCER

In addition to benefiting from preexisting policies, health social movements can benefit from the efforts of movements or groups that came before them. The EBCM, for example, has reaped the benefits of successes by the broader breast cancer movement. The overall breast cancer movement is also an example of how movement success can result in functional differentiation. As a result of this differentiation, the mainstream breast cancer movement continues to focus on policies aimed at diagnosis and treatment, whereas the EBCM focuses more on securing funding to investigate possible environmental causes of breast cancer. I discuss both efforts.

RESEARCH FUNDING POLICY Following early successes in requiring informed consent for breast cancer surgery in order to prevent unnecessary mastectomies, the general breast cancer movement shifted its attention to securing funding for additional research, mainly for detection and treatments rather than for etiology. Thirty years later the EBCM has taken up the effort to secure research funding, but this time to study the potential environmental causes of breast cancer. The EBCM, though national, has been shaped by the strength of Long Island, Bay Area, and Massachusetts activists, who have obtained funding for locally based research. Yet although federal breast cancer research has increased dramatically as a result of the broader breast cancer movement, less than 3 percent of the money (a total of $800 million in 2001) has been directed toward finding environmental connections to breast cancer.[2]

At the national level, Congress directed the DOD's involvement in a breast cancer research program beginning in 1992. Congress has appropriated more than $1.5 billion to this program since its inception. In 2001, the congressional appropriation for the program was $175 million dollars, but it was cut to $124 million in 2002 and has remained at $150 million since 2002. The DOD program is unusual in the extent to which social activists have successfully shaped the process, not only in terms of level of funding, but also in the overall structure of the system and the allocation of funds (e.g., more than 90 percent of the funds went directly to research grants).

The Stamp Out Breast Cancer Program, another successful policy effort, is centered around a breast cancer stamp whose net proceeds of eight cents above the cost of postage are designated for research purposes. The stamp also demonstrates the tension within the diversified breast cancer movement. For example, EBCM activists have attempted to get a portion of revenue from the stamp moved from the NCI to the NIEHS, where it would be put toward more research into environmental causes.

More than $3 billion has been spent on intramural breast cancer research at the NIH. Breast cancer activists applaud the importance of research, but they also question whether the money is being invested wisely. They are calling for external monitoring of the NIH and believe the public should design and participate in an oversight process that will track how the money is being spent and whether it is being spent well.[3]

One last important development in research funding is the Breast Cancer and Environmental Research Act of 2001, which established multidisciplinary, multi-institutional research Centers of Excellence to study environmental factors that may be related to the development of breast cancer.

COMPENSATION/TREATMENT As recently as the 1970s, physicians relied on nearly eighty-year-old research in justifying radical mastectomies as the most effective treatment of breast cancer. Given the complete absence of any alternative research, the breast cancer movement first passed legislation requiring women's informed consent before mastectomies could be performed. Changes to broaden the number of women receiving treatment occurred at both the state and the national level. EBCM activists, along with activists from the mainstream movement, were active in pushing California to take important steps toward guaranteeing treatment to everyone diagnosed with breast cancer. California's 2000–2001 budget earmarked $20 million for breast cancer treatment for low-income uninsured and underinsured individuals. Shortly thereafter, President Clinton signed into law the Breast and Cervical Cancer Prevention and Treatment Act of 2000. This law gives states the option of making medical assistance for breast and cervical cancer–related treatment services available to certain uninsured low-income women screened and found to have breast or cervical cancer under a federally funded screening program. As of May 2005, all fifty states, four U.S. territories, the District of Columbia, and thirteen American Indian and Alaska Native organizations had opted into the program or had passed legislation adopting the new Medicaid option to expand eligibility to uninsured women.

REGULATORY POLICY Few regulatory policies exist in the area of breast cancer and environmental risk factors, perhaps in part because of the equivocal state of research on breast cancer and the environment. One important exception came in 1996. At that time, provisions of both the 1996 Food Quality Protection Act and the reauthorization of the Safe Drinking Water Act called for the creation of a screening and testing program at the EPA that would provide information about possible endocrine effects of certain chemicals on human health, including pesticides, industrial chemicals, and environmental contaminants. The Endocrine Disrupter Screening and Testing Advisory Committee brought together representatives from industry, government, environmental and public-health groups (including the NBCC), and academia to deliberate for two years to design this program. Although the committee's 1998 recommendations were a large achievement, they still failed to include low-dose testing and a focus on early developmental disorders. Many observers believe the committee's recommendations have not been implemented at the level necessary to study the many known endocrine disrupters.[4]

The CDC recently collected the third national data set on human exposure to environmental chemicals, which eventually may lead to an improved understanding of environmental health effects and even to new regulation.[5] Although this step is important, exposures that occur prior to adulthood or that leave no biological markers may also be very important. For now, the paucity of regulatory policy is at least in part a reflection of the underdeveloped scientific base in this area.

CITIZEN PARTICIPATION Citizen participation is very much a part of research on environmental factors in breast cancer. The federal government implemented citizen participation requirements for the Long Island study and for the DOD breast cancer research program. Silent Spring Institute bases all its activities on a citizen participation model, and California activists achieved mandated citizen participation in state-funded research. Citizen participation has recently been mandated in the new Breast Cancer Centers of Excellence, multidisciplinary institutes studying breast cancer with advising by activists.

Another program that institutionalizes public involvement in research is the NIH's Consumer Advocates in Research and Related Activities program. The NIH is recruiting 150 advocates who can serve multiple functions, such as advocating for and setting research priorities, participating in early design efforts of clinical trials, advocating and evaluating clinical trials, identifying research gaps, peer reviewing research, and participating in other programmatic activities.

GWRIs

Gulf War veterans, unlike breast cancer or asthma activists, have struggled within a policymaking context that is heavily shaped by the power wielded by the DOD. However, early in their fight to get the government to treat and compensate them, they benefited from the status accorded to military veterans. This status and the congressional support it generated have resulted in quite a few policies both to provide compensation for Gulf War veterans and to protect soldiers in future wars. Yet this support has failed to produce the scientific evidence that would justify further or stronger policies.

RESEARCH FUNDING POLICY Research on GWRIs has been supported through executive orders, congressional legislation, and internal policy decisions. Between 1994 and 2000, the federal government spent approximately $155 million for Gulf War research. This money from the VA and DOD funded 192 basic and applied research projects. In addition, the VA contracted with private research firms such as RAND to conduct reviews of scientific literature. Federal funding has declined steadily since 2000. Veterans have not pushed for additional research dollars, but in some cases they have applied pressure to ensure that certain researchers would be the recipients of that money, including some "dissident scientists" such as Robert Haley.

Intramural research has also been supported. For example, in 1997 the VA requested that the Institute of Medicine conduct a review of the peer-reviewed scientific literature on depleted uranium used in munitions, sarin nerve gas, and pyridostigmine bromide, a nerve gas antidote. This request was later turned into legislative mandate when Congress passed the Veterans Programs Enhancement Act of 1998 and the Persian Gulf War Veterans Act of 1998. The Institute of Medicine study was released in 2000 and has been used to justify future research priorities.[6]

Intramural research also resulted in the 2001 study by the VA and the DOD that linked service in the Gulf to ALS, also known as Lou Gehrig's disease. This study was launched after veterans voiced concerns about a perceived disproportionate number of cases of ALS among them. The VA initially denied that veterans were suffering from ALS at a higher rate than the general population, but the preliminary results of an epidemiological analysis now suggest otherwise, and the VA has promised to offer immediate compensation and benefits to any Gulf War veteran diagnosed with ALS.[7]

Distrust of the government, however, has led veterans to be suspicious of much of the research conducted by the VA and DOD. The relationship is better, though, with the three Environmental Hazards Research Centers, located in VA Medical Centers in Boston, East Orange, New Jersey, and Portland, Oregon, and created by the VA in 1995, although the VA later retooled them as general occupational and environmental medicine centers rather than centers dedicated to GWRIs.

COMPENSATION/TREATMENT An important issue regarding Gulf War illnesses has been the attempt to guarantee compensation and benefits for veterans with undiagnosed symptoms. Due to a VA policy that requires a service-linked diagnosis in order to receive compensation and benefits, veterans whose symptoms have gone undiagnosed have not received the care they would like. The Persian Gulf War Veterans' Benefits Act of 1994 was intended to provide for the payment of compensation to veterans suffering from a chronic disability resulting from an undiagnosed illness, but the VA interpreted the legislation to exclude undiagnosed illness. Under the Persian Gulf War Illness Compensation Act of 2001 (HR 612), *undiagnosed illness* was redefined so that veterans can still receive compensation even if the etiology of their symptoms cannot be identified. HR 612 also extends the presumptive period for undiagnosed illnesses to 2011, which guarantees veterans continued coverage during the intervening years even if their illnesses are not connected to their military service. HR 612 was incorporated into and abbreviated in the Veterans Benefits Act of 2001 (HR 2540), which passed the House in July 2001. As passed by the Senate and enacted into law as Public Law 107-94 in December 2001, the bill became the Veterans' Compensation Rate Amendments of 2001 (S 1090), which provided cost-of-living adjustments for veterans but lost some of the benefits proposed in earlier versions of both HR 612 and HR 2540, ultimately making no mention of undiagnosed illnesses.

In 1999, the Veterans Millennium Health Care and Benefits Act extended for four years the requirement that the VA operate a program to evaluate the health status of Gulf War veterans' dependents and assured that those who are sick but not in the military will maintain eligibility for VA health care. Other legislation has included the Veterans Programs Enhancement Act of 1998. The December 2001 findings on ALS were so prominent, representing the first known causal connection to disease and Gulf War exposure, that even prior to the scientific publication of the data, the VA immediately offered full disability status to any Gulf War veteran with the disease. Some observers familiar with GWRI research raise the possibility that this apparently munificent policy, although a victory in one sense, was an easy decision to make because ALS affects very few veterans and would therefore not be costly to the VA.

REGULATORY POLICY Because research has failed to provide a clear understanding of the etiology of GWRIs, few regulatory policies have resulted. Probably most important is a DOD policy to improve predeployment health monitoring. But despite having been in place after the Gulf War, this policy was not implemented successfully in the Balkan peacekeeping action or in the Afghanistan War. In June 2004, with soldiers returning to Iraq, the Senate unanimously passed the Armed Forces Personnel Medical Readiness and Tracking Act of 2004 (S 2400) to establish a new system of health screenings for active-duty soldiers and reservists. Also important is a policy against the DOD's using military personnel as research subjects without their knowledge.

CITIZEN PARTICIPATION It took veterans some time before realizing the importance of participating in the review of research proposals and in research design itself. In 1993, VA secretary Jesse Brown established the Persian Gulf Expert Scientific Committee, consisting of scientific experts and representatives from VSOs. The committee advises the VA on clinical, research, and compensation issues for Persian Gulf veterans. One outcome of this committee's work is the recommendation that all research efforts receiving federal money include veteran representatives on their community advisory boards. But this recommendation has not always been followed, and veterans have failed to participate in collaborative work to the extent that breast cancer activists have, despite the high level of expertise many veterans have acquired with respect to the science underlying much research.

EXPLANATIONS OF VARIATION IN POLICY OUTCOMES

From this review of the policy outcomes for each of the three diseases, it becomes clear that the level and types of policies vary quite drastically. These differences are apparent in the summary of the existing policies in each of the four policy-type categories in table 5.1. Four factors help to explain this variation, as summarized in table 5.2.

Table 5.2 includes the strength of health social movements as one of the factors that explain the variation in policies for the three diseases. But the movements themselves are shaped by the first three factors in the table. Consequently, I describe these three factors in terms of the challenges a health social movement might face in legitimating its claims of environmental causation. To gain both public and medical acceptance of an environmental causation hypothesis, strong supporting scientific evidence must be produced. Therefore, the science base for environmental causes is the first factor I discuss. Because the science base for environmentally induced diseases is typically not strong, the next step for a movement is to generate additional scientific evidence to warrant new research. With an adequate science base, the movement might focus more on regulatory policies, as in the case of asthma, or on treatment and compensation policies, as in the case of GWRIs.

Health social movements may struggle to find institutional support or even media or public support if the second factor, the consequences of the disease, do not appear serious enough. For example, affected populations that are not particularly vulnerable may not get public empathy, and affected populations that lack adequate wealth and resources to mobilize may not be able to raise public awareness and perception of risk. The more public awareness there is and the greater the perception that there is a widespread risk, the more likely a movement is to marshal the resources necessary to push through new health policies. Finally, the movement may be more successful if it gets support for environmental causation hypotheses from well-respected sources. Activists alone cannot produce scientific evidence in support of an environmental causation hypothesis. They need the support of scientific researchers, physicians, and politicians.

At the same time, the strength of health social movements is dependent on the same factors. Weakness in one of the factors, however, is often compensated for by strengths in other factors. For example, there is a relatively weak science base for the environmental causation hypothesis for breast cancer. But due to the mortality rate of the disease, especially compared to asthma, for example, and the perceived vulnerability of the afflicted population (i.e., women and, more specifically, mothers), breast cancer activists have had many policy

TABLE 5.2 Factors Shaping Policy Outcomes for Asthma, Breast Cancer, and Gulf War Illnesses

	ASTHMA	BREAST CANCER	GULF WAR ILLNESSES
Science Base for Environmental Causes	• strong for environmental triggers (e.g., PM$_{2.5}$, allergens) • weak for environmental causes (though pollution causes of other respiratory diseases are confirmed)	• weak for environmental causes (but strong theoretical reasons to suspect environment)	• some evidence of increased ALS, but no GWRI symptoms conclusively linked to environment
Prevalence and Public Perception of Risks	• widespread and increasing among children (disproportionately impacting inner-city children) • higher rates of hospitalization and days of work or school missed in inner-city populations • seen as condition that can afflict anyone • high prevalence and high public awareness	• highest rates in white women of high socioeconomic status • increasing morbidity but decreasing mortality due to improved detection and treatments • highest mortality in black women • seen as disease that can affect any woman • family history perceived as primary risk factor	• symptoms emerged over time, but limited to men and women who served in Gulf War • high morbidity among those who served, but low mortality • condition seen as isolated to war • decreasing awareness due to isolated occurrence in 1991 Gulf War
Sources of Support for Environmental Causation Hypotheses	• scientists, EPA • some physicians and public-health professionals • environmental justice activists • substantial academic research and foundation support (ALA; American Academy of Allergy, Asthma, and Immunology; Pew Health-Track), though little research on environmental causes	• scientists • activists • some Congress members • substantial academic research and foundation support in general (e.g., ACS), but little research on environmental causes	• less so scientists • little academic research or foundation support • symptomology like chronic fatigue syndrome, fibromyalgia, and multiple chemical sensitivity (few resources for study of those diseases)
Strength of Health Social Movement	• strong, though not exclusive to health-related issues (related to environmental justice more broadly)	• strong, in both broader breast cancer movement and EBCM	• once moderately strong, but now waning

successes in this area. Gulf War veterans, like asthma sufferers, experience low mortality. But veterans have benefited from extensive support from a small group of politicians. This support has strengthened the movement and its ability to push through legislation guaranteeing treatment and compensation. In the following sections, I discuss impacts on policy and social movements by the first three factors from table 5.2: strength of the science base, public awareness and perception of risk, and support for environmental causation hypotheses.

SCIENCE BASE FOR ENVIRONMENTAL CAUSATION

The first limiting factor in creating environmental health policy is the strength of the science base for environmental causes. Though some scientists may find evidence of that link, the relationship between the environment and human health is highly contested in most scientific forums. Policymakers are therefore unable to create strong policies in the absence of firm scientific evidence. Even when evidence exists, policymakers often use the mere presence of debate as a reason to avoid regulation and enforcement. Health social movements with a strong scientific base are more likely to advocate for regulations, whereas movements without that strong scientific base must first work to establish one.

Asthma activists have benefited from a strong science base that demonstrates that environmental factors, such as particulate matter and allergens, can trigger asthma attacks. There is much less evidence that the same factors cause asthma, though ozone has recently been shown to cause it, and air pollution is known to cause other respiratory diseases. The strength of this science base has legitimized more research in the effects of particulate matter on health, and more asthma-specific research in the past couple of years has solidified the base even more. Despite the strength of science here, industry and government have fought very hard against it, as evidenced in the trucking industry lawsuit and the EPA's backing down on stronger regulation of air pollution.

In the case of breast cancer, there is a strong science base in terms of detection and treatment, but less is known of causes in general and even less in terms of environmental factors. In the face of this weak science base, the mainstream movement continues to push for policies that address detection and treatment and to some extent policies that will provide funding for research into genetic and lifestyle factors. The EBCM, however, focuses on policies that will provide funding for research into environmental causes. Activists have been successful in getting more research done, and though it has not yet provided them with the results they hope for, some studies are showing environmental effects. Activists also hold up the larger body of endocrine disrupter research as related to potential causes of breast cancer, and they find some support in the recognition of estrogenic hormone replacement therapy as a breast carcinogen.

Gulf War veterans also deal with a lack of significant scientific evidence linking their symptoms to environmental causes. Like EBCM activists, they have relied on political pressure to demand the funding necessary to investigate the environmental factors they suspect are causing their symptoms. Unlike the EBCM, veteran activists have not been successful at establishing positive relationships with scientists who allow them to participate in the scientific process. Only in 2004 did research find some evidence of a lung cancer connection to Gulf War environmental causes. The science base here is therefore weakest among the three diseases. Veteran activists, like EBCM activists, tend to be highly knowledgeable about the science behind their disease, but in this case their knowledge has resulted in a contentious relationship with scientists instead of the cooperative one that many in the breast cancer movement enjoy.

PUBLIC AWARENESS AND PERCEPTION OF RISKS

Even without a strong science base or institutional support, health social movements may be able to mobilize scientific research and even policy if the morbidity and mortality rates of the disease seriously impact particular populations. For example, asthma is currently a widespread condition that is increasing among children, especially among inner-city children. But because it has a low mortality rate, it is not always perceived as a serious condition that warrants swift and broad policy changes. In fact, if it were a greater problem among suburban middle- and upper-middle-class children, there might be greater attention paid to the causes. The fact that inner-city populations are more likely to be hospitalized and to miss days of work or school suggests that higher socioeconomic status improves a person's ability to manage asthma by purchasing allergy control supplies and equipment and by having a healthier diet and better living conditions.

Breast cancer, though not as pervasive as asthma, is the most common form of cancer for women, with a one-in-eight lifetime prevalence rate, with its occurrence rising steadily for half a century. Women face this disease with much fear and attention, and its high mortality rate makes it a disease that policymakers see as warranting new policy. The breast cancer movement has benefited from the perception that innocent women and, more particularly, mothers are being taken by this indiscriminate disease. Although morbidity is increasing for all individuals, the rates of mortality are decreasing due to improvements in detection and treatment. Although the highest rates of breast cancer occur in white women of high socioeconomic status, African American women experience higher rates of mortality due to the disease. That gap is narrowing, with the excess of incidence among white women being not much larger than the rate for black women, and with the mortality rate for blacks rising even more relative to the rate for whites. This change may make environmental justice activists more aware of breast cancer as an issue, which is something I have seen in my current research. GWRIs, in contrast, have a relatively high rate of morbidity among those who served in the Persian Gulf, but a low mortality rate. The initial high morbidity drew attention to GWRIs, but their low mortality has hampered veterans' abilities to keep public attention focused on their condition.

Public awareness of a disease, including perception of its relative risk, affect whether health social movements can tap into populations beyond the affected population when mobilizing supporters. Asthma, for example, is seen as a disease that might afflict anyone. Most people know someone with asthma and are aware of the triggers and approaches to managing an asthma attack. But asthma is also a disease that people believe they have little risk of getting if they did not get it during childhood. Some people perceive the risk of asthma to be greater for those who live in "dirty conditions." To the extent that such a perspective places blame on the individual, there is less sympathy for asthma sufferers. Breast cancer shares the widespread awareness held by asthma and is likewise seen as a disease that can afflict any woman. Nevertheless, in order to get funding for research into environmental causes, the EBCM has had to overcome the perception that genetics and lifestyle choices are the primary risk factors for breast cancer. Gulf War veterans have struggled the most in this arena because their condition is seen as isolated to the war environment. Veterans have been unable to keep GWRIs in the media and thereby in the public consciousness.

SOURCES OF SUPPORT FOR ENVIRONMENTAL CAUSATION HYPOTHESES

Finally, health social movements must have allies who support the environmental causation hypothesis. Asthma activists benefit from a scientific community that has largely embraced the plausibility of air pollution as a trigger and perhaps a cause of asthma, and most definitely as a cause of other respiratory and cardiac diseases. Even physicians and public-health officials share this perspective. As previously mentioned, the Clean Air Act mandates that scientific evidence exist to justify updates in the NAAQS. Thus, asthma activists can shift attention from research-related policies to regulatory policies—for example, those achieved by ACE.

Breast cancer activists have some allies in the scientific community, but the research priorities of genetic and lifestyle factors create an obstacle. Even those scientists who are investigating environmental causes of breast cancer, such as those at Silent Spring Institute, report that such work, especially when it involves lay advocates, is not always respected by their peers. As mentioned earlier, breast cancer activists have relied heavily on political support. They have been able to mobilize political allies—from female politicians to male politicians whose wives and other family members have experienced breast cancer—to work toward passing legislation that can begin to create the infrastructure to support environmental causation research.

Gulf War veterans have had the fewest allies. Initially, veterans succeeded in getting President Clinton to create the President's Special Oversight Board, in receiving invitations to testify before Congress, and in obtaining the unyielding support of several representatives, including Christopher Shays and Bernie Sanders. But government officials in the DOD, VA, and other agencies have been less likely to offer vocal support of veterans. A few of the scientists I interviewed, who have spent many years researching GWRIs, remarked that their commitment to finding the cause of GWRIs is related to their sense of injustice for the veterans. Nevertheless, even these sympathetic scientists experienced tension working with veterans who want quick answers and do not respect the slow process of accumulating scientific knowledge.

Another source of allies is health voluntaries and other organizations. Several organizations focus much attention and resources on asthma—for example, the ALA and the American Academy of Allergy, Asthma, and Immunology. Though these organizations fund much research on asthma, little of that money goes specifically to the investigation of environmental causes of asthma. Nevertheless, the ALA has been a more progressive force, in large part due to its decades of struggle against the tobacco industry and a medical establishment that refused to take on the industry. The breast cancer movement gets some support from organizations such as the ACS for overall breast cancer research, but the ACS is highly critical of the search for environmental factors. General breast cancer groups, such as the Komen Foundation, have moved to support some environmental research, and the NBCC has responded well in that area, especially because of pressure from state affiliates. Gulf War veterans, in contrast, have not benefited much at all from such health voluntary organizations, and even the mainstream VSOs have been reluctant to take on GWRIs for fear that the uncertainty about whether people really are sick will tarnish their wider success. Organizations that support sufferers of chronic fatigue syndrome and multiple chemical sensitivity have attempted to support veterans, but these groups lack a strong institutional infrastructure and are already dealing with stigmatized groups, so that they have little to offer Gulf War veterans.

These three factors—the science base supporting the environmental causation hypothesis, public awareness and perception of risk of the disease, and the sources of support for the environmental causation hypothesis—combine in a variety of ways. Health social movements' policy successes are not dependent on any of them, but the movements certainly stand to benefit to the extent that they can manipulate these factors to their advantage. In table 5.2, I summarize some of the impacts the three factors have had on the three diseases I have been discussing.

THE RELATIVE IMPORTANCE OF HEALTH SOCIAL MOVEMENTS

Overall, the power of a social movement can trump the other factors. Some of the movement's power stems from social structural features. Race and gender politics are very important in the successes of the asthma movement and the EBCM. Asthma activism has taken on an environmental justice frame, linking the disease to racism in housing, education, land use, health care, and transportation. The EBCM and its larger allied breast cancer movement stem in large part from the women's health movement, probably the country's most powerful and widespread health social movement. Gulf War veterans, though largely from the working class, do not put forward an explicit class-based politics, and such a politics does not go very far in the United States anyway.

Yet there is no simple algorithm for how health social movements and other factors interact to yield diverse policy outcomes. If anything, we must be attuned to the necessity of evaluating the complexity of each disease on an individual basis. For asthma, we see a strong regulatory policy in the form of the Clean Air Act, but this legislation existed before asthma activism emerged. Considerable research and a high degree of mandated and actual citizen participation in that research occur for asthma. Yet despite general coverage of asthma treatment, there is little provision of hypoallergenic products. These policy outcomes result from a strong science base, firm institutional support, and widespread public awareness. Support for environmental causation (including triggers, not necessarily causes) has grown, yet the low mortality has not put asthma high on many agendas. The social movement for asthma education and prevention is local, often based in environmental justice groups that are frequently allied with university research centers. It is a growing social movement, but not nearly as powerful as the nationally and locally centered breast cancer movement, so that support for it has not yielded significantly additional policy benefits. Also, it is easier to achieve research policy than regulatory policy because the latter has powerful economic implications.

For breast cancer, we see no major regulatory policy, but considerable research funding growth, a high degree of citizen participation, and major effects on detection and treatment. These elements occur in the context of strong institutional support, despite a weak science base for environmental causation. The high morbidity and mortality and the broad public awareness have been powerful features in bringing breast cancer issues to the nation's attention. And there is growing support for the environmental causation hypothesis. Like all other policy outcomes, this result has been due to an extraordinarily large and successful social movement, perhaps one of the strongest in all areas of health. To the extent that support for environmental causation is growing, despite broad scientific skepticism, we see how the strength of the social movement trumps the science base in this case.

For GWRIs, several pieces of regulatory policy have been developed, but none that directly affects sufferers. There has been extensive research funding and, despite initial government reticence, a growth in medical treatment and benefits. Citizen participation in research is relatively weak, both in mandate and practice. In looking at the factors responsible for these outcomes, we find a very weak science base and hardly any institutional support. High self-reported morbidity and extremely low mortality are matched by relatively weak public awareness. Though many activists support environmental causation, hardly any scientists do so. Social movement organizing for GWRIs is the weakest among the three diseases, despite some articulate leaders. Any policy successes we see in this case are more likely tied to congressional and public support for veterans as a class of people rather than to the power of a social movement.

The centrality of scientific evidence for any movement dealing with a known or supposed environmentally induced diseases can be explained by figure 5.2. Drawing from the three factors listed in table 5.2, we can see how the preexisting science base for environmental causation shapes movement approaches. The diagram does not suggest that a strong science base is a necessary condition for policy. Indeed, this is not the case in the many successes of the breast cancer movement. Rather, figure 5.2 illustrates how the absence of a strong science base leads health social movements to mobilize whatever resources possible to generate scientific evidence that will justify future policies. In turn, their policy priorities begin with funding for further research

Weak science base	→	HSMs rely on: 1. institutional support (e.g., academic, government, health voluntaries) to generate scientific evidence 2. support for the environmental causation hypothesis from key sources (prominent physicians and scientists)	→	Policy priorities of HSMs: 1. research 2. citizen participation 3. compensation/treatment 4. regulatory action
Strong science base	→	HSMs rely on: 1. morbidity and mortality rates of the disease or condition to attract 2. public awareness and perception of the risk related to the disease	→	Policy priorities of HSMs: 1. regulatory action 2. compensation and treatment 3. research 4. citizen participation

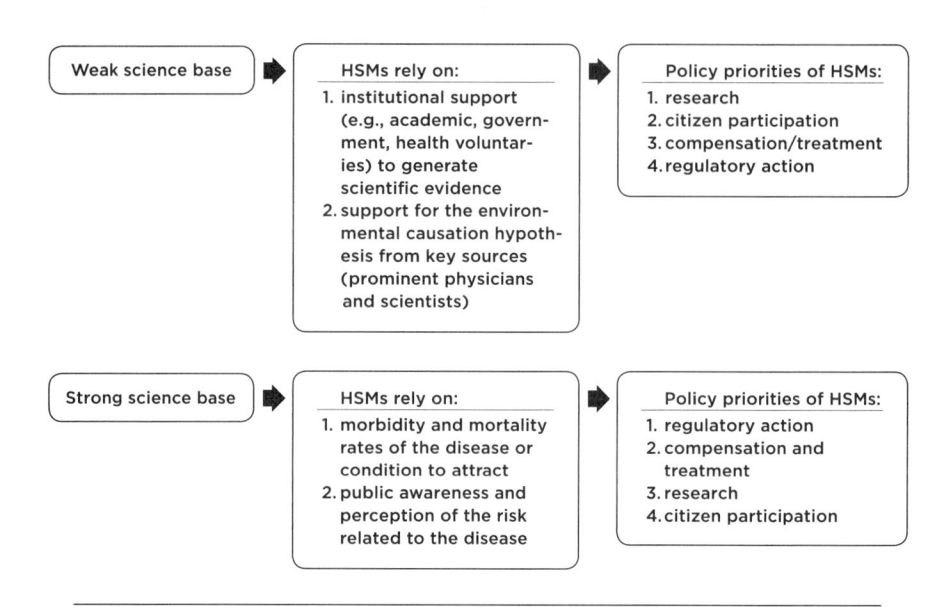

FIGURE 5.2 How science shapes health social movements' (HSMs') policy priorities.

and citizen participation. As these goals are achieved or as the movement differentiates, compensation and treatment become priorities. As the scientific evidence warrants, the focus eventually shifts to regulatory policies. When the science base is strong, a movement mobilizes public support by emphasizing the morbidity and mortality of a disease in order to increase public awareness and perception of the risk related to that disease. These efforts are geared toward regulatory policies, compensation, and treatment. In the event that the existing science is challenged, further research and citizen participation may also become priorities.

Activists whose work stands on a weak science base will often attempt to create a stronger one to legitimate their concerns. This point requires some clarification. First, activists generally do not believe that science by itself is sufficient; rather, they view it as one of several factors that help advance their efforts. Public support, political connections, media attention, and social movement organizing are crucial as well, so that science alone is not as important as it was in the past. Second, when we examine activists' search for scientific evidence, we often see that they challenge the DEP and take innovative approaches that transcend "normal science." For example, EBCM activists pursue the endocrine disrupter hypothesis, which is indeed a challenge to the bulk of mainstream breast cancer research. Gulf War veterans push for research on synergistic relationships of substances, in contrast to the common focus on single-source toxics. Environmental justice activists concerned with asthma put much energy on particulate emissions, which, although important to air-quality researchers, is generally not dealt with much by asthma researchers and clinicians. These activists also push for cumulative risk assessment rather than risk assessment of one substance at a time. And in all these cases, activists push for a strong community initiative in deciding what scientific issues to address and for a community-academic partnership in pursuing those issues. In this sense, the health social movement and the pursuit of science are highly intertwined in various forms of citizen-science alliances that are engaging in public-paradigm production and dissemination.

Hence, both in the type of scientific evidence and in the process used to develop it, environmental health activists challenge the traditional canons of science. Further, we see that science alone is not the highest criterion for activists. They see science as not separate from moral values, but rather as integrated with those values; societies and interest groups within those societies make science decisions based on what they want to know and do, not on a completely value-free knowledge base.

By focusing on health social movements that deal with diseases believed to have environmental causes, I have illustrated how the strength of the science base, public awareness and perceptions of risk, and the sources of support for the environmental causation hypothesis shape movements' strategies. Comparing these three diseases does not provide a sufficient base for generalizing about all diseases in terms of policy outcomes, but in the absence of much comparative research it can provide the beginnings of a contested illnesses model for examining the dynamic relationship between health social movements and health policies.

|6|

THE NEW PRECAUTIONARY APPROACH

A PUBLIC PARADIGM IN PROGRESS

MOVING FARTHER UPSTREAM: THE PRECAUTIONARY PRINCIPLE

Sandra Steingraber's book *Living Downstream* takes its title from the oft-used metaphor about preventive health. Villagers living along a river notice people floating downstream and regularly pull them out, developing new rescue techniques as time goes by. But no one thinks to look upstream to see why the people are in the river in the first place. As a society, we cannot pinpoint the exact location upstream where the toxic threats to public health and the environment originate because many of the suspected hazards are ubiquitous, so we have to address these hazards farther upstream until we are able to remove them altogether. The metaphor may be a little timeworn, but there is nothing more apt when we talk about environmental health. After all, we all live downstream from someone.

We also live upstream from other people whom we may be contaminating, and it is always possible that the direction of flow can change. Put otherwise, we cannot always tell in what direction toxins will flow, and we know they cannot be readily contained. What we release may indeed come back to poison us. Our contamination can harm our own future generations. Therefore, in protecting ourselves, we also protect others. Going farther upstream is a way to prevent harm, and this approach has become synonymous with the precautionary principle.

This chapter follows the development of the precautionary principle in the United States, showing how it exemplifies the notion of a public paradigm. I discuss how this seemingly new approach has deep roots in U.S. history and public policy, even if it has only recently been formulated as a powerful alternative vision. This chapter looks at the roles of the Massachusetts state agency TURI and the activist Precautionary Principle Project that later became the Alliance for a Healthy Tomorrow. It also examines the appeal of the precautionary principle to various sectors and groups and observes how an increasing variety of environmental health and environmental justice organizations have found it to be very applicable to their missions.

This precautionary perspective has led scientists, academics, and activists to shape a far-ranging model of environmental protection and environmental health. Environmentalists believe that the government's environmental and public-health agencies have largely failed to regulate and protect the population, so social movement organizations have stepped up to put the public back in public health. The environmental health movement spreading from this precautionary approach is a powerful contribution to American democracy, with ties to many other social movements. The precautionary principle has been a key discourse for much of this development and is worth considerable attention in this book.

When I chose TURI as a research site, I did so because it was advocating a method of reducing toxics, including those suspected of being in part responsible for asthma, breast cancer, and Gulf War illnesses. I was not aware at the time that some staff members of the Lowell Center for Sustainable Production, TURI's neighbor at the University of Massachusetts–Lowell, were working with activist groups throughout Massachusetts to develop a statewide network to foster the precautionary principle. I was fortunate to see this effort start at a very basic level and mushroom into a major initiative that coalesces with related actions throughout the nation and the world. We rarely have the opportunity to observe any paradigm in development from its origin, and this case offered the potential to see the origins and diffusion of a *public paradigm* such as I described in the first chapter. A public paradigm is a pattern of public action and attitudes not just for a specific scientific argument, but also for a broad perspective underlying current scientific and societal processes. A new public paradigm leads people to alter fundamentally how they view the world, poses deep levels of criticism as a route to an alternative model of social life, and occurs as a result of large-scale public participation.

The precautionary principle is giving a new name to a long-standing belief in environmental protection. Advocates for environmental health have been following a precautionary path since the time of Rachel Carson. In its U.S. version, the precautionary principle was codified in January 1998, when an interdisciplinary group of creative thinkers gathered at the Wingspread Conference Center in Wisconsin to discuss the precautionary principle and how to understand the environmental and public-health threats facing the world. The conference produced the Wingspread Statement, which situated uncertainty at the heart of the need for precaution: "Uncertainty becomes the reason for taking action to prevent harm and for shifting the benefit of the doubt to those beings and systems that might suffer harm."[1] The Wingspread Statement highlighted threats to human health and made a direct link between the failures of risk assessment in protecting public health and the importance of implementing the precautionary principle as an alternative.

THE RAPID CREATION AND DIFFUSION OF THE PRECAUTIONARY PRINCIPLE

The precautionary principle and other related approaches using different terminology have spread very rapidly. The precautionary principle is a very broad-based public paradigm that has diffused so quickly because it has an encompassing vision and because it meets the goals and needs of many different parties. That spread has been due largely to very effective organizing by the Precautionary Principle Project (now the Alliance for a Healthy Tomorrow, to be discussed in detail later), which not only organized extensively in Massachusetts, but also sent talented speakers around the country to teach about the precautionary principle at universities, research institutes, and environmental organizations. Leading proponents also wrote extensively about the precautionary principle and convinced key journals to publish material on it, including special issues and sections of *Public Health Reports* and the *American Journal of Public Health*. Leaders also conducted conferences that brought together many people, including European regulators who had already implemented precautionary principle methods. Furthermore, many environmental groups, including EBCM organizations, realized that the precautionary principle addressed their long-standing concerns and provided a vehicle for developing a firmer scientific underpinning and a catchy public frame.

For environmentalists, especially those dealing with human health, this precautionary approach has been implicit and often explicit ever since the publication of Rachel Carson's *Silent Spring* in 1962. Even before environmental epidemiology became as developed as it now is, scientists and environmental advocates knew they needed to apply caution regarding the many chemicals they saw proliferating. Toxic waste activists learned that health studies were often too rare, took too long, and were often unable to link exposures directly with health effects because of the way they were designed.

For environmental justice activists, the precautionary principle offers a new vocabulary and practice that bolsters their desire to prevent overburdened minority and poor neighborhoods from bearing additional environmental hazards. Environmental justice groups' focus on environmental racism has already educated many people about inequities and has triggered considerable public policy, but the addition of the precautionary principle offers a new framework that may make more people understand the problem of *preventing* unequal burden.

For occupational safety and health advocates, precaution has been around as long as there has been any organized concern for worker safety, easily longer than a century. Workers, unions, and occupational health professionals knew that companies routinely took the cheapest possible route to production, forsaking what were often very simple safety precautions. They well understood that this cost-benefit thinking drove unsafe practices, despite a clear awareness that precautions could save health and lives.

The precautionary approach has offered many scientists a way to challenge many of the old-fashioned canons of the scientific establishment by means of notions of causality and a holistic approach linking together many areas of science. The fact that biologists, ecologists, epidemiologists, and physicians were playing a strong role in the precautionary principle organizing gave credible support to an alternative vision of new scientific models.

For those regulatory officials and scientists who were somewhat sympathetic to environmentalist goals, the precautionary approach early on provided some of the legitimation they needed to convince colleagues, superiors, and political powers in legislative and executive branches. These regulators were able to point to the sensible examples of "late lessons from early warnings" and to show that thoughtful regulatory agencies in other countries were already taking action based on the precautionary principle.[2]

All these parties just mentioned were conducting discussions on precaution in a very public light. Their challenges to traditional ways of knowing routine courses of action were very visible, leading to the kind of public deliberation that is the hallmark of a public paradigm.

Above all, a movement was necessary to spark the collaboration of all these parties. This movement was expressly led by activists pushing the precautionary principle, including the Precautionary Principle Project/Alliance for a Healthy Tomorrow, various EBCM groups, and some environmental groups. Although raising the precautionary principle as a public organizing slogan or strategy in the 1990s might have seemed awkward, by the turn of the twenty-first century it was showing up in many articles in peer-reviewed publications and was being addressed in many papers presented at scientific and public-health conferences. And June 2006 saw the First National Conference on Precaution, organized by the Center for Health, Environment, and

Justice and designed to use the banner of the precautionary principle to bring together groups working on conservation, disease prevention, environmental justice, "green" purchasing, health, pesticides, toxic and nuclear pollution prevention, and worker safety and health. Due to the growth of approaches and local regulation based on the precautionary principle, the center also launched a clearinghouse with details on precautionary-based laws, policies, local ordinances, and industry agreements on a range of issues, including conservation, environmental precaution, genetically modified organisms, hazardous-waste transportation, PVC, pollution prevention, purchasing of toxic chemicals, corporate purchasing, mercury, and pesticides. This confluence of interests and activities accounts for the extensive diffusion of the precautionary principle in such a short time.

Although informed by European examples of more stringent control of toxic substances, especially by regulatory bodies of individual nations and the European Union, the precautionary principle in the United States has developed out of three overlapping pathways: toxics reduction, toxic waste activism, and EBCM activism, as shown in figure 6.1.

The toxics reduction approach—which aims to prevent toxics from entering the environment in the first place, rather than engage in command-and-control practices in order to remove them after the fact—arose largely from the movement toward pollution prevention in the 1980s. Incorporated into policies such as the Toxic Use Reduction Act passed by the Massachusetts legislature, toxics reduction led naturally to the precautionary principle in its incorporation of the idea of reducing the use of potentially harmful substances, its challenges to assumptions of safety in industrial products, as well as its encouragement to industry to implement prevention strategies in order to bypass later regulatory enforcement. Toxic waste activism, which arose in response to Love Canal in 1978 and the discovery that a substance's known

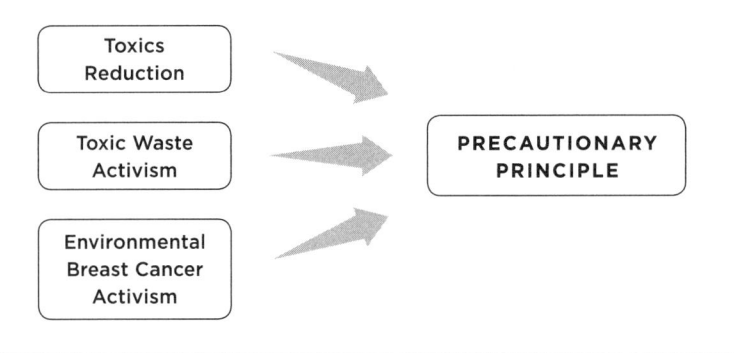

FIGURE 6.1 Three pathways to the precautionary principle.

toxicity did not always result in its removal from the public sphere, also supported the use of precaution and prevention given scientific uncertainty and began to reject the notion of "acceptable risk."[3] These pathways represented preventive approaches wherein traditional conceptions of the "innocence" of environmental pollutants were rejected, and they laid much of the groundwork for the development of the precautionary principle.

Although environmental justice was not an earlier pathway into the precautionary principle, it now represents a new impetus to further development of the principle because a growing number of environmental justice groups have begun to employ the principle as a description of their efforts to create safer, healthier communities.

The current emergence of a new name, however, is more than mere recognition of a continuing belief and practice; it helps coalesce a wide array of people and organizations under a banner that unites disparate activities, with a forward-looking vision toward a safer future. In other words, a breast cancer activist concerned with organochlorines can find common ground more easily with a person seeking nonpesticide approaches to handling West Nile virus or with a hospital worker seeking to avoid harmful phthalates and other substances in his or her workplace. By recognizing European countries' leadership in the precautionary principle, U.S. supporters are highlighting the harmful stances taken by the United States on many environmental issues, while also pushing for a more cooperative, internationalist perspective. On a political and philosophical level, by invoking the idea of a larger *principle*, supporters of this approach frame it as a code for organizing a better and healthier society. A healthy environment and a reduction in health hazards are no longer seen as "extra" policy matters, but as fundamental to a healthy and equitable social structure. Further, the internationalist element recognizes the cross-national production, use, and disposal of chemicals, and it fosters a collaboration that transcends narrow national interests. This new underlying principle's breadth of coverage and its location in a worldwide web of knowledge and action make it the kind of deep and public engagement that defines a public paradigm.

Quijano notes that in the use of the precautionary principle, we can

1. Take preventive action in the face of uncertainty
2. Shift the burden of proof to the proponents of an activity (reverse the onus)
3. Eliminate those substances about which we have sufficient evidence of harm
4. Orient policy toward helping communities

5. Explore a wide range of alternative products and processes
6. Consider uncertainty as a potential threat, rather than something to be ignored; always ask what we *do not* know
7. Make sure production and disposal processes are technically and scientifically sound
8. Make all information publicly accessible
9. Use an open democratic process with public participation in decision making
10. Ask whether we really need many of the products that we manufacture and use because many are unnecessary elements of an overconsuming society.[4]

These ten ways of acting are derived in part from European countries' recent laws, international treaties, and protocols, but they are also a direct outgrowth of decades of general environmental action. People concerned with environmental degradation and human disease are impatient with a process that puts the burden on real or potential victims to show the hazards of chemicals and other substances. They know that standard scientific procedure makes it hard to demonstrate hazards and that such efforts often take too long. Further, they replace the question "How bad is it?" with the more precautionary question "How much can we prevent it?" Similarly, instead of "What can an ecosystem absorb before it is damaged?" they ask, "What safer alternatives are there?"

Although the precautionary principle is not explicitly cited in U.S. legislation, activists argue that there is implicit reference to the use of precaution in protecting health in several major policies. Environmental, worker safety and health, and food quality laws often imply erring on the side of caution. For instance, the Occupational Safety and Health Act of 1970 requires employers to "furnish to each of his employees employment and a place of employment which are free from recognized hazards." An important component of the precautionary principle, the reversal of the burden of proof of safety from the consumer to the producer, is reflected in the federal Food, Drug, and Cosmetics Act of 1938. This act requires pharmaceutical manufacturers to demonstrate the safety of a product, rather than rely on limited government testing or public-health officials to demonstrate negative health effects. The Delaney Clause, included in the Food Additives Amendment of 1958, was another precautionary piece of legislation. It states that no food additive (including pesticides) should be considered safe if found to induce cancer when ingested by people or animals. Perhaps the most fundamental application of precaution in federal legislation can be found in the National Environmental Policy Act of 1969, which implicitly advances the precautionary principle's focus on exploring alternatives through the provision of information prior to making decisions.[5]

The precautionary principle has recently been adopted by the community of environmental, community, and worker organizations in attempting to improve chemical policy through the Louisville Charter for Safer Chemicals, "a platform for creating a safe and healthy environment through innovation." The Louisville Charter is a national environmental platform designed to outline broad social principles for protecting human health and the environment from the threat of toxic exposures. Drafted following the convening of leaders in the environmental health and justice movement in May 2004, it implicitly promotes a precautionary approach to reducing toxic hazards through the logic of "green" innovation in production processes. It calls for fundamental reform to the U.S. chemical policy by following six steps: (1) requiring safer substitutes and solutions to hazardous chemical use; (2) phasing out persistent, bioaccumulative, highly toxic chemicals; (3) giving the public and workers the full right to know by providing public access to data on hazardous chemicals; (4) acting on early warnings and taking immediate action when exposure occurs; (5) requiring comprehensive safety data for all chemicals; and (6) taking immediate action to protect communities and workers at risk. Hundreds of individuals and organizations have signed onto the Louisville Charter, marking the growing support for a precautionary approach to managing toxic chemicals and substances.[6]

Precautionary principle proponents offer their perspective as an antidote to many problems in how our society currently examines hazards. First, as indicated by Quijano's ten considerations for action noted earlier, mainstream risk assessment practices are narrow and short-sighted, are typically focused on single substances rather than on cumulative burden and synergistic effects, err on the side of allowing too many poorly understood chemicals into use, and protect corporations more than people. Precautionary activists contest traditional risk assessment with *cumulative risk assessment.* Considerable impetus for this cumulative risk assessment comes from environmental justice groups who argue that the multiple assaults on their communities cannot be understood if government and science focus on isolated, individual chemical risks. They argue that chemicals, particulates, and other hazards work in tandem with each other. They have so consistently brought up this issue to regulators and policymakers that those audiences have had to take it seriously. As a result, this new approach has diffused from activists and their scientist allies to some EPA offices, as evidenced at a 2004 conference held in Boston by EPA Region 1, which environmental justice activists helped organize and where they presented many papers.

Second, even traditional risk assessment approaches are being subverted; new laws and policies increasingly require regulators to show, according to their official calculations, that regulation will not be costly. According to the Clean Air Act of 1970, safety is supposed to come first, through using health criteria to set standards. But the process of *meeting* those standards is subject to cost-benefit calculations, allowing the corporate sector and the antiregulatory political administration to increase their attacks on adequate implementation. Even the health-based setting of standards is now being violated all the time, as the George W. Bush administration cut back on safety measures for such well-known dangers as mercury, arsenic, and particulate matter emissions. Clean water regulations allow standards to be set on feasibility and cost consideration. With arsenic, what many experts considered a safe level of 1 part per million (ppm) was problematic because laboratories had technical difficulty in detecting that level; at the same time, municipal water departments have trouble lowering the level below 3 ppm. As a result, the level was set at a rather high 10 ppm, yet the current administration fought against the arsenic regulation even at this more lenient standard.

Opposition to stricter regulation of arsenic is but one of the many efforts made by the Bush administration to subvert science. These efforts are so brazen that it is easy for many people to develop distrust in the objectivity of the science of risk assessment. Many actors both from the activist community and from within science itself are becoming increasingly vocal about the influence of corporate interests in the outcome of supposedly objective risk assessments. In addition, there has been a much publicized shift away from science-based decision making within the federal government under the leadership of the Bush administration. Given these critical flaws in the public trust in risk assessment, the precautionary principle is filling a new space to reshape current policies in hazardous substance management.

Third, even the pro-corporate, cost-benefit analyses that drive current policy are flawed. Extensive revelations of conflict of interest in science and medicine have been made, as I explore in chapter 7, and they call into question the quality of the data used to make public policy on science matters.

One example of successful, cost-effective implementation of the precautionary principle in Europe is found in the Danish government's ban on phthalates (plasticizers) in children's toys, implemented in 1999. This ban came not as a result of traditional quantitative risk assessment techniques as used in the United States, but rather from qualitative considerations of the toxicity of phthalates, the susceptibility of children, and the availability of reasonable alternatives.[7] Despite scientific uncertainty, the Danish government banned the use of phthalates in all toys intended for children under the age of three. Meanwhile, in the United States the risks associated with such exposure were determined to be relatively low, so no ban was implemented. However, given the action on the part of European nations and consumer warnings announced by advocacy groups, officials have suggested a voluntary removal of phthalates from such toys.

A recent example from Sweden provides a glimpse into how the precautionary principle can be applied and can achieve quick results. PBDEs, a class of chemicals similar to PCBs, have been widely used since the 1960s as flame retardants in clothing, furniture upholstery, computers, and many other products. In the 1980s, Swedish researchers found high concentrations of PBDEs in breast milk. Although there is still uncertainty about the effects of PBDEs on health, the Swedish government quickly worked with industry to remove these substances from the production chain, which resulted in a noticeable decline in breast milk burden. The reduction in body burden was sufficient evidence for the success of the new policy in Sweden, in contrast to how things happen in the United States, where policy action requires years of contested scientific studies. The European Union acted on this evidence as well, banning several types of PBDEs in 2003.[8] This example shows that a productive, advanced capitalist society can conduct rapid evaluation of health hazards and implement quick regulation without hurting the economy.[9] In contrast, in the United States the action of private organizations and the media was required to get government attention regarding the hazards of PBDEs. Even without a formally legislated ban, the EPA, in combination with state actions and the activism of environmental health groups, was able to convince major sources of PBDE, such as the flame retardant manufacturer Great Lakes Chemical Corp., to phase out the use of the chemical by 2004.[10]

The European Union formulated a broad policy to protect human health and the environment through Registration, Evaluation, and Authorisation of Chemicals, which was introduced in October 2003 and required ratification by all member nations, a process still under way in 2006. This policy puts the precautionary principle into more widespread action by requiring various levels of testing for all chemicals used in quantities of one ton or more annually and by placing restrictions on the most dangerous chemicals on the market. It puts the burden of proof on companies rather than on government and seeks to harmonize rules across the European Union. In stark contrast to current practice, chemicals already in use would not be exempted from testing and registration. In keeping with the precautionary approach, companies would need to show that chemicals of high concern, like drugs, can be used safely, that no safer alternative existed, or that the benefits of the chemical to society outweigh the risks. Even in the cases that meet these standards, "authorization" for use is provided only for a limited time. The U.S. chemical industry and the Bush administration opposed the proposal, but more than ten thousand Americans, including representatives of sixty-one environmental, health, and trade groups, signed the "U.S. Declaration of Independence from Hazardous Chemicals," which supports the European Union reforms.[11]

Considerable corporate opposition centers on the fear that precautionary approaches will lead to higher costs for business. Opponents fear that a more radical interpretation of safety will neglect considerations of cost-effectiveness in implementing precautionary policies, thus forcing businesses to engage in expensive practices despite scientific uncertainty. Furthermore, under this perspective, the precautionary principle is seen as "antitechnology" and "antiscience," working to halt all potential innovations through a "zero risk" stance, thereby leaving industry powerless.[12] However, European experience has demonstrated that precautionary policies actually promote industrial innovation and create jobs. Even though there is limited ability to study this result in the United States, there is substantial evidence about the relationship between the environment, jobs, and the economy. Environmental spending creates more manufacturing and construction jobs than the general economy's average. Pollution prevention and control jobs are one of the nation's highest growth sectors. States with stronger environmental regulation have greater job growth than other states; although this statistic does not mean such regulation produces jobs, it certainly shows that regulation does not reduce jobs. Environmental protections at present generally cost about 2–3 percent of companies' expenses, thus not making them a reason for moving overseas, and those industries with the most overseas flight are not the ones with high environmental costs. Many fearful estimates of high compliance costs are exaggerated; oftentimes there are little or no net costs because new technologies save costs in other ways. Much new technology development is made possible only when firms are shown the environmental problems they face, which then spur them to innovation.[13]

Forward-looking businesses may realize that they can be affected by expensive cleanup, costly liability, higher insurance premiums, potentially more onerous regulation, and public pressure. Such companies not only can be models of change, but can pressure other firms. The Swedish construction firm Skanska, the third largest in the world, suffered a major industrial disaster when using a water sealant in constructing a rail tunnel in 1997. Without the benefit of protective gear, 223 workers suffered acrylamide poisoning, with nausea, dizziness, and prickling sensations in their fingers.[14] In response, Skanska put together a list of chemicals it wanted replaced, then gave this list to all its suppliers, demanding they provide safer substitutes.[15]

It is not enough to apply precaution to regulation; it also needs to be applied to research. Alternatives must be readily available in order for precautionary regulation to succeed. For example, scientists need to work on "green" chemistry, alternative materials, and substitute processes to find safer paths for industrial production and commercial practice.[16] Exploratory research is also needed to characterize the many chemicals about which we know little or nothing; an excellent example is Silent Spring Institute's household exposure study, which found concentrations of thirty chemicals never before measured in homes.[17] As Levins remarks, we need rapid detection of actual and potential harmful effects, just as animals and plants in nature respond to threats. In particular, we need to be aware of systems at the boundaries of tolerance because they are the most susceptible. This awareness entails looking closely at the most vulnerable groups of people, such as the poor, race minorities, children, and people with chronic illnesses.[18]

APPLYING THE PRECAUTIONARY PRINCIPLE

Beginning in the late 1990s, a growing number of environmental organizations and health groups found the precautionary principle to be a useful foundation for extending environmental health concerns. The role of TURI and some related units at the University of Massachusetts–Lowell has been central in these activities. TURI was created by the 1989 Massachusetts Toxics Use Reduction Act. During a high point of environmental activism in this state, the Massachusetts Public Interest Group gathered enough public support potentially to pass a ballot initiative that would set strict state standards for the use of toxic substances. To avoid the possibility of such strict regulation, industry leaders and state officials met with environmental leaders to negotiate a compromise. The legislature voted unanimously to pass the Toxics Use Reduction Act, creating TURI and the Office of Technical Assistance, which work with the Massachusetts Department of Environmental Protection. Together these agencies work to reduce toxics throughout Massachusetts by providing technical assistance to industry.[19] TURI reduced toxic emissions by 58 percent in its first decade, 1990–2000, and saved Massachusetts firms more than $15 million while doing so.[20]

The Toxics Use Reduction Act program is funded through a structured fee system for the use of a selected list of toxic chemicals by industrial firms. Each company must submit biannual plans to the Massachusetts Department of Environmental Protection that must include pollution prevention opportunities addressing each toxic chemical used. These toxics use reduction plans must be certified and approved by a toxics use reduction planner. One of TURI's biggest roles is the training of these planners. It offers training seminars and certification tests, along with extension classes to keep planners informed about new pollution prevention technologies and strategies. TURI also acts as a research laboratory and often designs and tests alternative cleaning processes at the request of industrial firms. Its function as a research institution and its technical assistance role exist to challenge the notion that pollution prevention is too costly and to fund the development of innovative production methods.[21] Because the University of Massachusetts–Lowell had extensive capacity in research and training in occupational safety and health, alternative production and disposal, and environmental planning, TURI was located there. It made sense that there would be significant overlap of people working in TURI and in two university units, the Lowell Center for Sustainable Production and the Department of Work Environment.

Key individuals in these three units wore multiple hats, allowing them to use their talents in varied arenas. Ken Geiser, TURI's director, and Joel Tickner of the Lowell Center for Sustainable Production and the Department of Work Environment held leadership roles in the Precautionary Principle Project and later in the Alliance for a Healthy Tomorrow. While working at the Lowell Center for Sustainable Production, Joel Tickner earned a doctorate in the Department of Work Environment under the guidance of Ken Geiser and became one of the key leaders of the Precautionary Principle Project. This configuration of personnel produced a continual cross-cutting interaction between the different organizations, allowing each to contribute parts to the larger framework of the developing precautionary paradigm.

Beyond the Lowell-based connections, many interorganizational connections developed stemming from two sources: the transformation of toxics reduction activities into precautionary principle activities and the participation of many people there in a long legacy of other environmental activism in the state. In Massachusetts, the precautionary principle is centered in the work of the Alliance for a Healthy Tomorrow. The project was initially a joint effort of the Massachusetts Breast Cancer Coalition, the Clean Water Fund, the Lowell Center for Sustainable Production, and the Science and Environmental Health Network, but it now has a much larger organizational core, with more than one hundred member organizations. The original Precautionary Principle Project was the only statewide organization designed to spread knowledge about and implement the precautionary principle as a guide to health policy. It developed an organizational structure of its own, including statewide conferences, the most recent of which attracted approximately two hundred people. It conducted a variety of public educational forums, held meetings with scientists and scholars on university campuses, encouraged state and local public-health officials to be more cautious on spraying to kill mosquitoes that spread West Nile virus, and worked with state legislators and government officials to develop model policies with a focus on the problems associated with spraying and potential solutions to address broad public-health concerns.

In the fall of 2001, the Precautionary Principle Project held an international conference focused mainly on building the scientific base for the precautionary principle. Even though the conference was held only weeks after the 9/11 terrorist attacks, virtually all preregistrants attended, despite the difficulty of travel from abroad for many of them. Following the success of the conference, the Precautionary Principle Project transformed itself into the Alliance for a Healthy Tomorrow in order to foster broader public participation.

The Alliance works toward a wide range of goals: implementing safe cleaning products and practices in schools, hospitals, and public buildings; getting hospitals to avoid the use of PVC and to phase out mercury-containing products; requiring that safe materials be used in new school construction and school renovations; pursuing safer cosmetics and personal care products; and promoting integrated pest management at both residential and municipal levels. By the summer of 2005, the Alliance had three bills under consideration in the Massachusetts legislature, sponsored by senators and representatives who were friends and members of the Alliance. The Safer Alternatives to Toxics Program would require companies to replace toxic chemicals with safer alternatives wherever feasible, starting with ten of the worst toxic chemicals.

The Safer Cleaning Products bill would require that only healthy cleaning products (products screened to avoid health impact) be used in various public areas. A bill on Safer Alternatives to Mercury Products would phase out the use of mercury-containing products and keep them out of incinerators, landfills, and the environment. The mercury bill was unanimously passed in both the Massachusetts House and Senate in February and March 2006, respectively.

The toxics reduction infrastructure created through TURI is unique in the United States, serving as a model for others who seek statewide toxics reduction policy. Despite TURI's success in toxics reductions without an attendant cost to business, there was significant opposition in 2000 to renewing the time-limited Toxics Use Reduction Act; business and pro-business legislators found the regulations to be burdensome and disadvantageous for Massachusetts industry with regard to competitors from other states. Industry leaders voiced their concerns at the hearing to renew the act held by the Committee on Natural Resources and Agriculture, but their opposition failed, and the act's structure remained intact. Indeed, TURI now has the enforcement responsibilities previously held by the Department of Environmental Protection. This position, however, holds out the potential for TURI's becoming less friendly to industry because of its emphasis on enforcement rather than on industrial retooling and cost savings.

It is also important to recognize that TURI's strong basis in science led the Alliance for a Healthy Tomorrow to emphasize a science base in its organizing activities. For example, Ken Geiser wrote the book *Materials Matter: Toward a Sustainable Materials Policy,* which provided strong support for industrial reengineering. Alliance conferences always feature scientists who provide strong support for both the study of hazards and the design of alternative chemicals and processes. Position papers and organizing materials always put forth substantial evidence from the scientific literature, giving credibility to many of the Alliance's programs.

All these components of the precautionary principle organizing program fit into the notion of a public paradigm. As I noted in chapter 1, the public paradigm represents a complex set of public actions and attitudes that challenge the DEP underlying current scientific and societal processes, and it offers an overall approach to environmental causation. The precautionary principle, as an exemplar of the public paradigm, orients many people to a dramatically new perspective. It seeks to involve the public in participation in science, rather than keeping the public on the outside, and to give laypeople some influence over the formation of new scientific knowledge.

I now briefly describe some of the directions taken by precautionary principle activism on the local, state, and national levels.

WEST NILE VIRUS ACTIVITIES

In 1999, seven people in the New York City area died from the mosquito-borne West Nile virus. In 2000, West Nile virus was confirmed in twenty-one people, two of whom died. As a result, state and local governments sprayed insecticides widely. On the basis of the precautionary principle, however, environmental activists and scientists argued against widespread pesticide spraying. Many believed that the virus is a cyclical event that might not be controllable through spraying. They also considered the government response to be exaggerated, founded on a poor knowledge base, and lacking sufficient regard to the potential health effects of the spraying.[22]

Scientists supportive of the precautionary principle and activists throughout the East Coast formed local and state groups aimed at educating citizens and public-health officials about the issues. The Lowell Center for Sustainable

Production, a mainstay of the Massachusetts Precautionary Principle Project, convened a meeting attended by the state deputy public-health commissioner and several major city commissioners and deputy commissioners, who were willing to engage in such discussions and to reevaluate the hazards. Similar meetings were held at local levels throughout the state. These meetings provided a neutral place for scientists, officials, and advocates to talk about the possible public-health trade-offs in insecticide spraying. Some officials admitted that the spraying had been done because of political pressure from mayors rather than because of a calculated public-health assessment.

Although malathion—an organophosphate classified by the EPA as a class III toxin whose potential ill effects include organ toxicity and damage to the nervous system in humans and widespread ecological toxicity—was never sprayed in Massachusetts, there was widespread use elsewhere. Antispraying activists pointed out that malathion was the second leading cause of hospitalization for occupational pesticide poisoning in the United States during the period 1977–1982. Prior research indicated that malathion may compromise the immune system, cause reproductive harm and genetic mutations, or interfere with normal cell replication. Resmethrin and d-phenothrin (Sumithrin), the other pesticides used, are synthetic pyrethroids. Though less toxic than malathion, they can affect the nervous system, and allergic responses have been reported.[23] Activists were especially concerned with the synergist included in the spraying, piperonyl butoxide, which has been classified by the EPA as a possible human carcinogen. Importantly, many local and state health departments neglected to discuss potential risk from piperonyl butoxide, addressing only malathion and resmethrin. Without capacity to do their own research, many local officials used the CDC's pro-spraying guidelines.

Activist efforts were also quite successful at a national level. In April 2001, a sixty-five-page draft by the CDC outlined new guidelines for addressing West Nile virus. Despite the expectation of more cases in summer 2001, the CDC suggested pesticide spraying only as a last resort, instead urging a preventive approach aimed at eliminating mosquito breeding grounds, spreading larvicide to kill mosquitoes before they emerge as adults, and educating the public on mosquito avoidance. The CDC acknowledged flaws in its previous approach, when it suggested spraying in a two-mile radius each time health officials found an infected bird, mammal, or mosquito. CDC action is critical because many state and local health agencies base their programs on CDC guidelines.[24]

There is no way to credit precautionary principle advocates conclusively with the success in changing the approach to this issue. However, they were extremely visible and active in Massachusetts under the banner of the precautionary principle. Activists realized that the public-health officials had their own form of precaution in spraying because indeed there were real risks to those contracting West Nile virus. Some activists in other parts of the country used an explicit precautionary principle approach, and those who did not clearly aligned their work with the precautionary principle, even when only implicitly. The example of West Nile virus shows the complexity of how competing frames of risk management can develop. Although some of the typical struggles over risks and hazards involve laypeople struggling against corporate profit-seeking, there can be overt significant inputs to decision making that raise legitimate concerns.

CHILDREN'S ENVIRONMENTAL HEALTH ACTIVITIES

Proponents of the precautionary principle have devoted considerable energy to children's health. Childhood is the most susceptible time for environmental agents to damage development, especially neurological development. Also, children's slower metabolic rates can leave the toxic agents in their bodies for longer periods of time. Relative to body weight, children are exposed to a greater amount of toxic materials than are adults.[25] Recent rises in asthma, developmental disorders, and cancer in children have focused attention on children's environmental health, including the EPA's 1996 National Agenda to Protect Children's Health, the Food Quality Protection Act of 1996, and the 1997 Executive Order 13045 on Children's Environmental Health. The EPA, the CDC, and the NIEHS funded eight university-based centers for children's environmental health, and HUD's Healthy Homes project has supported many urban efforts. Despite the number of bills and attention paid to children's health, however, most of the funding is directed toward screening and quantifying risk rather than toward examining, banning, and restricting chemicals or toward building alternative, substitutive production approaches.[26]

Activists implicitly fought for the application of the precautionary principle when trying to pass the Massachusetts Children's and Families' Protection Act in 1996. The act, finally passed in 2000, prohibits use of the most toxic pesticides at schools and day care centers, and it requires these institutions to develop integrated pest management solutions to reduce the amount of pesticides through a combination of enhanced monitoring, biological controls, and limited chemical usage. Senator Resor's announcement of this bill was a highlight of the December 2000 statewide conference of the Precautionary Principle Project. The bill did not use the term *precautionary principle* because it had become a flashpoint, triggering instant corporate attacks. By framing their approach in terms of protecting child health, precautionary principle supporters could avoid fighting over words and instead focus on convincing legislators that it was valuable to protect the state's youth.

PRECAUTIONARY PRINCIPLE ACTIONS BY OTHER MOVEMENTS

Interestingly, even as the term *precautionary principle* became a liability for some groups, such as the Alliance for a Healthy Tomorrow, it became increasingly widespread in the perspectives and writings of environmental health and environmental justice organizations. Some of these groups could afford to use the term because they were not trying to win over a legislature or industrial producers. At the same time, some groups retained the term more explicitly in their efforts to seek government approval for precautionary policies; the San Francisco alliance that succeeded in getting the city to adopt the precautionary principle is one example of this approach.

BREAST CANCER AND OTHER WOMEN'S CANCER ACTIVISTS Activists in the EBCM employ the precautionary principle in their organizing. The Massachusetts Breast Cancer Coalition is one of the original four groups that started the Precautionary Principle Project, and its breast cancer organizing is based on the precautionary principle. Using it as a strategic frame means advocating for the reduction and eventual elimination of chemicals suspected of causing breast cancer. The coalition started the Silent Spring Institute, the nation's only research institute specifically designed to investigate environmental factors in breast cancer. Activist groups elsewhere, such as Breast Cancer Action in San Francisco, use the precautionary principle in their campaign for a research agenda that focuses on identifying and eradicating the causes of breast cancer. Breast Cancer Action evokes the precautionary principle to argue against the preventive use of drugs such as Tamoxifen and Raloxifene. Although the media portrays these drugs as "prevention pills," Breast Cancer Action argues that the drugs have serious side effects; in particular, Tamoxifen is a known uterine carcinogen. Within the context of the precautionary principle, this group argues that the medical community interested in preventing breast cancer should "first do no harm." It "encourages the use of environmentally-safe alternatives to ways of doing business that we know—or have reason to believe—are harmful."[27] As discussed more fully later, the Breast Cancer Fund and other Bay Area breast cancer groups were central players in getting the San Francisco City Council to adopt the precautionary principle as a guide for municipal policy.

One cancer activist group, the Women's Community Cancer Project in Cambridge, Massachusetts, provides support for women with all forms of cancer and utilizes the precautionary principle in its critique of traditional risk assessment. It echoes the principle's tenet of action based on suspicion rather than proof and advances the precautionary principle as an alternative to risk assessment. It cites toxics use reduction as an example of the precautionary principle because it "instructs firms to identify ways to reduce their waste and, subsequently, use of those chemicals—any amount of use is considered too much."[28] The Women's Community Cancer Project's focus on alternatives to toxic chemicals reflects its critical approach to traditional cancer policy that typically blames genetic and lifestyle factors. Instead, it emphasizes socioeconomic risk factors and chemicals, areas overlooked by most medical researchers.[29] This group is so certain that the precautionary principle is a fundamental way to look at the world that it produced an extensive mural in Harvard Square, painted by Be Sargent in 1998, that portrays twelve women cancer victims, including Rachel Carson and Audre Lord, and the words "Rachel Carson was right." It considers the mural to be a testament to the precautionary principle, and the beautiful booklet describing the mural and the women portrayed in it has emblazoned on its cover, "Indication of harm, not proof of harm, is our call to action." Be Sargent later produced another mural on the precautionary principle in Central Square, less than a mile down the road.

ENVIRONMENTAL JUSTICE GROUPS Environmental justice and the precautionary principle have begun to transform traditional approaches to environmental policy, research, and community organizing, though few people have provided a framework to link them. Environmental justice and the precautionary principle have some important overlapping goals, as Morello-Frosch, Pastor, and Sadd point out.[30] First, both approaches emphasize disease prevention and public health, but eschew the need to wait for studies to show health effects after the fact of exposure. Second, both seek to shift the burden of proof away from affected groups. For environmental justice groups, this approach entails rejecting the need to "prove" that environmental inequality stems only from discriminatory intent and disparate impact. For precautionary principle advocates, it means that proponents of a proposed new product or chemical must demonstrate that it is as safe as possible. Third, both approaches emphasize procedural justice and democratic decision making. For environmental justice groups, this emphasis means building community capacity to shape environmental policy and enforcement through community partnerships in research and action. For precautionary principle advocates, it means seeking procedural justice by promoting alternatives assessment in a milieu of democratic and open deliberation.

These complementary approaches are now showing up in environmental justice groups' literature and speeches, even though only a few years earlier the precautionary principle may have seemed too theoretical and distant for use in community organizing. Groups have used the paradigm of preventive action as the basis for programs and campaigns, such as WE ACT's Community-Based Initiative to Protect Infants in Northern Manhattan from Environmental Risks, which promotes precautionary actions as tantamount to protecting vulnerable populations. The California-based environmental justice group Greenaction has also actively made the connection between mitigating disparate impacts in minority communities and the precautionary approach, successfully advocating for the incorporation of the precautionary principle into San Francisco's environmental code.

In California, the Environmental Justice Coalition for Water was among many environmental justice groups that successfully pushed the California EPA to develop a program explicitly based on the precautionary principle. The Ecology Center in Ann Arbor, Michigan, has also been supporting the integration of environmental justice and the precautionary principle in its Community Water Rights Project.[31] Aimed at educating decision makers and the public about California's public-water rights and responsible water management, the project emphasizes the disproportionate burden on certain communities created by current water policies, and it advocates for an adherence to the Public Trust Doctrine, a legal doctrine that places navigable waters in the trust of the state for the good of future generations and that is described as "argu[ing] for a precautionary principle."[32] Thus, a more equitable situation might be created by applying this doctrine that demands precaution and prevention.

Given the successful integration of the goals of the precautionary principle and environmental justice in a growing number of programs, the precautionary principle is now seen as one of the crucial steps in eradicating environmental health disparities. In *Building Healthy Communities from the Ground Up: Environmental Justice in California,* a publication based on the research and experiences of five environmental justice organizations in California, the use of the precautionary principle in environmental laws, regulations, and decision-making processes is given as the first policy recommendation for achieving environmental justice in California. It argues that "a precautionary approach should be utilized to address existing environmental injustices and to prevent the creation of new ones."[33]

THE BE SAFE CAMPAIGN

The Center for Health, Environment, and Justice, started and led by Lois Gibbs (former head of the Love Canal Homeowners Association), has taken up the precautionary principle in many ways, including the Be Safe Campaign with its multiple projects. One focuses on PVC, a ubiquitous plastic used in construction materials, consumer products, packaging, electronics, and medical products. When this widely used material is incinerated in waste disposal, it releases dioxin. If not incinerated, it leaches component chemicals into soil and groundwater. PVC is made from vinyl chloride, widely known for causing angiosarcoma, a form of liver cancer among workers who manufacture it, though industry has sought to deny this risk.[34] Some of the major environmental justice struggles in Louisiana have centered around existing or proposed PVC factories, making this concern very prominent. Local environmental justice groups dealing with specific factories and national organizational efforts by Greenpeace laid the groundwork for the more recent campaign.

By 2004, the Center for Health, Environment, and Justice and other environmental groups had already convinced more than forty producers to phase out PVC, including many major multinationals. On December 7, 2004, they launched a campaign targeting two major firms to eliminate PVC: Microsoft (which uses PVC in blister pack packaging for software) and Johnson & Johnson (which uses PVC bottles for health and beauty products).[35] By February 2005, the center had received responses from both Microsoft and Johnson & Johnson stating that, given reasonable alternatives, they hoped to phase out the use of PVC in their products in the near future. A Johnson & Johnson spokesman stated that his company was "actively engaged with suppliers to identify alternatives to replace our existing PVC packaging, and is avoiding PVC use in future products," and Microsoft intended to find an alternative to PVC in blister packs by the end of 2005.[36]

As mentioned earlier, the center also organized the First National Conference on Precaution as part of its Be Safe Campaign. It also applied its Be Safe approach to healthy schools, nuclear waste, and dioxin. Given the center's savvy approach to organizing with many toxics groups that might not respond to a distant-sounding terminology, the extension of the precautionary principle into a broad reach of American political organizing is a sign of its appeal.

THE CAMPAIGN FOR SAFE COSMETICS

Similar to the PVC campaign, many environmental organizations are taking up the precautionary campaign to make cosmetics safer via a coalition: the Campaign for Safe Cosmetics. The Alliance for a Healthy Tomorrow, the Breast Cancer Fund, Commonweal, EWG, Friends of the Earth, National Black Environmental Justice Network, National Environmental Trust, and Women's Voices for the Earth started the group in 2002, and more than fifty other groups have signed on since then.

The Safe Cosmetics Campaign began with the release of a report, *Not Too Pretty: Phthalates, Beauty Products, and the FDA*. Environmental and public-health groups engaged a laboratory to test seventy-two brand-name beauty products for the presence of phthalates, a family of chemicals linked to birth defects and possibly to breast cancer and recently found to have an association with genital deformities in male babies when the mother is exposed to them during pregnancy.[37] The lab found phthalates in nearly three-quarters of the products tested, even though the labels listed none of them. In a second report, *Pretty Nasty*, the coalition documented similar lab results in Europe.[38]

In February 2003, the European Union passed a new amendment to its Cosmetics Directive that prohibits the use of known or suspected carcinogens, mutagens, and reproductive toxins in cosmetics. This amendment went into force in September 2004. It is one more example of a case in which European environmental protection is far more advanced than U.S. protection, and it served as the jumping off point for the Campaign for Safe Cosmetics, which seeks to convince U.S. manufacturers to adhere to those European standards. More than one hundred companies, including some major ones, have signed on in the first year of organizing.[39] Direct action with corporations has become important to precautionary principle advocates and other environmental health groups because the government is very antagonistic to such regulation. Faced with consumer pressure, however, companies can understand the need to change their products. Moreover, they cannot sell such products in Europe unless they reformulate them to be safer, sparking a growing awareness on the part of corporations that do business with more regulatory regimes. We can think of this action as an "anti-NAFTA" approach: the North American Free Trade Agreement and other similar trade agreements such as CAFTA for Central America force pact members to accept the *lowest* level of environmental regulation, whereas meeting Europe's precautionary regulation forces U.S. firms to accept a *higher* regulatory level.

WE ACT's youth group has taken up this campaign, and their teenage activists speak to high school audiences about such dangers. So, too, have Asian American girls in the Oakland area's Sisters in Action for Reproductive Empowerment.[40] The campaign is one example where the precautionary principle has attracted environmental justice organizations and has proven to be an effective organizing approach.

SAN FRANCISCO AS A MODEL FOR PRECAUTIONARY POLICY IN LOCAL GOVERNMENT

San Francisco became the first city in the country to adopt the precautionary principle as a policy on June 17, 2003. Breast cancer activists, led by the Breast

Cancer Fund, were in the forefront of this activity, working not only with other environmental groups, but with sympathetic city officials. The policy requires all city agencies to implement the precautionary principle in a public participatory fashion in the selection of products, services, operations, and planning. Although the ordinance does not require any specific actions, it promotes the precautionary approach in all areas of public life.[41] Following up on this policy, the San Francisco Board of Supervisors approved in June 2005 the Precautionary Purchasing Ordinance, which requires that the city buy the most sustainable, healthy commodities possible. For example, if the city needs to buy cleaning products for its janitorial staff, it will have to purchase the least toxic products available within a reasonable cost.[42]

Environmental activists and local officials are working in many other locations to implement similar policies, some of them very broad and some only at the level of goods purchasing. In the fall of 2004, the Portland City Council and the Multnomah County Board of Commissioners unanimously passed a joint resolution to direct a toxics reduction strategy for the city and county, specifically utilizing the precautionary principle. The Center for Health, Environment, and Justice and the Science and Environmental Health Network provided resources and guidance to the Oregon Center for Environmental Health to do relevant research and to shepherd the process through government channels. The project will identify short- and long-term goals to reduce use of toxic products in all city and county agencies.[43] A proposed program in Seattle would go as far as to measure benefits to health by tracking outcomes of asthma, diabetes, and obesity.[44] As noted in several places in this chapter, much attention is being devoted to school systems, with plans for integrated pest management, safer cleaning products, removal of carpets, and improved food quality.

DEMOCRATIZATION OF INFORMATION AND SCIENCE

Public participation in matters concerning scientific decision making is an integral part of the precautionary principle and a cornerstone of a public paradigm. Community members and activists are often excluded from traditional risk assessment processes that devalue the lay understanding of threats to public health. Through the education of community members and the democratization of science, the precautionary principle encourages public participation. Undemocratic science maintains that technical matters be addressed solely by experts. Democratizing science means including laypeople in the scientific process, whether it be in establishing research priorities or in producing scientific knowledge.[45] Though TURI does not explicitly frame public participation in toxics use reduction within the precautionary principle, its Toxic Use Reduction Networking grants increase public participation by funding local projects, many of which are designed by environmental groups and carried out by laypeople and sometimes in lay-professional collaboration. This approach is a significant step toward creating public support for implementing the precautionary principle through providing it with technical support and project funding.

The Toxic Use Reduction Network grants enable the implementation of toxics use reduction at the level of community organization or municipal government, offering lay people access to information and resources. Grant recipients receive pollution prevention education so that they can train other individuals in their community to reduce toxics both in the workplace and at home. For example, ACE received funding for the Healthy Hair Show project that taught hairdressers how to use less toxic products, and a Marblehead gardening group was supported to organize residents to reduce lawn and garden pesticides. One project, Green Cleaners in Boston Schools, supports the Massachusetts Coalition for Occupational Safety and Health in its Healthy Boston Schools Janitorial Project, which engages custodians, Boston public schools, and the larger school community in a project to reduce the use of toxic cleaning chemicals in schools and to promote this project as a model throughout the state and nation. I have been fortunate to observe the work of this project in my study of labor-environment coalitions.

Recent collaborations between TURI and the Massachusetts DPH seek to increase availability of scientific data to the public and to teach people how to utilize the data in their toxics reduction efforts. This initiative, targeted specifically at aiding parents in minimizing their children's exposure to toxins, includes outreach programs on how to understand labels on household products, where to find safer alternatives, and what the connections are between products in the home and adverse health effects.[46] In this instance, the effort reflects a recognition that laypeople are more likely to lean toward the side of public health in the face of uncertainty. In other words, citizens will be less willing to accept potentially dangerous substances in their environment merely because those substances have not yet been proven to hazardous.

By encouraging community members to become involved in the scientific community, the social ownership of risk assessment becomes less an elitist tool and more a public capacity. Traditional approaches to risk assessment assume that the generation of scientific knowledge required to conduct objective research need not involve the public and that the public is not even capable of understanding scientific knowledge. This arrogance fails to understand that laypeople have access to information that might otherwise be unavailable to experts. It also ignores people's right to be part of decisions that affect them. Hence, alternative ways of understanding risk and acting on it are excluded. Informed citizens are capable of intelligent participation in matters of scientific decision making. By creating the opportunities for citizens to become informed, precautionary principle advocates transform the role of community members from victims of pollution to empowered activists capable of combating corporate and state scientific authority. This public empowerment, widely found in literature by precautionary principle supporters, reinforces the basic human right to life, which is often overridden by the right to private property.[47] The incorporation of community activists and community leaders into scientific decision making increases the significance of the public understanding, or public paradigm, of the risks to public health.

SUMMARY

I have argued that a new public paradigm has been evolving since the publication of Rachel Carson's *Silent Spring* and is continually developing through current activism concerning toxic chemicals. Whereas early environmental protection focused on control of toxic emissions already being produced, later efforts sought to prevent pollution by encouraging business, government, and the public to reduce consumption of toxic chemicals. Even then, however, pollution prevention strategies were limited because they excluded the public and environmental health activists and instead focused on cost-saving arguments to entice corporate buy-in. In working to move beyond toxics use reduction, toxics awareness advocates in Massachusetts initiated the Precautionary Principle Project (later the Alliance for a Healthy Tomorrow), which, through its focus on alternatives to risk assessment, has moved farther upstream than pollution prevention. More broadly, the environmentalism that sparked the state's toxics reduction approach has expanded to new and broader areas of environmental protection.

Today we see the precautionary principle as a very sensible outcome of current social problems such as toxic pollution, environmental threats to children's health, and the growing breast cancer epidemic. The precautionary principle, in the forms we presently see in the United States, has developed out of three overlapping pathways: toxics reduction, toxic waste activism, and the EBCM. The democratization of science is an integral part of the precautionary principle, empowering environmental activists and encouraging community participation in decisions regarding public health. In the process, informed citizens are continually demonstrating their ability to employ informed arguments against toxic exposure and to pursue creative alternatives. Examples of activists and scientists working together through citizen-scientist alliances, as in the West Nile virus and environmental breast cancer issues, demonstrate the successful integration of social responsibility with science, cemented by the rubric of the precautionary principle. Such successes pave the way for further implementation of the precautionary principle, as exemplified by the great strides in applying the alternative paradigm in Europe. The precautionary principle also leads to a significant policy direction in health tracking and monitoring, which is discussed in chapter 7 as one of many implications of my contested illnesses perspective.

It is exciting to watch a public paradigm—the precautionary principle—develop and spread so rapidly in a short period. This growth is ample testimony to the fact that the precautionary principle represents for so many people, groups, and sectors a viable way of looking at how we consider risks and hazards. Because precautionary principle organizers emphasize the science behind their policy orientation, they add to the legitimation process of environmental activism and offer empowerment to those activists. In the years that I have been developing the contested illnesses perspective, the precautionary principle has definitely shaped my approach. It highlights so many of the issues that come up when laypeople and sympathetic professionals engage in contestation over environmental factors in disease, as developed in chapter 1.

Richard Levins, in closing the 2001 international precautionary principle conference, eloquently placed the precautionary approach in the framework of a general model for a people-oriented science: "The conflict is not between science and antiscience, but between different pathways for science and technology; between a commodified science-for-profit and a gentle science for humane goals; between the sciences of the smallest part and the sciences of dynamic wholes."[48]

|7|

IMPLICATIONS OF THE CONTESTED ILLNESSES PERSPECTIVE

THE CONTESTED illnesses perspective goes beyond traditional notions of lay involvement to argue that the role of organized social movements and the organizations in them is crucial for recognizing and acting on diseases and conditions of known or potential environmental causation. This perspective places much emphasis on both political-economic and ideological factors as determinants of the contestation around these diseases and conditions. By viewing this contestation through the lens of the DEP, my perspective focuses on the structure and alterations of public understanding, scientific knowledge, and public policy as acted on in the spheres of government, science, and public life. It also emphasizes the role of health social movements and highlights the new environmental health movement that is battling for the truth about toxins and their health effects. It is a battle of many components, involving government policy, corporate practices, production and control of scientific information, and the public right to know what communities are exposed to in daily life.

In this chapter, I examine the many implications of this perspective. Implications for public policy and for science are very often intertwined, and it is hard to parcel them out. So in some cases I briefly mention a particular issue in the first section on policy implications and then delve into it more in the section on science implications. In doing so, I make a number of prescriptive statements about needed changes and additions in all these spheres. Given the contested illnesses perspective I am coming from, it is not logical to separate such suggestions from research and analysis of the issues. First, though, I make some points about ways in which contested illnesses are very much a part of our present society.

CONTESTED ILLNESSES AS COMMON FEATURES OF THE CURRENT ERA

Contested illnesses are complex combinations of diseases and conditions. For some, like asthma, environmental factors are not the sole cause, yet they are a significant contribution, with evidence of the connection growing all the time. For others, such as breast cancer, environmental factors are harder to demonstrate, but there is some evidence and much potential knowledge of causation. Many scientists and activists are convinced that the recent increases in incidence of asthma and breast cancer are very likely attributable to environmental factors. For other conditions, such as GWRIs, it has been hard to find evidence for environmental factors, despite the involvement of various toxic exposures, though even in that case some recent developments offer a little evidence for environmental factors. In all three examples, the disputes over environmental causation are very contested, with laypeople and a small set of scientific and governmental allies challenging a DEP.

Contested illnesses have always been with us, but the stakes are higher with the newer ones, especially breast cancer, because of the scientific challenges and policy implications. For instance, struggles over black lung from coal mining and over asbestosis and mesothelioma from asbestos are examples of strong contestation that arose from lay discovery and had powerful economic and political impacts. Unions, occupational safety and health groups, and sympathetic scientists had to fight to prove the existence of diseases that corporate doctors argued did not exist, and then they had to struggle to show the path of causation. Sufferers and their allies had a hard time getting government regulation and financial compensation, which is still mired in controversy. The economic costs of protecting health and caring for victims may have seemed high, but they fell primarily on the involved industries. In contrast, the environmental causes implicated in asthma and breast cancer go to the heart of the entire economic system and require massive policy shifts. In addition, the strong certainty of signature causes in black lung disease, asbestosis, and mesothelioma are not matched in asthma, breast cancer, and GWRIs, or in many of the other currently contested illnesses.

It has long been a standard perspective in public health to view the current era as one in which chronic diseases predominate, having replaced an earlier era of infectious diseases as the main scourges. From a contested illnesses approach, this view is too simplistic. First, we already understand the continued threat of infectious diseases in many parts of the world, as has been thoroughly demonstrated by the AIDS epidemic and by the resurgence of other infectious diseases such as cholera linked to climate change. Second, there are now a number of diseases for which the etiology is so unclear that we cannot necessarily distinguish between whether they are infectious or chronic. This is true both for AIDS, with which many people live for long periods, and for diseases or conditions with MUPS for which we simply do not know the causes. The current era is full of mysterious diseases and conditions and unclear etiologies: lupus and other autoimmune diseases, a growing number of allergies, unexplained increases in autism and neurological and developmental disorders. Although we are continually finding more evidence for environmental causation of many diseases, the links are sparse for others. As we develop new avenues of research, some of the mysteries may become clearer: GIS mapping may provide geographic and historical patterns of exposure that are not yet known; toxicogenomics may allows us to determine complex interactions of genes and environment.[1] But scientific advances alone will not do the job; we also need dedication by science and government to new outlooks. Environmental health activism will continue to be the pressure point in developing such dedication.

Asthma, breast cancer, and GWRIs are only a few among the contested diseases and conditions of our current era, mainly since World War II (what some call postmodern diseases). Their status as contested illnesses of this era arises from several sources. First, some of them stem from the production, use, and disposal practices of the past half-century. Second, they reflect a growing uncertainty over the specific causes and expressions of symptoms. And third, popular participation in science and politics has made the identification of illness and its causes much more public.

The environmental components for different disease are varied. For example, both asthma and breast cancer are affected by post–World War II developments, but for different reasons. Asthma exacerbation and perhaps incidence are very prevalent biological responses to air pollution, much of it from increased power plant, incineration, and vehicle exhausts in the past half-century. The environmental factors in breast cancer are more linked to post–World War II chemical production and a society that enjoys a highly disposable consumption of plastics. As pointed out in the previous chapter, resistance to looking for environmental factors comes from a variety of sources: corporations whose production, distribution, and disposal practices depend on dangerous substances; governmental agencies with an antiregulatory position; and scientists who have a firm opposition to environmental causation explanations. Ultimately, all forms of resistance to environmental causation are linked; they stem from people, corporations, and institutions that are integral parts of a social order based on profit rather than on well-being. If we add to this mixture the frequent media attacks on environmental causation, we observe a many-sided opposition to environmental factors that support the dominant social structure.

This widespread resistance to environmental causation makes it hard to achieve support for adequate research and theory. Yet once that research gets a solid footing and is accompanied by a perspective that explains the political-economic context for our problems, the far-reaching implications of targeting environmental factors can make people more critical overall of their society. That is why so many disease activists expand their intellectual sphere and their political worldview to incorporate a larger array of critical perspectives on social inequalities and power dynamics. We have, for instance, observed how asthma activists take up a broader civil rights and environmental justice perspective and how breast cancer activists have a wider awareness of sexism, corporate control of medical research, and the dangers of unfettered production of contaminants. These activists, along with people challenging a variety of environmental health effects, have become part of a multifaceted environmental health movement that brings together citizen groups, disease-based organizations, other social movement organizations, and scientists. Even some government officials and scientists are part of this movement, through the process of what Wolfson calls "interpenetration."[2] Examples include program officers and other officials involved in NIEHS's support of environmental justice collaborative research and in the CDC's body burden studies. Such people may not define themselves necessarily as activists or as allies of activists, but by merely doing their jobs they are serving in such roles.

Decades of activism by health social movements have led to a growing democratization of science and medicine, bringing citizens more into the process of disease detection and investigation. Occupational safety and health activism, toxic waste organizing, AIDS activism, the women's health movement, environmental justice activism, and related mobilizations have thoroughly transformed popular expectations about science and medicine. They have also altered the worldviews of many scientists and health providers who have come to exercise a more humanitarian approach centered on the needs of people with actual or potential sicknesses. In this light, mysterious and contested diseases can no longer remain hidden from public view, but are indeed discussed and debated in a very public forum. In C. Wright Mills's classic formulation, *individual troubles* have been transformed into *social problems*.[3] The primary vehicles for that transformation have been health social movements, which are now central to all environmental health concerns.

With these points in mind, this chapter examines the implications of the contested illness perspective for many areas of society and for diseases and conditions other than the few on which I have focused in this book.

For clinical medicine, implications include more patient-centered health care that takes seriously the complaints of the sufferer, broader conceptions of the impact of diagnosis, better informed consent, and closer connections between providers and researchers. Implications for the corporate sector revolve around more responsible practices in the area of production, use, and disposal of toxic substances, as well as around drug development, clinical trials, and corporate involvement in advocacy work. For public policy, they include implementation of the precautionary principle, improved health surveillance and monitoring, stronger environmental regulation, changes in the amount and focus of federal research dollars, an end to censorship and political control of science, and improvements in military and VA health care, research, and entitlements. In terms of scientific research, implications involve the way we look at courageous scientists who buck the tide of normal science, developments in critical epidemiology and citizen-science alliances, and new approaches to scientific methods.

My view of these implications encompasses both what I think they mean for their respective social sectors and my specific recommendations for action on them. This discussion of implications thus frames a broad, holistic perspective on health and the environment that challenges dominant ideas and practices in science, health care, government responsibility, and citizen involvement.

IMPLICATIONS FOR CLINICAL MEDICINE

LINKING CLINICAL PRACTICE WITH RESEARCH AND AWARENESS OF ENVIRONMENTAL FACTORS

It is important that environmental activists are on the forefront of identifying health effects of contaminants and promoting appropriate public policy, but we clearly need action on the part of health professionals as well. A clear goal for the environmental health movement must be to make clinical medicine more responsive to environmental health concerns.

To get a sense of how much this change is needed, let us take a look at the current state of knowledge about and practice of environmental health among physicians. Despite the increasing public interest in environmental health issues, there has been a critical shortage in the number of physicians who have been trained to recognize and treat environmental health effects, with a need for between 1,600 and 3,500 more occupational and environmental medicine specialists, according to a 1991 Institute of Medicine report.[4] Case studies of toxic waste sites where community action created high public visibility have found a lack of physician involvement.[5] Local physicians living and practicing at or near toxic waste sites—often rural areas, small towns, or small cities—are more likely to be tied to sources of political, economic, and social power.

In 1989, the Agency for Toxic Substances and Disease Registry awarded cooperative agreements to eleven states for environmental educational activities for physicians and other health professionals. Using data from one project funded by the agency, I examined 1,609 responses to a survey of Massachusetts physicians on the role of physicians in the detection and treatment of environmentally induced illness. I discovered that physicians whose practices are located within communities that have EPA National Priority List hazardous waste sites (Superfund sites) or other highly publicized toxic waste sites within their borders are no more knowledgeable about environmental health hazards than physicians whose practices are not located in communities with National Priority List sites.[6] We might expect this discrepancy, given my point about the lack of physician support in toxic waste crises. Nevertheless, it is a troublesome finding because one would ideally expect physicians to have awareness of their communities' major health hazards and to learn more about environmental health so as to be able to include such concern in examinations and to watch for potential environmentally induced diseases.

Women with breast cancer commonly report that their doctors do not ask about any environmental exposures, and those women who raise environmental factors find that their doctors are very skeptical. People with asthma fare better because there has been a stronger acceptance of environmental factors in asthma. Yet even in this case physicians have a narrower understanding of the role of the environment, focusing on indoor exposures such as dust mites, mold, tobacco smoke, and pests rather than on outdoor exposures such as particulate matter from air pollution. And as we saw with GWRIs, the military and VA medical establishments were very unwilling to consider environmental factors. Interestingly, many of the physicians called upon to conduct research on GWRIs and to run special clinics came not from occupational and environmental medicine, but from backgrounds involving the study and treatment of MUPS.

Physicians for Social Responsibility has been in the forefront of teaching doctors, nurses, and other health professionals about environmental factors in disease. It is a national organization of more than thirty-one thousand physicians, health care professionals, and supporters originally focused on the health consequences of nuclear and other weapons of mass destruction, but more recently emphasizing the health consequences of environmental pollution. The group holds workshops using its curriculum "In Harms' Way: Toxic Threats to Child Development." Health Care Without Harm is another organization of health professionals with an emphasis on preventing harm in hospitals from toxic contaminants, especially mercury and PVC plastic. It also seeks to eliminate incineration of medical waste, encourage food-purchasing systems that support sustainable food production and distribution, provide healthy food at health care facilities, secure a safe workplace for all health care workers, provide patients with access to information about chemicals used in health care, and promote human rights and environmental justice for communities impacted by the health care industry. The environmental health centers supported by the NIEHS also provide much education to health professionals, though there are only a handful of these centers. More of this work can be done by environmental organizations, some of which now provide needed awareness of health effects that health professionals do not get from mainstream medical sources. Physician organizations, such as medical societies and specialty boards, can be important players here, as can medical schools and their academic associations. Environmental health easily fits into so many areas of medical education—for example, particulate matter effects on heart and lung function and endocrine disrupter effects on reproduction. This topic should be integrated into routine education, not shunted off to an occupational and environmental medicine elective. Moreover, further federal support needs to be developed, going beyond the limited success of the Agency for Toxic Substances and Disease Registry program and various NIEHS efforts.

The incorporation of environmental health into mainstream medical practice has financial benefits as well. Managed care organizations can learn much from environmental perspectives on disease. For asthma, which is a huge expense for these organizations, attention to environmental triggers can lead to reduced emergency room visits and hospitalization. But most major managed care organizations do not even follow the established guidelines of the National Heart, Lung, and Blood Institute in whole or part.[7] Having a community orientation, even if not directly aimed at air quality regulation, can be immensely helpful in clinical outcomes. The focus of so many asthma programs on environmental justice, or at least on air pollution, has led to a diffusion of more community-based programs with direct involvement of parents and children in self-monitoring and self-management. I have had the pleasure of working with Draw-A-Breath (subsequently the Providence School Asthma Partnership), a community-based asthma program that uses intense school-based recruitment, prevention, and treatment in tandem with parent support groups (including one in Spanish for Latinos), special training for nurses, health fairs, and a summer asthma camp. Draw-A-Breath's community focus includes an awareness of air pollution, voiced in involved parents' group letters to politicians expressing their demands for better air pollution regulation. In its first year alone, Draw-A-Breath made remarkable gains in reducing symptoms, school days missed, hospital days, and emergency room episodes. Similar programs are at work across the country, and they need more support.

A MORE PATIENT-CENTERED MEDICINE: LISTENING TO THE PATIENT AND HAVING ALTERNATIVE CONCEPTIONS OF DIAGNOSIS

By taking seriously the environmental factors in health, medical practitioners can be part of an overall trend toward a more humane, patient-centered medicine. Health activism since the 1960s has challenged paternalistic medical practices in demanding access to personal health information, general information on treatments and procedures, a share in health care decisions, and openness in clinician-patient discourse. My discussion of health social movements in chapter 1 showed how powerful these elements have been in altering the landscape of medical belief and practice. What Packard and colleagues refer to as "communities of suffering" have become crucial in the medicalization and recognition of personal sets of symptoms as illnesses.[8] A wide variety of actors—including the media, medical practitioners, and political leaders—and factors such as the etiology of a disease and its epidemiological presence play roles in how a condition or disease is portrayed and addressed. Yet strong activism by communities of suffering can be instrumental in placing a disease on the public health agenda.

Such collective action indicates that a more humane, patient-centered medicine is not merely about better communication strategies in dyadic relationships, which is what dominates the agenda in clinical interaction reform. Rather, it is about taking on the larger public health concerns, such as how social structural and neighborhood characteristics affect health. Health professionals' activism on issues such as fast food, sexual violence, gun control, and drunken driving are examples of how patient-centered medicine operates at larger societal levels. In this light, a better medical practice necessitates transforming medical practice into a combined medical and public health practice. That is why health professionals' involvement in environmental health issues is so crucial.

More profound listening to the patient requires asking appropriate questions that will elicit possible toxic exposures in the workplace, school, and home. It also requires knowing about symptoms and syndromes that are indicators of known or likely toxicity. Because these symptoms and syndromes are not always generally accepted, health professionals need to be willing not to limit themselves to traditional diagnoses. For example, if something is not clearly a regular and known disease, it is not necessarily the somaticization of an underlying psychiatric disorder. Patients need to be taken seriously in their suffering, and providers need to be open to the fact that they may be seeing something new that is still not understood, as was certainly an issue with Gulf War veterans.

BETTER INFORMED CONSENT, ESPECIALLY FOR THE MILITARY

In an earlier section, I pointed to the need for extending informed consent to the community level and for allowing individuals to receive personal exposure data. These approaches, as part of a broader community-based participatory research approach, offer not just research protections, but what I think of as *democratic participation protections*. Especially with regard to GWRIs, we need to take up the special issues of informed consent in the military.

Some military personnel wanted to refuse anthrax vaccinations before being sent to the Gulf War and now the Iraq War, but were not permitted to do so under military law. It seems very unfair that the hard-fought gains in informed consent, a result of repudiation of the Nazi medical experiments and the Tuskegee syphilis study, should be denied to those supposedly serving to protect freedom. The military approach to informed consent must be seen in light of the tradition of the "atomic veterans," soldiers and sailors who were forced to watch nuclear weapons tests very close to ground zero but were not told that they were being exposed to huge radiation doses for experimental purposes. Decades later these veterans had cancer at rates far above local or national averages and had to fight bitterly to get those cancers recognized as service related and hence as eligible for treatment and compensation. The military's obstinate position was one of its worst blemishes and, tied together with the more recent Agent Orange situation, makes very suspect its claims that it adequately cares for the health of servicemen and servicewomen.

Those serving in the military should have the same rights to informed consent as civilians, but it will likely require special legislation to circumvent the military's likely objection that giving them these rights will jeopardize capacity and effectiveness. Even then, effective informed consent will be hampered by the problem that military physicians, the ultimate gatekeepers of informed consent, do not have individual service members as their clients; rather, their client is the military. They are what Daniels calls "captive professionals," much like company doctors, so that this situation requires extensive reallocation of responsibility and retraining in order to protect people's confidentiality, informed consent, and right to know.[9]

IMPLICATIONS FOR CORPORATE PRACTICES

Tracking and monitoring health, developing sustainable production techniques, using the least toxic alternative, and taking precaution with potentially dangerous substances are elements of a growing public paradigm, stemming from the precautionary principle, that helps us to realize that we all live downstream and that we must deal with what is happening upstream. The business community has been reluctant to accept such an outlook, but some recent changes have occurred in this sector. Corporate reform in environmental matters typically stems from activist organizing and from government regulation. During parts of the post-1970 period, when the EPA was established, some administrations were able to wrest important concessions from industry. But in the current era of antiregulatory policy, it will mainly be up to state government and environmental groups to press for changes in corporate practices.

Some companies voluntarily choose safer practices. For example, the Zoots Dry Cleaners chain got rid of perc (tetrachloroethylene), the dangerous chemical used in traditional dry cleaning, and replaced it with a safer "wet cleaning" product. For this company, the change is part of a belief in safer business practices, and its CEO even testified in the Massachusetts legislature in support of toxics reduction laws. The change is also good business in that increasing numbers of consumers will seek nonperc cleaners. Further, it will likely lead to reduced costs of disposal and insurance. As TURI in Massachusetts has shown, businesses can do well economically while also reducing toxics, and more state agencies of this sort can be enormously helpful in spreading that awareness. Indeed, business would be wise to press states to initiate TURI-type agencies and to suggest a coupling of tax incentives for forms that succeed in toxic reduction. As other more enlightened countries implement healthier environmental policies, U.S. companies will have an incentive to meet those standards. For instance, Massachusetts' exports of industrial and technology products are largely to western European nations that have increased their precautionary policies for clean production. Without meeting such standards, Massachusetts firms will stand to lose much business.[10] A growing "green chemistry" approach has shown that many production processes can be both safer and cheaper, while pioneering exciting research developments and yielding new patents.[11] At the Lowell Center for Sustainable Production, associated with TURI and the Department of Work Environment at the University of Massachusetts–Lowell, scientists and educators work with industry to develop safer chemicals and materials, to redesign goods so there is less threat to health over the life course of products, to minimize waste in production, and generally to boost sustainable production and consumption. The Lowell Center has been well received in many business circles and shows the potential for more such efforts.

Even without the previously mentioned prompt by foreign regulations, companies are finding that public desire for goods created by safer production and disposal methods can be a strong market pressure. Much as corporations have moved into selling previously marginalized products, especially organic foods and "green" cleaning supplies, firms will find an ever-expanding number of niches for marketing safe goods and practices—for example, nonchlorine swimming pool disinfectants, nontoxic building supplies, and cosmetics without endocrine-disrupting parabens. Overall, few of these changes will stem from a socially conscious mindset, but there is hope that a business world guided by opportunistic reform will generate a wider ideological reform attitude as well.

Activists for safe cosmetics are finding that they can influence corporations in safer production. In March 2004, the Breast Cancer Fund requested that many cosmetics manufacturers halt the use of phthalates (plasticizers found in children's toys, cosmetics, and medical devices). An October 2004 European Union ban of two types of phthalates, dibutyl phthalate and diethyl hexyl phthalate, provided some impetus for companies to agree. Procter & Gamble and Estée Lauder agreed last year to reformulate several lines of nail polish to eliminate phthalates in the United States. In January 2005, Revlon, Groupe L'Oreal, and Unilever said they no longer are using phthalates in nail polish, fragrances, and hair sprays.[12]

In the wake of increasing globalization and the failure of governments to regulate multinational corporations adequately, many activists have found it more possible to achieve success by targeting business rather than government regulators. In our interconnected world, new modes of regulation will become increasingly necessary in a globalized production network where a single corporation has manufacturing contracts with as many as four thousand factories in fifty-five countries, and those may have a multitude of regulatory regimes. Groups such as the Fair Labor Association, Social Accountability International, and Worldwide Responsible Apparel Production have worked across a variety of arenas, such as child labor, health and safety, work hours and conditions, and right to organize. In one example of success, such activism was able to effect a 98 percent reduction in organic solvents and glues in Nike's Vietnam factory. "Industrial ecology" is expanding, a concept that integrates multiple approaches to yield safer and healthier methods of production, consumption, and disposal. It may lead to experimental technologies such as "socially responsible barcodes" that a consumer can scan with a camera phone to learn detailed information on the environmental and human rights records of the company that manufactured the product.[13]

IMPLICATIONS FOR PUBLIC POLICY

IMPLEMENTING THE PRECAUTIONARY PRINCIPLE

We have witnessed the rapid growth of the precautionary principle, especially in the case of breast cancer. A growing number of environmental justice groups and the broader environmental health community have made it a cornerstone of belief and practice and have worked hard to make it a key part of public policy.

San Francisco became the first city in the country to adopt the precautionary principle as a policy, as I discussed in chapter 6. Other such municipal legislation seems likely because cities have many concerns that logically fit under a precautionary approach, including toxic hazards in public schools, hazards of pesticides used for mosquito spraying, municipal waste disposal, and toxic substances in public hospitals and other health facilities. It is likely that we will see activists pushing for such local policies because lobbying and organizing can be easier at the local level and because a number of cities with progressive governments are more receptive. Many of the concerns are local by definition, such as public schools, an area where there has much recent concern in matters such as the use of hazardous chemicals for pest control and cleaning, the use of hazardous building and decorative materials, and the building of new schools on abandoned toxic sites.

Not all implementations of precaution will openly use the term *precautionary principle*, even though they may substantively be based on the principle. For example, much of TURI's work occurs within the parameters of the precautionary principle without naming it as such. And, as we have seen, some of the people working at TURI provided an incubator of precautionary principle activism that led to new organizational forms. It is likely that other government agencies can be formed or transformed along similar lines, so supporters should be prepared to take that tack as well. TURI's marked success can be a strong incentive to other states. Similarly, the Alliance for a Healthy

Tomorrow's programs, as ensconced in legislative bills, often avoid the term *precautionary principle* out of fear that it is a lightning rod for many people. The accumulation of city policies can become a major push for state and federal policy, though the EPA and other regulatory agencies have been so pro-corporate under the George W. Bush administration that it is difficult to imagine much success right now. Nevertheless, it is to be hoped that the groundwork is being laid for more sympathetic administrations in the future.

RESEARCH AND SURVEILLANCE ON ENDOCRINE DISRUPTERS

Much of the impetus behind the precautionary principle initially came from researchers and advocates who were examining the many effects of endocrine disrupters. This effort coincided with the work of some EPA officials who were supportive of a stronger regulatory regime and who were aided by congressional mandate. Provisions of both the 1996 Food Quality Protection Act and the reauthorization of the Safe Drinking Water Act called for the creation of a screening and testing program at the EPA that would provide information about the possible endocrine effects of certain chemicals on human health, including pesticides, industrial chemicals, and environmental contaminants.

The Endocrine Disruptor Screening and Testing Advisory Committee brought together representatives from industry, government, environmental and public-health groups (including the NBCC), and academia to deliberate for two years and to design this EPA program. Although the committee's 1998 recommendations were a large achievement, they still failed to include low-dose testing and a focus on early developmental disorders. Many observers believe the recommendations have not been implemented at the level necessary to study the many known endocrine disrupters.[14] Critics of the current chemical-by-chemical regulatory approach might have predicted this outcome, noting that the size of the task is too large and that the resources required are too great. Moreover, the task grows harder every year as two to three thousand new chemicals appear on the market. Preventive action will require policies that address chemical classes, not just individual chemicals, and that are focused on the sources of chemical contamination. Future agency directives may be more successfully implemented by focusing on what we know of the characteristics of entire chemical classes, rather than on health effects of individual chemicals, because health studies can take a long time and are often inconclusive.[15]

The original Screening and Testing Advisory Committee mission included only research, with no regulatory power. Any coherent approach to studying endocrine disrupters must have a regulatory component. Even that much is no longer feasible in the current Bush administration, and it will likely require a new Democrat government to reintroduce policy on endocrine disrupters. It is to be hoped that the European Registration, Evaluation, and Authorisation of Chemicals program for chemical regulation will provide some legitimation for the United States to pursue this goal.

In particular, we need more attention to endocrine-disrupting compounds. Much early attention to such compounds stemmed from the diethylstilbestrol (known widely as DES) problem. DES was given to pregnant and postmenopausal women but is now widely known to cause vaginal cancer in users' daughters. It is frequently cited as another "late lesson from early warnings" that bolsters the need to expand the precautionary principle.[16] Colborn, Dumanoski, and Meyers's widely read book *Our Stolen Future* unified three major research elements to develop the endocrine disrupter hypothesis: (1) the malformative properties of diethylstilbestrol, (2) many findings of wildlife reproductive disorders, and (3) research on the global decline in quality and quantity of human sperm.

In the meantime, increased attention to endocrine-disrupting compounds may occur because of the recent findings that hormone replacement therapy for postmenopausal women has been very hazardous. Hormone replacement therapy initially consisted of estrogen, but after earlier research on estrogen's carcinogenic properties, a supposedly less dangerous treatment combining estrogen and progestin was devised. The quick rush to prescribe this therapy was a collective problem caused by the error of overeager scientists who failed to design proper studies, the financial incentives for pharmaceutical firms that pressed the use of these drugs in spite of early warnings of both side effects and lack of benefit, and the laxity of government regulators who did not adequately investigate the medical evidence. In 2002 and 2003, key studies found hormone replacement therapy to cause excess cancers and heart disease, leading to a major shift in treatment.[17] As rapidly as this therapy was initiated, it has been reduced in clinical practice through recommendations of medical associations and journals, but after severe costs to many people. But unlike DES, it has not been withdrawn by the FDA. Many women's health activists and breast cancer activists have long pointed to the dangers of hormone replacement therapy, and much of the environmental health concern for endocrine disrupters cite it as a key example.[18] Even in light of this development, many researchers maintain that the high estrogenic doses in the therapy are not matched by other endocrine disrupter exposures. Still, it is likely that other researchers will shift their beliefs in the other direction. This case, too, now appears as one of the many "late lessons from early warnings" that precautionary principle proponents mention.

IMPROVED MONITORING AND SURVEILLANCE

Much contestation over environmental factors stems from lack of information. One way to deal with this lack is health tracking and monitoring, something generally absent in public health up to now. However, some promising moves have been made in this direction. In March 2001, the CDC released the first-ever report on toxic chemicals present in our bodies. For chemicals with known toxicity levels, the CDC can show how currently accepted toxic levels are exceeded in many people. For all chemicals, the project can show differential exposure by age, sex, and racial or ethnic group. The CDC can use this data to track exposure over time and to set research agendas for health effects. The tests measured 27 substances, including mercury, organophosphates (found in pesticides), and phthalates. Although some of these chemicals had been tested in soil, air, and water, levels in humans had not previously been directly measured. The CDC continued testing, adding 89 chemicals to the screen, as shown in their 2003 report. It points out that the data detail exposure of the U.S. population to toxins; it does not present new data on health risks caused by different exposure levels.[19] The CDC's third report, released in 2005, included blood and urine levels for 148 chemicals, adding several PCBs, certain phthalate metabolites, the pyrethroid insecticide permethrin, and the organochlorine pesticides aldrin, endrin, and dieldrin.[20] The CDC will continue to increase the number of chemicals monitored, including 309 and 473 chemicals in 2007 and 2009, respectively.[21] These reports serve as tools for other researchers to begin to assess environmental health threats and, as trends become increasingly clear, will help to guide public-health decisions.[22]

This screening process is a step in the right direction, and the information collected can and should be used to promote more research into environmental links to diseases, as proposed by the Trust for America's Health (previously Health-Track). This organization's mission, supported by the Pew Charitable Trusts, is to identify and track the links between environmental hazards and illnesses and to provide researchers and public health officials with the necessary tools to prevent disease. Beyond the process of health tracking, of course, is the application of its findings to reducing or eliminating toxic substances. In the later section on implications for scientific research, I cover a new phenomenon of "grassroots body burden testing" in which activists conduct human biomonitoring.

The National Children's Study, authorized by Congress in 2000 and launched in 2004, offers another longitudinal approach to monitoring. Led by a collaboration of federal agencies including HHS, EPA, CDC, and NIEHS, the study examines the influences of environmental factors on the health of children by following more than one hundred thousand children from before birth to the age of twenty-one by enrolling pregnant women, couples planning pregnancy, and women of child-bearing age not planning pregnancy to participate in a minimum of fifteen visits with a local research team.[23] Several factors differentiate the National Children's Study from other studies being done in the United States, one of the most influential being its definition of *environment*, which includes genetics, cultural influences, physical surroundings, diet, and biological, chemical, and social factors.[24] The study's results will be made continually available two to three years after its launch. Furthermore, the study is aimed at examining a variety of different issues that arise during childhood development, from attention deficit disorder, schizophrenia, autism, asthma, obesity, and heart disease to injuries resulting from poor neighborhood planning and violent behavior in teenagers.[25] However, despite indications that the results of the study may reduce annual health care costs by an amount greater than the costs of the study, the federal budget for the study was cut from $27 million to $12 million in the fiscal year 2005.[26]

The EPA has taken a different approach to systematizing the information available regarding children's environmental exposures in its America's Children and the Environment project, which aims to bring together information from a variety of sources in reports regarding trends in environmental contaminants, concentration of contaminants in children, and trends in childhood illnesses possibly associated with environmental exposures. The most recent report, *America's Children and the Environment: Measures of Contaminants, Body Burdens, and Illnesses*, released in 2003, explicitly points out the critical differences in the ways in which children and adults are exposed to environmental contaminants.[27] Although America's Children and the Environment is not a tracking project, it shares many of the same goals with a project such as the National Children's Study in attempting to make information readily available, to track trends in environmental health, and to analyze data in a larger environmental context with hopes of informing policymakers and the public.

The government can learn much from Trust for America's Health and should take seriously its push for more federal activity in health tracking. In 2002, members of Congress introduced the Nationwide Health Tracking Act, largely tied to antiterrorist biomonitoring, which appropriated $17.5 million for building a health-tracking network, with another $28 million added to this appropriation in 2003 in order to expand the program.[28] As a result, the CDC was able to award health-tracking grants to several states in order to build environmental public health capacity, improve collaborations between environmental health agencies, evaluate existing tacking systems, build community partnerships, and develop model systems.[29] Grants were awarded to states that worked in collaboration with CDC-funded schools of public health that were deemed Centers of Excellence in Environmental Public Health Tracking and that offered expertise to the CDC and state grantees. In collaboration with the Centers of Excellence, state grantees worked to develop plans, assess the status of existent programs, and begin to evaluate environmental triggers. For example, northeastern states have worked with Johns Hopkins University to assess air quality and its relationship to chronic diseases.[30]

However, despite the success of grants originally appropriated through legislation targeting terrorism, it would be a mistake to tie routine health monitoring to antiterrorism. Many public health professionals have been concerned that state public health departments have curtailed many of their routine activities in favor of terrorism preparedness. This result has been very widespread at the CDC, which has historically provided some of the best health monitoring and surveillance, but now has largely transformed its mission into terrorism preparation. Better health monitoring will not only help environmental health, but can move our society further along to overall benefits from monitoring for many health concerns.

It is also important to provide surveillance on various diseases and conditions that are known or thought to be environmentally caused. For example, the Association for Birth Defect Children (now Birth Defect Research for Children) runs its own register of birth defects, hoping to find causes of those defects, because there is no national registry for doing so. Registries are also needed to track autism, Parkinson's disease, and other diseases that are increasingly thought to involve some environmental causation. Perhaps state and federal agencies will take cognizance of these efforts and emulate them in official practices.

For the armed forces, the failure to have adequate predeployment health data for the 1991 Gulf War should have motivated them to do better in the future. Following the war, an effort was made to improve data collection through issuing brief predeployment health surveys, creating a centralized health-tracking database in order to track information on hospitalized soldiers in the DOD medical system, and, last, initiating the Millennium Cohort Study, a longitudinal study targeted at gauging the long-term impacts of deployment by capturing information about soldiers both during and after their time in the service. According to the 2003 testimony of Assistant Secretary of Defense for Health Affairs William Winkenwerder, the deployment process now includes a health assessment, a medical record review, collection of a blood serum sample, and a health care provider review, all of which are archived electronically along with postdeployment health assessments in the Defense Medical Surveillance System.[31]

However, despite the implementation of these programs in the Iraq War, starting in 2003, many problems remain in collecting and organizing predeployment health data. As the Institute of Medicine recognized in its 1999 report *Strategies to Protect the Health of Deployed US Forces*, even with these measures in place, the lack of central authority for military public health and the disjunction between the VA and DOD make achieving these public health goals throughout the military difficult.[32] Efforts continue, however, in both pre- and postdeployment health tracking, with a strong focus on monitoring the mental health of active soldiers and veterans in order to better understand rates of posttraumatic stress disorder and depression.[33]

PRESSURE FOR ADDITIONAL RESEARCH FUNDS

As mentioned earlier in this chapter, much resistance to environmental research is strongly ideological, especially when that research goes against the cost-effectiveness of science. The Bush administration's refusal to allow stem cell research is a clear example of this situation. Therefore, simply pressing for research funds is not necessarily a winning goal. Nevertheless, the longer-range implications of the contested illnesses perspective are that people need to push for increased federal funding for a wide range of environmental health matters.

Within the NCI, the best-funded NIH unit, virtually no funds go to environmental factors research. Overall, in 2002 and 2003 only one out of every nine federal research dollars was spent on examining environmental links.[34] Federally funded institutes such as the NCI focus primarily on the molecular biology of cancer and the role of genes in the treatment and development of cancer. Furthermore, the bulk of the environmental research focuses on gene-environment interactions.[35] The NIEHS, which supports most of the environmental health research, has a far smaller budget: $645 million for 2005 compared to the NCI's $4.8 billion. Activists will therefore need to put pressure on the NCI to work more in environmental directions. EPA programs, such as its Science to Achieve Results (STAR) grants, have been either ended or sharply reduced. Activists have been pushing for a long time to get more support for environmental research, including pressure for some of the money from the breast cancer stamp (sold at an eight-cent premium above regular postage, to fund $37 million in breast cancer research since its initiation in July 1998) to be shifted to the NIEHS for environmental factors research.

Funding needs to be rethought so that it includes increased research on environmental factors in the other NIH institutes. Toxics affect neurological, developmental, reproductive, and autoimmune health, and the relevant institutes should be contributing to these research directions. Until this larger framework for environmental health effects is taken into account, research on environmental health will be both underfunded and misunderstood.

It is clear that without substantial increases in funding, we will not be able to learn what is necessary. When we remember that most chemicals in use have not been evaluated for human safety, when we realize how many Superfund sites remain unremediated, when we see how little goes toward alternative energy and sustainable production, and when we notice how few funds are available for ecosystem protection, we begin to see the extent of the underfunding of research on environmental issues. Even among these issues, environmental health is often one of the poorest funded because advocates for environmental health tend to challenge the status quo by focusing on the effects of core features of the nation's economic structure.

SUPPORT FOR COMMUNITY-BASED PARTICIPATORY RESEARCH

As noted previously, for those involved in contestation over environmental factors in disease, research support is often difficult to obtain. Activists and sympathetic scientists and policymakers have frequently turned to community-based participatory research as the approach most likely to advance both science and democracy. This approach is widely used by environmental health and environmental justice groups and their academic partners and is one of the primary areas in need of support.

Over the course of the NIEHS's program in community-based participatory research, the most expansive ever, evaluators have found that collaborative projects are very successful. They generate new research questions, build trust between participating groups, increase relevancy of research question and data, increase quality of data, expand dissemination of findings, increase the likelihood of creating effective policy, and create and sustain new infrastructure to support additional community-based projects. These benefits extend beyond citizen participation in research to participation in advisory boards that make siting decisions for facilities such as waste transfer stations, bus depots, and nuclear energy plants.[36]

In one example, three university researchers and eight tribes of northeastern Oklahoma organized as Tribal Efforts Against Lead to study and reduce the body burden of lead in children associated with proximity to mines. The results showed that the primary source of lead was dust ingested through hand-to-mouth behaviors. These new data increased the knowledge base for lead sources, which had previously emphasized paint, and expanded the range of prevention strategies to reduce lead poisoning. Because community advocates could draw upon local data rather than prior studies emphasizing lead-based paint exposures, they were able to secure resources and funding for more relevant solutions to their problems. Novel interventions included vacuums equipped with HEPA filters for community members to better control lead-contaminated dust.[37]

Skill acquisition enables community members not only to participate in future research studies, but also to conduct their own studies as they see fit. We may think of this benefit as *social research capital,* whereby communities build their own arsenal of research skills and resources with which to effect change. This term expands on the popular "social capital" concept, which holds that collective ties and active communities can offer beneficial health outcomes. For example, Roxbury, Massachusetts, citizens have gained confidence, research skills, and a long-term air quality monitor through ACE's AirBeat project in Dudley Square, as mentioned in chapter 3 on asthma. These skills contribute to their collective social research capital. In this case, the "capital" is a tangible, quantifiable marker that the community has benefited from their participation. This example stands in contrast to a long history of research studies where communities give of themselves to researchers without gaining anything, be it useable knowledge, altered outcomes, or additional resources. Through collaboration in the AirBeat project, the Dudley Square neighborhood has become a healthier physical and social environment.[38]

ACE used its growing base and its alliances with other groups to expand its activism to other issues. When Boston University won a large federal grant to open a bioterrorism research laboratory, ACE took a leading role in opposing the location of a potentially hazardous facility full of chemical and biological agents in a dense, largely minority neighborhood. In December 2003, ACE issued a report citing as causes for action against the proposed National Biocontainment Laboratory the lack of transparency and community oversight in plans for the laboratory, community safety concerns given the lab's proposed location in a densely populated area, the fact that the lab would not be providing jobs or economic support to community members, and the failure to consult the community and address environmental justice concerns.[39]

As well as individual research grants, some community-based participatory research funding has been made available to set up ongoing centers. In 1998, the NIEHS, the EPA, and the CDC teamed up to develop the Centers for Children's Environmental Health and Disease Prevention Research. The goal of this program was to support research efforts that would translate basic research on children's environmentally related diseases into community-based prevention and intervention applications. Respiratory disease was one of two foci for the initial round of funding. Five centers currently focus on asthma, respiratory disease, and air pollution. Their activities include monitoring air pollution, tracking community trends in respiratory health, decreasing pollution, doing more educational outreach, increasing community participation, and increasing community capacity to conduct future research. One of these centers, the Columbia Center for Children's Environmental Health, works in collaboration with the activist group WE ACT, as mentioned in chapter 3.

Community-based participatory research has changed the process of reviewing proposals and conducting research. For the NIEHS, the main granting source, this change has occurred not merely through having members of community-based organizations on review panels, but from having entire panels committed to a thorough community-centered approach. This commitment has meant making sure that proposals are not written primarily by academic partners who then invite in community groups at the last moment. It has also meant equalizing resources among the partners because earlier proposals did not always provide adequate financing for community groups. The NIEHS has also supported the idea of community-based organizations serving as the principal investigator in order to put them more in charge of the process. Further, by reducing allowable indirect costs to 8 percent, it made sure that large chunks of grant money were not eaten up by regular federal indirect rates that often reach 60 percent. Annual conferences for grantees, largely designed by local grantees who host the conferences, emphasize the environmental justice activism that is so central to this program, including speakers who are local and national environmental justice leaders and featuring "toxic tours" of the area.

The NIEHS's approach is a model that may well be emulated by other NIH institutes and by other agencies where participatory research is weak or nonexistent, such as the NCI. We need to develop more interagency programs for such work because these programs cross boundaries and thus often yield innovation along with increased community participation. A few years ago the National Institute of Occupational Safety and Health joined with the NIEHS to coordinate efforts in environmental justice programs. Community-based participatory research will also necessitate agencies' acceptance of new forms of partnerships that appear less traditionally professionalized than they are used to. The new director of NIEHS, David Schwartz, who took office in 2005, has indicated plans to dismantle the dedicated participatory research and environmental justice programs, though a definitive decision had not yet been announced at the time of writing (2006), so activists and their scientist allies have had to spend time and effort in just pressing for the program to continue, which takes away from the actual community research they need to be doing.

Some organizations engage in community-based participatory research without specific funding from participatory programs. For instance, even before its recent NIEHS grant in 2004, Silent Spring Institute has always engaged in a citizen-science alliance approach toward its study of breast cancer and the environment, as described in chapter 2 on breast cancer. At Brown University, the Community Outreach Core of our NIEHS Superfund Basic Research Program engages in participatory research work even though the official requirements of the program do not specify that we do so. But universities and research facilities need more official mandates; without adequate funding across many agencies, we are losing out on an enormous amount of talent and dedication that thrives in collaborative research programs.

POLITICAL CONTROL AND CENSORSHIP OF SCIENCE

In the previous section, I discussed a very open, democratic approach to scientific research and science policy. From a contested illnesses perspective, scientific research and policymaking should be open and accessible to all, without allowing political and ideological factors to trump science. Because there has been so much political control and censorship of science, advocates for environmental health are compelled to combat them. Efforts to do so reverberate in the larger halls of science, where even nonenvironmental issues have been under similar attack. In this way, environmental health advocates play a cutting-edge role in overall health and social welfare policy.

One of the reasons I focused my research on asthma is the quality of the research that links air particulate matter to asthma, but even more clearly to other pulmonary and cardiac morbidity and mortality. The original particulate research (such as Dockery's powerful Six Cities Study) and the policies it led to were not centered on asthma, but much more major research on asthma followed from this line of inquiry. This body of research is among the strongest in all of environmental health, yet despite the science community's very broad agreement on this research, regulation lags behind. As discussed in the chapter on asthma, the EPA was forced to spend enormous amounts of energy combating an industry lawsuit that challenged its right to promulgate regulations based on particulate research.

The current antiregulatory stance of the Bush administration makes it unlikely that further air quality regulation will be produced during this term, especially considering the executive branch's claim that this science is far from conclusive. Instead, the administration has agreed to "study" regulations broken by polluting industry, has weakened health standards and air pollution regulation, and has delayed the deadlines for removing fine particles from the air.⁴⁰ Further, the Shelby Amendment of 1999 requires researchers to turn over primary data to any party, a practice that gives corporations a major weapon in opposing environmental health research and threatens the confidentiality of health data. Many activists view this amendment as a move intended to squash critical research that can affect clean air regulation and as a clear message from a conservative Congress that the EPA's research and regulatory approach to air pollution is not going to be tolerated. The 2002 Federal Data Quality Act allows actors, such as corporations or industry groups, to force the government to halt regulation if a minor error is found anywhere in the volumes of research documents used in support of regulatory processes.⁴¹ As a result, industry can successfully challenge scientific information that might support a case for further regulation.⁴²

Also in 2002, the Bush administration had the HHS secretary disband the National Human Research Protections Advisory Committee and the Advisory Committee on Genetic Testing because organizations on the religious right considered those committees' advice on genetic testing and research to be threats. In addition, fifteen of eighteen members of the Advisory Committee to the director of the National Center for Environmental Health at the CDC were replaced at once, mostly with scientists long associated with chemical and oil industries and with records of opposition to environmental protection. Within the federal government, the National Center for Environmental Health has been one of the strongest forces for environmental health research and stronger regulation.⁴³ Similar purges have occurred on the President's Council on Bioethics and on NIH study sections.⁴⁴ Similarly, pro-industry scientists were placed on the HHS Advisory Committee on Childhood Lead Poisoning Prevention in order to oppose a planned review of what are considered as elevated blood levels of lead.⁴⁵ Donald Kennedy, the editor of *Science* and previously the FDA commissioner in the Carter administration, noted that "loyalty tests" instead of scientific expertise are now required for members of federal science advisory panels.⁴⁶

Current practice on air pollution resembles the Bush administration's approach to global warming and climate change—deny the reality of scientific consensus, express concern about economic costs, and stall by calling for more research. In 2003, when the EPA published a report on the state of the environment, the administration had the agency remove any discussion of climate change, which is another issue on which the overwhelming body of scientific evidence agrees.⁴⁷ Advisors to the president even demanded that references to a National Research Council report indicating human contributions to global warming be removed, even though the report had previously been referenced by the president.⁴⁸ Despite public outcry and allegations of scientific censorship, the administration and EPA director Christine Todd Whitman defended the edited version of the report. A memorandum even circulated among EPA staff members stating that "E.P.A. will take responsibility and severe criticism from the science and environmental communities for poorly representing the science."[49] Leading up to that action, the administration had convened a conference to prime public opinion on the uncertainty of such research, repeatedly claiming that the it wanted to move toward a "science-based" policy that would require further investigation regarding the possibility of climate change.

Leading air pollution researchers have not backed down from the challenge, but it is an unfortunate waste of their time to have to defend knowledge that most colleagues know to be valid. These scholars have continued their work, expanding the knowledge base even more so that it can withstand attacks. The EPA has far to go in recouping legitimacy in this area and others. In 2002, Eric Schaeffer, director of the EPA's Office of Regulatory Enforcement, quit over this issue, going public with sharp criticism of the administration.[50] Even the conservative EPA administrator Whitman found it unpalatable to work in the Bush government and resigned in 2003.

Federal environmental regulation has been turned around in many other areas as well. The government has refused to sign the Kyoto Protocol on greenhouse emissions, retreated on clean air and clean water standards, allowed oil drilling in the Alaskan National Wildlife Refuge, opened logging and mineral exploration on many previously protected lands, relaxed permit regulations allowing commercial development on wetlands, exempted defense agencies from the Endangered Species and Marine Mammal Protection Acts, repealed the Roadless Area Conservation Rule, and repeatedly cut discretionary funding for the environment.[51] In view of this broad antiregulatory stance, environmental health supporters have to expand their critique and activism, while also realizing that the many positive implications of their work cannot be realized at present. By pursuing research and advocacy, however, these supporters can reach out to individuals who may later be in Congress and in the EPA, where they can pursue policies that give first priority to human health and environmental quality.

Indeed, environmental health supporters have continued to battle an administration that has been largely hostile to environmental regulation. Environmental groups have filed suits against the Bush administration's changes to Clean Air Act rules that resulted in weaker regulation and allowed plants to avoid mandatory improvements.[52] Such groups also actively contest the administration's claims regarding the costs and benefits of drilling in Alaska. Proponents of drilling in the Alaskan national Wildlife Refuge argue that it will reduce U.S. dependence on foreign oil, but environmental groups have continued to put forth information showing that arctic drilling will not reduce oil prices and will have a greater environmental impact than the administration has claimed, and at the same time that a number of renewable energy options are available.[53] Similarly, environmental activists have monitored the Bush administration's rollbacks concerning endangered species, and in August 2006 a federal judge overturned a new rule that would have made it easier for pesticide producers to discount the effects of their products on endangered plants and animals.[54] In other areas, environmentalists have had little success. In September 2006, the consequences of the Bush administration's failure to enforce many elements of the Clean Water Act became increasingly clear when male fish in the Potomac River basin were found to be bearing eggs as a result of dumping in the Chesapeake and Shenandoah rivers.[55] Likewise, the administration has successfully maintained weak fuel economy standards, issuing a new rule in March 2006 that allows most sport utility vehicles and pickups to remain exempt from fuel-saving requirements.[56]

Although there are clear economic reasons for the Bush regime to oppose environmental protection, it would be a mistake to view everything as based on straight economics. Some policies are less overtly corporate and more ideological. For example, Bush's No Child Left Behind policy has spent considerable money on standardized examinations, which in large part have forced the replacement of creative teaching with a mindless teaching-to-the-test approach. This policy has no immediate economic goal, but rather reinforces a conservative mindset and an education that avoids independent, critical thinking. Imagine if those funds were used to hire more teachers and to make schools more environmentally healthy. However, an administration that has attempted to unravel the whole history of modern environmental protection is not likely to be receptive to taking on environmental issues in schools.

The administration's general hostility to environmentalism is a hostility toward a sensibility that embraces cooperation between people and communities and that views people and their surroundings as part of a holistic system. Although there are some conservative forms of environmentalism, much of the environmental movement is sympathetic to other desires for social equality that the administration opposes.

Bush administration and corporate opposition to Clean Air Act regulations works along with the assault on science by conservative scientists, organizations, and think tanks. They often couch their position as one of supporting science over citizen pressure and use this approach to dispute virtually any connection between industrial activities and harm to people and the environment. This conservative onslaught has pressed opposition to class-action lawsuits on environmental matters (as well as on other health issues) and has called into question much evidence for routine environmental regulation. Furthermore, it has expanded "Daubert challenges" to scientific testimony in court, based on the 1993 Supreme Court decision in *Daubert v. Merrell Dow*, a case concerning side effects of medication, which effectively made judges into the gatekeepers of science. The Daubert ruling, which states that judges should allow only evidence that is relevant and reliable in the scientific community and that uses "generally accepted" scientific methods, has effectively barred many scientific experts from the courtroom and has been viewed as an attack on science with allegations that such experts' work is nothing more than "junk science."[57] Thus, although members of the administration are claiming that they support science over politics, their practice is directly the opposite—challenging science when it comes up against the dominant political and economic forces.

I have already discussed this challenge to science in detail in terms of air particulates, and I have mentioned in passing such efforts in terms of climate change. It is part of a broad denial of science that resembles what I think of as a "new Lysenkoism," similar to the Soviet Union's ideological opposition to scientific developments that challenged the state's political authority. Some alarming instances can be seen in recent purges and gag rules, as well as in resistance to generally accepted scientific knowledge.

Global warming is perhaps one of the best-known examples of how political and economic interests are being used to oppose science. Few ecologists, oceanographers, or other scientists with any respectable credentials fail to accept the strong evidence that global warming and other climate change effects have occurred due to human activities, in particular the emission of greenhouse gases such as carbon dioxide. Arguing against global warming is like arguing that HIV does not cause AIDS or that tobacco does not cause lung cancer. Recognition of the problems of global warming comes from the National Academy of Science, the Intergovernmental Panel on Climate Change, and the American Geophysical Union, the world's largest organization of earth scientists. But the Bush administration consistently denies this scientific consensus and censors scientists and agencies who try to address the problem. The White House even criticized its own State Department for discussing global warming in a report to the United Nations. Russell Train, who ran the EPA under two conservative Republican presidents, Nixon and Ford, claims that the removal of climate change from the annual air pollution report was the worst imaginable interference, something he had never encountered in his tenure. The administration even stopped its Agriculture

Department from reprinting a brochure that helped farmers reduce greenhouse gas emissions.

Mercury is one of the most well-studied neurotoxins, causing brain damage in children and harming reproduction in women. Most mercury emissions come from coal-fired power plants. When the EPA prepared a report on children's health and the environment, the administration tried to suppress publication. Only because EPA officials leaked the report in 2004 did it get out to the public. Now the administration plans to allow mercury trading, through which polluting power plants can buy the right to emit dangerous levels of mercury.

The administration plan to allow more pollution is called Clear Skies, truly an example of Orwellian doublespeak because its purpose is to increase the government's privileging of the oil and power industries, resulting in dirtier skies. In particular, it guts the Clinton-era NSR process, which required scrubbers and other pollution controls on power plants that engaged in substantial renovations or overhauls. The doublespeak language of Clear Skies is mirrored in the Healthy Forests plan that allows more clear-cutting of forests, while claiming the goal is to prevent forest fires.

The purges on federal advisory panels have been frightening. As we know, lead is one of the country's most serious health risks for children, causing brain damage and neurological disorders. Four hundred thousand children under age five have elevated lead levels in their blood, and the CDC has been crucial in reducing lead exposures. The current risk level of 10 micrograms per deciliter (μg/dl) is the definition of undue lead absorption, but as recently as 1975 lead poisoning was defined at 30μg. It only went to 25 in 1985 and to the current 10 in 1991. Considerable research now shows significant health effects even under 10. The routine process of evaluating hazard levels is a major public-health task. And so in 2002 the CDC's lead advisory committee set about to review whether it was time to change the number again because the onward march of science had given us better information on which to act. A few weeks before the deliberations, HHS secretary Tommy Thompson personally rejected new members of the advisory panel, the first time an HHS secretary ever did so. At the same time, Thompson filled the panel with people who had consulted for and testified for the lead paint industry, including one who argued that no health effects were found below 70 μg/dl. Similar removals have occurred in many other federal health panels, so many that I cannot even mention them in passing. The panel members considered acceptable are exemplified by one appointment to the FDA's reproductive health panel: an obstetrician-gynecologist who refuses to prescribe contraceptives to unmarried women and who recommends using Bible passages to treat premenstrual syndrome.

Major health research is being blocked for purely ideological reasons. Stem cell research has the potential to treat many disorders, some even prenatally. Yet the administration has virtually halted it, allowing the use of a small number of cell lines that meet its moralistic requirements. James Huff, a key scientist at the National Toxicology Program, was responsible for testing chemicals that might otherwise never be studied for their harmful effects. He apparently was too diligent, and because the National Toxicology Program published many data on toxics, he was told that he could not publish anything without prior approval (i.e., without censorship). Many scientists voiced their protest, and Huff's gag order was rescinded.

The Bush administration successfully put a gag order on the EPA from publicly discussing perchlorate pollution, although research shows high levels of the rocket fuel component in water and food. The EPA had recently recommended more stringent regulation in the face of widespread concern that concentrations greater than one part per billion were harmful to infants. Yet fears of liability for cleanups on the part of the Pentagon and defense contractors triumphed, as industry and military leaders argued that perchlorate is safe in drinking water at levels as much as two hundred times higher than EPA recommendations. The White House went so far as to propose a bill in Congress that would exempt the Pentagon and defense industry from most potential liability on the ground of military necessity. The White House Office of Management and Budget, which has increasingly stepped in to stop scientific research, halted regulatory action for perchlorate, pending further review by the National Academy of Sciences. In the interim, the EPA told staff not to speak about perchlorate, despite the clarity of the recent studies.[58]

This assault on science has been so extensive that the Union of Concerned Scientists published a widely cited report in February 2004, signed by twenty Nobel laureates and by former directors of the National Science Foundation, the NIH, and some of its units. *Scientific Integrity in Policymaking* (cited in chapter 3) documents a long legacy of the Bush administration's assaults on science, and union members have traveled around the country disseminating the results of their analysis.

In light of this discussion on political control and censorship of science, it may seem surprising to see my call for increased and better-placed federal funding. But there are always segments within the government that support a more humane agenda. Sometimes this means outposts of environmentalism in agencies existing in an antienvironmental government, and other times it means congressional support that can push environmental agendas despite general conservatism. Further, it is always necessary to keep alive proposals for more environmentally conscious programs because changes in government do occur.

IMPROVEMENTS IN MILITARY MEDICINE AND VA CARE

One last policy implication is very specific to the care of military personnel both in active service and as veterans. It is very clear from chapter 4 on GWRIs that the military and VA made many errors in handling those illnesses. Most commonly, GIs and veterans were considered to be exhibiting psychiatric symptoms, were often discredited, and had to fight to get treated. There is no way to overestimate how disheartening it is for people who served in a war to have to struggle to get medical professionals and administrators to recognize their suffering. Research was further hampered by a lack of communication between the VA and the DOD regarding predeployment data and potential sources of disease as well as by the VA's limited information sharing and collaboration with the Research Advisory Committee. For a time, the VA and DOD were even prohibited from communicating. Unfortunately, as they began to improve their collaboration, federal funding was on the decline, presenting yet another obstacle for such efforts.

Predeployment health information was scant, if at all available. Records of where troops served were very sketchy. Although the latter records are not medical records and hence not necessarily part of a medical issue, they become medical when the data in them are needed for treatment. One of the strongest recommendations made after the Gulf War was to ensure that proper predeployment health exams were carried out and made available, yet this plan has never been implemented in a serious manner.

The extent of GWRIs was such that it should have pushed military and VA medical policymakers to come up with a thorough alternative approach. That approach might include training of medical personnel by other professionals who have more experience with and sympathy toward the unexplained symptoms and illnesses. It might also include a network of centers dedicated to serving the medical and psychosocial needs of GIs and veterans reporting such symptoms, similar to the Specialized Care Program, in which Gulf War veterans who have inadequately explained symptoms, as judged by the DOD's Comprehensive Clinical Evaluation Program, participate in a three-week outpatient program with other veterans specifically designed to meet their needs.[59] The VA's 2001 creation of the two War-Related Illness and Injury Study Centers in East Orange, New Jersey, and Washington, D.C., which focus on addressing causes and treatments for unconventional chronic symptoms and develop specialized care programs tailored to veterans individual needs, is a step in the right direction in devoting attention to these difficult-to-diagnose illnesses.[60] Such centers might also include follow-up to see the effects of treatment on veterans who might otherwise disappear from institutional records if they do not return for more treatment. They would also offer veterans a larger role as allies in the research process. Although the research center I studied at the Boston VA Medical Center was very geared to a citizen-science alliance with veterans-activists, that approach was not typical policy, and at national research conferences sponsored by the federal government veterans repeatedly expressed disappointment in not being more central to the research process. The lessons of community-based participatory research have not taken root in the DOD and the VA, even though the DOD has played a positive role in allowing breast cancer activists to serve on review panels.

Military and VA officials have not usually learned from past errors. The many years they spent fighting treatment and compensation for Agent Orange–induced diseases should have alerted them to the problems they would face in the future. Only after Institute of Medicine reports clearly showed health effects from Agent Orange did the VA cover victims. A precautionary approach would have highlighted the uncertainty of current knowledge and the failure to learn from "late lessons" and thus would have assumed the likelihood that environmental factors played a part in at least some of the complaints. As always, late lessons seem more sensible in hindsight, as with recent developments: in 2004 a General Accounting Office report revised sharply upwards the estimate of service members exposed to sarin gas, and other research indicated a likelihood of increased lung cancer as one Gulf War outcome.

ENERGY POLICY

Asthma activism and other organized efforts on air pollution can have many ramifications far beyond asthma. Diesel fuel accounts for 70 percent of the total air toxics burden, according to the California Air Resources Board, a major source of air pollution research. There is a rapidly growing awareness that diesel is a major cause of many diseases and a high number of deaths.[61] Asthma rates are higher in neighborhoods bordering on major truck routes.[62] Activists have grasped the importance of this research, and it is understandable that they place so much emphasis on diesel and other sources of pollution. Although they have placed much of their organizing on transportation policy, it is also necessary to take on national energy policy approaches. We need overall reduction of auto use, increased hybrid production, hydrogen fuel cell research, stricter emissions controls, development of lower-sulfur diesel formulations, more rapid transit, and community zoning and development based on reduced commuting. These changes will help reduce the incidence of asthma and other cardiovascular and pulmonary diseases, while also making us less dependent on Middle East oil and the dangerous politics that stems from it.

Multisectoral approaches that bring together activists, researchers, and providers in collaborative efforts bring about action in a range of social sectors, including the public-health sphere, transportation, education, sanitation, and housing. These intersectoral efforts take into account the multiplicity of factors influenced by policy that contribute to environmental health. These collaborations often incorporate environmental justice concerns that arise as the underlying causes of health disparities are revealed, ultimately calling into question societal structures that are seemingly distant from initial environmental concerns. As seen in ACE's transportation campaign, discussed in chapter 3, the health problems and justice issues eventually become inextricably tied. Although justice concerns had been the initial focus of ACE's campaigns, asthma was found to be community members' primary concern and thus became a focal point of ACE's actions.

IMPLICATIONS FOR SCIENTIFIC RESEARCH

Searching for the distribution, effects, and causes of diseases and conditions can be a challenging and exciting scientific activity. From a contested illnesses perspective and with the community-based participatory research models discussed earlier, scientists can be of great social use by serving people's needs and by working in collaborative projects. Yet as scientists have long understood, it is often dangerous to engage in discovery, treatment, and prevention of most environmentally induced diseases. My contested illnesses perspective views this problem as very central, while highlighting many scientists' efforts to carry on in spite of such obstacles. I also see the growing importance of popular epidemiology and citizen-science alliances as laypeople continue to make advances in disease detection and prevention in tandem with professionals who take a critical epidemiology approach. The environmental health struggles I have written of in this book have pushed people to make significant innovations in measurement, method, theory, and organizing. This section focuses on those contributions to science.

DISSIDENT SCIENTISTS

Epidemiology begins for most of us with John Snow's 1854 discovery of the relationship between water contamination and a cholera epidemic. Snow listened to other people's anecdotal evidence about connections between water and cholera and then made his own observations of such connections. He determined a localized area where the epidemic was most severe and then surmised that the water pump at Broad Street was the cause. Snow's famous removal of the handle from the Broad Street pump actually occurred after the epidemic had peaked, though that fact by no means diminishes the significance of his action. Snow was quite radical in his recommendations to prevent cholera. Because he feared government obstruction, he wrote: "The communicability of cholera ought not to be disguised from the people, under the idea that the knowledge of it would cause a panic, or occasion the sick to be deserted."[63] Such withholding of information is a major problem in present-day relations between contaminated communities and government.

Withholding the highly charged political and economic ramifications of epidemiological investigations has long been a concern. In his 1882 play *An Enemy of the People,* Henrik Ibsen portrayed a remarkably modern perspective on epidemiological discovery. Dr. Stockmann, physician of the town's famous and profitable mineral baths, discovers toxic contamination from nearby tanneries. He refuses town government and business leaders' demands that he hide the evidence and hence is fired, evicted, beaten, stoned, stigmatized as an enemy of the people, and driven to mental breakdown, and his children are attacked and thrown out of school. Still, he vows to fight on for justice.

Stockmann's struggles are the same kind in which modern-day epidemiologists and other scientists engage. Rachel Carson, the inspiration for the modern environmental movement, was ruthlessly victimized by industry, whose attacks were personal as well. Chemical companies enlisted scientists to write negative reviews of *Silent Spring,* threatened to remove advertisements from journals that favorably referred to it, and continually made efforts to fill the public with a fear of a world without chemicals. One company, Monsanto Chemical, even published a parody of Carson's work entitled "The Desolate Year" in which the United States is invaded by insects because a year went by when no one used pesticides.[64] Carson was portrayed as a poetic dissident in the war against insects in which industry made itself into the hero. Further attacks undermined Carson's credibility, as seen in Congressman Jamie L. Whitten's accusations that "Miss Carson's" works were little more than "sentiment and nostalgia" and that she had ignored scientific evidence.[65] We thus have what I call *dissident scientists* who go against the grain of both established science and public policy.

More recently, epidemiologist Devra Davis has been ostracized for her pursuit of the connection between cancer and the environment. She is one of the leading scholars arguing that there is a real increase in cancer, but critics claim that the increase is due to better detection. Davis argues that there is a large increase in unspecified and ill-defined cancers, which are generally not subject to better detection. She has been attacked by some leading cancer epidemiologists and treated like a whistleblower. This story was big enough that the *New York Times Magazine* featured her in an article and documented the attacks made on her by people such as Richard Doll, known for his role in linking tobacco and lung cancer.

Ironically, one of the themes in Davis's 2002 National Book Award–nominated book *When Smoke Ran Like Water: Tales of Environmental Deception and the Battle Against Pollution,* is "heroic scientists'" daring to make environmental linkages. Some of them were hounded for such efforts, and others were muzzled and had their careers truncated. Mary Amdur, a pioneer in the effect of air pollution on the lungs in the 1950s and 1960s, was fired from her Harvard position for pursuing her research after corporate sponsors and senior faculty with whom she was working found it threatening.[66] In the 1960s and 1970s, Herbert Needleman, whom Sheldon Krimsky profiles as an exemplary scientist who set the standard for examining the effects of lead on children, was relentlessly pursued by lead industry–hired scientists who filed complaints with the NIH accusing him of misconduct.[67] Though finally exonerated by a scientific panel, Needleman was stigmatized by the University of Pittsburgh, which sealed his research files, allowed him to use them only in the presence of a university official, and refused an open hearing until the university faculty voted unanimously to demand it.

More recently, Steve Wing, an epidemiologist at University of North Carolina, faced similar harassment because of his research on environmental health effects and environmental justice issues stemming from massive pig farms, a key industry in the state. Working on an NIEHS-funded community-based participatory research project with Concerned Citizens of Tillery, and putting his technical skills in the service of poor and mostly African American rural residents from 1995 to 1999, he showed the extensive health effects and hazards of this industry. Pressed by the Pork Council, the state legislature forced the university to require Wing to share his data with the industry despite the potential that people's confidentiality would be compromised. Rather than defend Wing's academic freedom and the research subjects' right to confidentiality, the university attacked him. One administrator told him that if he refused to turn over documents as directed by the university attorney, they would have him arrested for stealing state property. A compromise was reached that allowed Wing to remove certain identifying data, but it was still a major threat to research ethics and integrity.[68]

Although not all environmental health scientists face such harassment, most realize that they will face stronger obstacles than other scientists. They are, by definition, dissident scientists who go against the grain of a corporate-dominate research enterprise wherein companies, trade associations, and industry lobbying groups increasingly dictate research programs at the nation's major universities. They rebel against the supposed objectivity that distances research from the affected people and communities. They fight against what Sandra Steingraber aptly calls "cigarette science" (named after tobacco industry scientists who conspired to hide the well-established role of tobacco in causing lung cancer), taking up instead "advocacy science."[69] Dissident scientists raise their voices against the many silences that keep at bay scientific knowledge and policies for prevention and remediation of environmentally induced diseases.

As I noted earlier, many scientists who participate in these debates are not targeted or harassed. Indeed, the more dissident scientists have made their mark, the more they have made it possible for others to do this kind of work with the support of a growing community of like-minded people. In addition, many scientists do alternative, critical work without defining themselves as "activists" or without calling attention to themselves. Thus, scientists can take other possible positions in working to engage in progressive, community-oriented environmental health work.

RESEARCH SILENCES

Obstacles are not only about acts done; they are often about things undone. Connections between environmental exposures and ill health have often been met with what I call *research silences* by government and science. Research silences are the questions we do not ask, the kinds of studies we do not do, the kinds of records we do not keep, the kind of implications we worry about making, the defenses of beleaguered critical scientists that we fail to mount or join. David Hess similarly speaks of "undone science."[70] The lack of effective disease registries is one example of research silence. Some European countries have disease registries that are useful in understanding both social and biological causes of disease and death. The United States has avoided registries, and not all states have even cancer registries.

Another form of research silence is the paucity of federally funded studies on disease clusters and environmental epidemiology. State and federal agencies carry out some of their own studies, but they are often reluctant to do so. Much of the pioneering work is done by environmental health activists, and it is great that they can provide such leadership. For example, the Collaborative on Health and the Environment, started by Commonweal, continues to grow, with hundreds of member organizations and hundreds more individual members. Its frequent national Internet conferences bring well-respected scientists and major government health and environment officials as speakers and participants. Yet the existence of such an organization also highlights the failure of most government agencies and mainstream academic centers to develop these new directions themselves.

As I wrote in chapter 6 on the precautionary principle, European countries and the European Union have gone much further in alternatives to traditional risk assessment and have implemented many precautionary approaches, including banning some chemicals. Silence in the United States on the same issues exhibits a fearful hesitancy to confront environmental hazards.

Academic science is largely silent on environmental factors. Much of that silence is the paucity of research on environmental factors, particularly when there is controversy. Environmental epidemiologists report much fear for their career prospects and find it hard to take risks. University-based scientists, a potential source of aid, frequently consider applied community research to be outside the regular academic reward structure. Universities' dependency on corporate and government support makes it harder for scholars to challenge established authority. Silence is also found in the failure of many journals to publish articles and editorials on environmental factors in disease, either because they are deliberately censoring the material or because they fear mainstream scientists will consider such articles unprofessional. When scientists have been unfairly challenged by industry, as noted previously, colleagues have too often been silent in their defense.

Extensive focus on cancer and insufficient study of other diseases is another form of research silence. Cancer incidence ought not be the main reason for people to seek environmental safety. As the rapidly growing environmental epidemiology literature indicates, toxic substances cause many other kinds of illness, not always cancer and not always leading to death. But not enough scientists have taken up these new directions. I do not mean in any way to minimize the important work of scientists who study known and suspected toxic links to cancer; otherwise, I would not have spent so much time on the importance of endocrine disrupters and breast cancer. Moreover, links between many toxics and various cancers are very important, especially given recent increases in many cancers that cannot be attributed to personal habits, genetic predispositions, or age.[71] What I am concerned about, however, is the common notion that if something does not cause cancer, it is not worth studying in terms of environmental impacts.

What do epidemiologists and other scientists offer in place of research silences? Similar to the popular epidemiology of lay activists, we see a growing number of professionals who engage in what I term *critical epidemiology*.

CRITICAL EPIDEMIOLOGY

In chapter 1, I described the overall perspective of critical epidemiology, which looks at the historical context of epidemiologic investigations, takes a social structural and health inequalities approach, and applies the discipline's epistemology for social justice by working with laypeople or by seeing themselves as tied to social movements or by doing both. Some epidemiologists make very clear these radical elements, such as Steve Wing, whose list of criteria for a humane epidemiology was already discussed. Others are drawn in to this approach without a prior political interest, but because they find environmental epidemiology to offer intellectual challenges that they want to take up without fear of consequences. Some nonprofit research and training organizations employ a critical epidemiology approach in assisting community groups—examples include John Snow Inc. and Tellus Institute in Boston. The scientific work of activist organizations such as the EWG (to be discussed later) also fits into this framework in that such groups employ epidemiology and toxicology to advance the environmental health movement and to change scientific practice.

In some cases, entire academic departments emphasize such an approach, as with the Environmental Health Department in the Boston University School of Public Health, which has for a long time worked with many community groups around toxic hazards. At present, several faculty and a doctoral student there are collaborating with the Toxics Action Center (a New England–wide group) and two local organizations, Health-Link and Haverhill Environmental League, on an NIEHS-funded study of how community groups deal with health studies that either they or government agencies have initiated.

Indeed, the NIEHS's environmental justice and community-based participatory research program has helped to legitimize epidemiologists who work with community groups. In dozens of projects under that program, epidemiologists, toxicologists, and other scientists have been able to focus their professional efforts on matters that would otherwise be difficult, if not impossible, to fund. In the process, they are learning more about the complexities of academic-community collaboration and becoming more attuned to the many changes necessary in such work.

Scholars working in this area have emulated Rachel Carson's holistic, integrative approach that ties together many strands of science and other areas and presents them in a humane, sensitive prose accessible to the average reader. Environmental scientists such as Theo Colborn, in her work with John Myers and Diane Dumanoski on endocrine disrupters, *Our Stolen Future,* pieced together evidence from multiple disciplines to provide the first major synthesis of how endocrine disrupters are so profoundly affecting human and animal health. Sandra Steingraber's *Living Downstream: An Ecologist Looks at the Environment* and *Having Faith: An Ecologist's Journey to Motherhood*, as well as Devra Davis's *When Smoke Ran Like Water* combine biography, narrative, biology, ecology, epidemiology, public policy, and ethics to make compelling cases for precaution, prevention, and environmental health activism.[72]

The combined effect of these disparate innovations in environmental health science is a strong challenge to the various DEPs currently being used to explain disease. Created from that challenge is a vibrant and progressive alternative vision of science in the service of bettering people's lives and communities, rather than at the beck and call of private profit and government control.

CITIZEN-SCIENCE ALLIANCES

In chapter 1, I spoke of citizen-science alliances as a central part of the contested illnesses perspective, and in the chapters on asthma and breast cancer we saw numerous examples of such alliances. Critical epidemiology is, of course, crucial to citizen-science alliances; without scientist allies, community groups would be less able to challenge corporate pollution and government regulatory shortcomings. But we must also focus our attention on how activists themselves affect science by being part of such alliances.

If we return to the Love Canal incident, which took Rachel Carson's legacy and pushed forward the modern environmental health movement, we see the beginnings of popular epidemiology, where citizens discover disease clusters and develop community organizations to investigate the problem. The Love Canal crisis taught laypeople that they could initiate health studies, and the Woburn, Massachusetts, toxic waste crisis showed them they could participate in the data gathering itself. In just over a quarter-century since Love Canal, community-driven and community-involved health research has thrived, with many hundreds of communities doing it.

Citizen-scientists often come up with methods that professionals have not. For instance, as I wrote in the asthma chapter, in the midst of an air-monitoring program set up by experts, teenage environmental justice volunteers suggested that monitoring focus on peak travel times rather than on daily averages and that particulates be measured at the height of small children rather than from rooftops. In California, Communities for a Better Environment devised "bucket brigades" where residents could use inexpensive, low-technology air-monitoring devices to capture local emissions. Subsequently, other communities have used this technique, especially in Louisiana.[73] Indeed, the Louisiana Bucket Brigade's presence in that state, where it monitored emissions from plastics factories and oil refinery, enabled it to play a significant role in recovery efforts after Hurricane Katrina in 2005.

Some activists become scientists, as with the case of some Woburn activists who became inspired to get graduate training and work for health and environmental nonprofits. Others develop strong organization skills and become leaders and staff people at environmental justice groups, where even if they are not scientists, they are quite conversant in scientific issues. Many environmental health and environmental justice activists who do not seek further formal education and who do not take on professional jobs still learn a great deal of science in order to carry out their organizing work. Throughout my career, I have seen many such people at toxic waste activist and environmental health conferences, and they are impressive in their command of technical matters. More than that, they bring to the table an embodied sense of the science: the hazards and the actual health effects have a deep resonance for themselves, their families, and their neighbors that drives them in their efforts.

While these community activists learn science, their professional allies learn community organizing. The result is a multitude of citizen-science alliances where there is a blurring between resident activist and scientist, so we have hybrids where you cannot tell where the activist part ends and the scientist part begins.

GRASSROOTS BODY BURDEN TESTING

One kind of new science that derives from critical epidemiology and citizen-science alliance is what I term *grassroots body burden testing*. In light of the CDC's study, mentioned in a previous section, the EWG, an activist organization, decided to publicize the need for monitoring humans. Working with Mount Sinai School of Medicine in New York and Commonweal, a California environmental group, researchers examined a far larger number of chemicals than the CDC was monitoring. They found an average of 91 industrial compounds, pollutants, and other chemicals in the blood and urine of their volunteers, but a total of 167 chemicals in those nine people. Of those chemicals, 76 cause cancer in humans or animals, 94 are toxic to the brain and nervous system, and 79 cause birth defects or abnormal development. As with most chemicals, the dangers of exposure to these chemicals in combination have never been studied.[74]

The EWG's small-scale body burden study included nine well-known environmental activists and other public figures, who permitted themselves to be publicly identified in order to show people what it was like to discover the toxic substances inside one's body. Besides producing a grassroots version of the CDC study cited earlier, the group also had impetus from seeing the increase in many diseases, from realizing that chemicals are usually studied only after there is known harm, and from understanding that federal regulation, especially the Toxic Substances Control Act, was failing to regulate chemicals. It also wanted to promote awareness that many effects are now found below "safe doses" set by regulatory bodies. On the EWG Web site, the viewer can click on the thumbnail photo of each person to see what contaminants are in that person's body. A graphic resembling the periodic table of the elements shows up, and the chemicals in that person's body are highlighted. Clicking on any chemical box brings up information on that chemical.[75] The group had a larger agenda; it wanted to publicize how little health data there are on many of these substances, and it sought to draw attention to the fact that the federal government was now monitoring chemicals.

By advising participants of their personal burden and having them go public, the EWG was pursuing a radical approach of informed consent in which people receive such information. Their intent was not necessarily that all participants in all studies receive such information, but to show that this strategy is a good one for volunteers in landmark public studies. Emphasizing the ubiquity of these substances in *particular* individuals who could symbolize in the public's mind an "anybody" because of their celebrity can prompt people to think, "If it is in Bill Moyers, it could easily be in me."

In the breast cancer and environmental justice project I am involved with in collaboration with Silent Spring Institute and Communities for a Better Environment, mentioned in the chapter on breast cancer, a core concern is the ethical responsibility to inform all research participants of the results of their biomonitoring. This approach contrasts with the typical epidemiological study approach, in which people are informed only that their population (e.g., people in their census tract or town or county) had a lower, similar, or higher rate than the general population. Not giving people their personal data, however, is a violation of their right to know what is in their bodies. I term the process of informing people of their personal data the *research right to know* and consider it to be a major step forward.

Several decades of activism concerning democratization of science have led to major changes in conceptions of who has the legal and ethical right to be involved in scientific research on people. Informed consent is increasingly understood as much more than a clinical treatment or research subject protection. Rather, it incorporates many concerns about the rights of people and communities to decide if they should indeed be the focus of research, about the right to own their own data, and about reciprocal relationships between researchers and those being studied. The growing movement of community-based participatory research has been key in this transformation of the view of informed consent from a largely bureaucratic process to a major democratic right.[76] African American, Latino, and Native American populations have been among the most critical of academic research that is exclusively controlled by scientists who develop research questions without community input and that has little or no benefit to community needs. In some community-academic partnerships, the community partner requires the researchers to inform the community before they disseminate research results or give presentations on these results.[77]

Now that activists have extended informed consent to communities, some are bringing it back to the individual level in a new way. Just as communities should have the right to know about research affecting them and should have control of the resultant data, so too should individuals have the right to receive their data. Yet, as noted, the typical epidemiological study gathers data and then reports on an aggregate level—for example, "your town has a higher/lower/equal rate than the state (or national or other average)." People do not usually learn their own body burden or household contaminant level, except for certain well-known testing protocols for chemicals with well-established health effects, such as lead levels in blood. As I wrote in the breast cancer chapter, my colleagues and I highlight that particular issue as we report back to individuals their household exposures and give them the opportunity to discuss the implications of those data.

The CDC's body burden study was a great advance, yet it is only the first step. As the EWG study showed, it is important to go beyond "individual chemicals in a multitude of people" to "individual people for a multitude of chemicals." And it is important to let people know their individual burdens. Many more such studies, with larger samples, are necessary to understand what levels of toxics are in our bodies and to correlate that burden with actual health status. Some researchers are increasingly turning attention to breast milk monitoring, given the high concentrations of toxics in it. This topic is a fruitful avenue for advocates to press because human breast milk has many toxins, often at levels that surpass EPA limits on manufactured food products.[78] Breast milk, perhaps more so than blood or urine, is a highly socially significant medium through which to announce the presence of toxics.

The EWG went further in its grassroots body burden testing with a breast milk–monitoring study in 2003, examining the flame retardant PBDE. The average level in the milk of twenty first-time mothers was seventy-five times the average found in recent European studies that led the European Union to ban a number of the substances. Milk from two women contained the highest levels of fire retardants ever reported in the United States, and milk from other mothers in the study had among the highest levels ever detected worldwide. The EWG is aware that the beneficial aspects of breast-feeding may outweigh the risks of PBDEs and other toxins in breast milk and so did not try to convince mothers to stop breast-feeding. Some environmental health groups are reluctant to support breast milk monitoring because of the problematic nature of such research, especially the fear that women will stop breast-feeding. Nevertheless, this kind of publicity can be helpful because it results in corporate action, for instance the Great Lakes Chemical Corporation's ceasing of production of the penta-PBDE mixture that is the largest source of PBDE exposure.[79]

Continuing its biomonitoring work, the EWG announced in July 2005 its results from a study of toxins in umbilical cord blood. The study detected 287 different chemicals in the umbilical cord blood of ten babies born in August and September 2004 in hospitals in the United States (averaging 200 chemicals and pollutants per infant). Of the 287 chemicals, many of which are found in consumer products such as stain repellants, fast food packaging, clothes, and textiles, 180 are known carcinogens, 217 are toxic to the brain and nervous system, and 208 cause birth defects and abnormal development in animal tests.[80] Among these chemicals, 212 were already banned or severely restricted by the federal government. The study results led the EWG to argue for more rigorous federal policies regarding industrial chemicals.

Another interesting approach to grassroots body burden testing occurred on October 20, 2004, at the annual meeting of the Society of Environmental Journalists in Pittsburgh. John Spengler, one of the best-known specialists in air pollution and other environmental health effects, was invited as a guest speaker. Spengler suggested that the journalists might learn a great deal if he took hair samples at the beginning of the conference, tested them for mercury, and returned the results by the end of the conference. Spengler's staff took 199 samples, flew back to Boston to analyze them, and brought the results back in two days. Aggregate data were presented, and each individual was able to receive his or her personal burden data. They found high levels of mercury in many people, mostly correlated with older age and greater fish consumption; combined, those two factors accounted for 45 percent of the variance, with fish consumption being the stronger predictor. The EPA's reference dose is 1 ppm for pregnant women, and one-fourth of the sample exceeded that level. A Greenpeace survey of 1,149 people found similar data, with 21 percent of women between sixteen and forty-nine years old exceeding that level. Projecting from those data, we would expect 300,000–630,000 births per year of children at risk of health effects, mostly developmental, from mercury.[81]

One might argue that Spengler's study was not a grassroots effort because it was conducted solely by a researcher, but I would hold that it was, even if a community group was not doing it. Spengler's approach is based on supporting grassroots groups, and the individual notification and rapid attention to personal data are hallmarks of a grassroots approach.

A growing number of nonprofit environmental groups are engaged in biomonitoring, including Commonweal, Washington Toxics Coalition, and Sightline (formerly NorthWest Environment Watch). The Pesticide Action Network of North America now owns a chromatograph to do its own analyses. It is interesting that communities are doing this work, too, at the request of community members. Some do testing under the auspices of federally funded community-based participatory research mechanisms, such as the NIEHS's Environmental Justice Partnerships for Communication program, which funds the work of Alaska Community Action on Toxics on body burdens of Arctic contaminants in people living near formerly used defense sites and the work on body burdens in the people of the Mohawks' Akwesasne community in northern New York. Commonweal's Community Biomonitoring Resource Web site is a fine example of how advocates are making resources available for communities thinking about biomonitoring. In addition to the goal of pushing for government policy and scientific research, one objective of these grassroots body burden studies is to personalize environmental issues. When people can examine their own burden of dangerous chemicals, they may be more likely to take both personal preventive action and broader social action for changes in corporate practices and public policy. Another objective is to show physicians the need to test people for dangerous chemicals. A powerful social policy impact, prodded by many of the activist groups I have discussed, is the passage and signing on September 29, 2006, of the California Environmental Contaminant Biomonitoring Program, the first statewide effort of this type.

But caution is necessary when dealing with this level of information. Unlike Spengler's study of mercury, where the many health effects are clearly understood, we do not always have clear evidence of the health effects associated with PBDEs and some other toxins. There is also danger is that we might be individualizing and medicalizing body burden to the point where the limited resources we have go into personal monitoring and detoxification treatments rather than into overall prevention and structural changes to reduce body burden in everyone. Hence, even some long-standing environmental health advocacy groups are concerned that providing information may scare people. In the case of breast milk monitoring for PBDEs, as noted, they worry that some women will choose not to breast-feed as a result. In fact, strong advocates of such monitoring do make public statements about the relative benefits of breast-feeding. But for this and other body burden testing, it is crucial to understand that we do not always know how to use all the information we can get. But this lack of knowledge is not necessarily a reason to hold off on such efforts. Community-based environmental health groups will mostly likely be the most prepared to grapple with the implications of such imperfect and potentially disturbing data. This willingness to deal with uncertainty by erring on the side of safety is emblematic of the precaution-infused framing by environmental health activists.

NEW APPROACHES FOR SCIENTISTS

The dissident, exemplary, and heroic scientists and the critical epidemiologists offer a beacon to other scientists to show them how scholars and researchers can go beyond the routine boundaries of science. We can hope that others will learn from them how to transcend narrow disciplinary constraints, how to keep humanitarian betterment in the forefront, and how to stand up against censorship and other obstacles. Although the ideal situation is for such scientists to join with laypeople in citizen-science alliances, it is also possible for them to contribute on their own outside of alliances.

I have already described a number of innovative methods and approaches taken in citizen-science alliances, such as air monitoring at peak traffic times and at children's heights. Scientists can continue to expand air monitoring to different settings, such as in the vicinity of large emitters that are otherwise escaping monitoring and adequate regulation. In chapter 2, I discussed the many endocrine disrupters and other chemicals that had not been tested until Silent Spring Institute did so, but there is still much to be done in this area. Because GIS mapping is so useful for identifying multiple hazards in a community, along with land use and demographics, many activist groups have pioneered the employment of this hot new technique. Scientists can learn a great deal from these applications, especially in terms of characterizing hazards that are not part of routine regulatory observation. Environmental health activists have also pursued ecological levels of analysis, whereby aggregate characteristics of communities, rather than individual exposures, are used to estimate health effects. Such measurements are often easier to obtain because they do not require individual sampling, and they are sensible because they refer to community-level problems. Epidemiologists and other public-health scientists have much to learn from this way of looking at environmental health.

So there is a growing array of methods and techniques often employed by community activists that can filter into general environmental science applications. Together with critical epidemiology, citizen-science alliances, and grassroots body burden testing, they make up a very large contribution to science overall. The dissemination of such methods exemplifies the way that lay approaches can be the impetus to serious scientific advancement.

We have seen a wide range of implications of the contested illnesses perspective in public awareness, public policy, science, medicine, and business. In these different arenas, people are pursuing a battle for truth that is hard to get and even harder to act on. Not only individuals and social movements are contesting illness, however. Industry, government, and right-wing political groups take part in contesting the existence and environmental causation of various illnesses. This rightist alliance seeks to maximize economic gain without regard to health, to prevent public input from affecting government policy, and to prevent the democratization of science that in itself can lead to a general resurgence of democratic rights. As progressive activists engage more effectively, industry, government, and conservative political groups ramp up their opposition, so that activists in turn have to be even more creative in developing new approaches to improve environmental health. The process of doing so is exciting because of the benefits for human health and increased social betterment and equality. This extensive potential for social change is very heartening, and many people and organizations will find it an attractive direction for a healthy future.

8

CONCLUSION

THE GROWING ENVIRONMENTAL HEALTH MOVEMENT

PERSONAL REFLECTIONS

I began this book with a personal recounting of how I arrived at my perspective on health and the environment and how and why I do the kind of work I do. It seems fitting to end in a largely personal vein. In chapter 7, I wrote about the implications of my contested illnesses perspective, so there is no need to repeat those points. Here I combine my personal reflections with a few notions about future directions for the new environmental health movement.

In truth, whenever people choose to do certain work, especially of a professional nature, they take into account the effects on others. Those of us engaged in careers that highlight community-based participatory research try to make that accounting more open and central. Further, we think of the work as collaborative rather than as a one-way service from professionals to lay recipients.

When I think of my combined work and activist life, I am pleased to link these two elements. It is gratifying to do work that has benefits for people living in contaminated communities, but there are always concerns. For instance, if in a traditional form of research, I find results that go against what I hoped for, I am the main person who "loses." *Lose* may be the wrong term because it may merely mean I have to take a different tack next time. Nevertheless, the outcome is something that affects me primarily. If I am involved with community partners in a community-based participatory research project, failures or negative results hurt many people and possibly an entire organization. This reason is indeed why some scientists and community groups avoid health studies: the results may be disappointing—often because of methodological issues, not because there is no connection between contamination and health.

Because others are affected by the research, they have the rights and responsibilities to be shaping the work and carrying it out, at least in part. Here's an interesting example of how collaborative work changes science. My research partners and I were prepared to do a random neighborhood sampling frame of households for air monitoring, drawing from all the eligible houses. Our Communities for a Better Environment colleagues noted that in their neighborhood organizing, when they informed residents that the sampling project was going to occur and that it would help residents in dealing with the huge refineries on their fence lines, many of them wanted to volunteer to be part of the sampling. These residents clearly believed they could play a positive role in scientific research that would help their community, in part because they trusted the organization for its past work, including a recent victory regarding refinery flaring that made national headlines.

We had some concern about the validity of a voluntary sampling frame, but we kept thinking about how to make it work. We came up with a method by which volunteers would still be eligible for random selection, and if they were selected in that way, there would be no threat to the validity of the sampling. But we also left it clear that we would keep up to half, twenty, of all households studied as volunteers. In writing up the study design, we would make it clear that this volunteer aspect was a crucial part of the way that community-based participatory research could be done, and perhaps without such volunteer enthusiasm we might not get access at all. We would, of course, analyze the data to see if there were any differences in the demographic characteristics and the data results from volunteer homes versus the results other homes. As it turns out, the volunteer homes are not representative, as we would have liked, and we are pressed to expand random recruitment to achieve the balance we want. But that result does not take away from our impulse to do things in a different way. I thought of this situation as a political equation: "20 + 20 = 100," in that twenty "random" and twenty "volunteer" homes would be a 100 percent solution. Everyone was pleased with the outcome, and it taught us much about how a community organizer can shape the scientific process.

In an average week—quite literally the week just ending on June 23, 2006—my research team and I are involved with a number of community groups and activist organizations. In our breast cancer and environmental justice project with Silent Spring Institute, Communities for a Better Environment, and Brown University, we interview and analyze responses from people who attended community forums we held recently in Cape Cod, Massachusetts, and in Richmond, California. We set up appointments to do air and dust sampling at additional homes in order to add to our knowledge of routine indoor contamination.

The Environmental Neighborhood Awareness Committee of Tiverton (ENACT) in Rhode Island is dealing with massive contamination from municipal gas waste that has led to a building and excavating moratorium. ENACT is an amazing community group that has done so much in a few short years on many different fronts, and it has been a major focus of our attention at Brown. Our team discusses the weekly ENACT meeting that we have just attended, thinking about the key issues currently facing the group. We engage in ongoing efforts to locate other contaminated communities dealing with similar municipal gas waste. We work together editing a brochure on raised-bed gardening and arranging a demonstration project to build one as part of a community-wide learning experience (because of contamination people cannot grow anything in their own soil). We work with state legislatures on housing legislation to provide home-equity loans for Tiverton residents through a state agency because banks will not lend to homeowners whose homes have literally no value on the marketplace. This effort has been successful, generating a bill in July 2006 to provide the nation's first focused home equity loan program for homes on contaminated sites.

With the Woonasquatucket River Watershed Council, we plan for meetings with EPA officials who are putting forth various remediation plans for a Superfund site laden with dioxin, and we work with the Watershed Council to locate technical expertise to make sense of the EPA proposals and to help organize community meetings so that locally affected residents can attend. The council's work is exciting because it combines traditional watershed protection with environmental justice concerns, such as parkland development on contaminated sites in poor neighborhoods populated largely by people of color.

With residents at Mashapaug Pond, we are organizing meetings, producing video documentaries, locating and archiving documents, teaching middle school students, and attending meetings to resist the construction of a public high school on the Gorham site, which is highly contaminated by what was once the nation's largest silverware manufacturer, employing 3,800 workers. This siting issue, following upon the recent siting of another school on contaminated land and with major disregard to neighborhood concern, has generated much activism in Providence for an environmental justice approach to schools and their safety. Activists have generated much engagement by the public, by courts, by legal advocates, as well as by local, state, and federal agencies.

I meet with the Toxics Action Center to coordinate our activities with a number of other groups in Massachusetts, including the ones mentioned earlier, as well as toxics groups in the small blue-collar towns of Alton and

Bradford that are suing textile companies for illegal dumping. We are also working on developing a statewide environmental health and justice organization, a joint effort over the past few years. It is exciting to see the state become a hotbed of toxics awareness activism, with some of this action visible on the national scene; for example, the Providence school siting activists have assisted the Center for Health, Environment, and Justice in a national effort for environmentally safe and clean schools.

With the Rhode Island Department of Environmental Management, we work on a new statewide stakeholders group to achieve better communication and remediation strategies for brownfields in Rhode Island.

I work with colleagues on a project at the Boston University School of Public Health that explores how communities deal with health studies. This week's tasks include editing a brochure for community groups in Louisiana and helping plan a consensus conference on biomonitoring.

Our Katrina research team carries out interviews in New Orleans with neighborhood rebuilding groups and with citywide and statewide environmental health and environmental justice groups in order to help those groups better combine efforts to achieve just and healthy reconstruction.

When I reflect on these efforts, I ask myself, Why can't we all be doing that? Yet at the same time, I have to ask, Why do we have to? Shouldn't the leading government, corporate, and philanthropic agencies and organizations be taking care of occupational and environmental health? Why should the affected communities—especially ones with few resources—always be on the defensive?

All over the United States, academics, scientists, and activists are engaging in similar collaborations to deal with many environmental health problems: asthma and lead in various cities; refinery flaring in Richmond, California; food security in Los Angeles public schools; occupational health for Chinese restaurant workers in New York, Vietnamese floor sanders in Boston, and Brazilian housecleaners in Somerville, Massachusetts; military toxics among Alaska Natives; hog farm waste in North Carolina's black rural areas; uranium mine tailings on Navajo lands; contaminated subsistence fish caught by the Yakama Tribe in Washington State.

Sometimes many of us get together in the same room, as we did at the NIEHS annual grantees conference for environmental justice and community-based participatory research projects. Then the synergy is powerful. We all may think we are familiar with the general issues of corporate and government pollution, with government failures to regulate and remediate, and with heroic efforts by affected people, but we are surprised as we listen to some of our colleagues and are moved to a deeper emotional and intellectual experience.

In the 2005 conference in Alaska, we were hosted in Anchorage and Talkeetna by Alaska Community Action for Toxics and Aleutian-Pribiloff Islands Association. Taking toxics tours from Anchorage to Talkeetna, we heard firsthand stories in small villages of military dumping and of corporate destruction of religious and cultural landscapes. Once at the main conference site, with the grandeur of Denali out the window, we heard vivid testimonies of the lived experience of many tribes of Alaska Natives, including Yupiks from St. Lawrence Island—so far off the coast of Alaska that half of it is underneath Russia. Yupik activists shared their concerns over contamination of fish and mammals that provide their physical and cultural sustenance. They spoke of their lives as subsistence hunters and of their work to avoid and remediate toxic contamination. They shared how they love natural beauty and how their lived experience belies any artificial distinction that others might make between preservation and environmental justice. These Yupiks, speaking about environmental justice organizing and dancing traditional dances in Talkeetna, were at the same time the cousins of my student, also a Yupik, and the activist collaborators of another student, a white Alaskan. This situation brought home the interconnections among us and solidified the point that we all are related somehow, that we struggle together for a safe and healthy world.

By including numerous undergraduate and graduate students in these projects, I join with other faculty members in developing a new corps of publicly engaged, community-based participatory researchers who have learned that personal satisfaction comes when socially conscious work is merged with respectable academic production. Some students are supported on research grants, a number are part of the Contested Illnesses Research Group that Rachel Morello-Frosch and I run, and others are engaged in service learning through courses I teach. We all are involved in a whirlwind of activity and a gratifying mix of teaching, scholarship, community service, and professional development. I work to share with my students the lineage of decades of social movements that have cogenerated with students and scholars an extensive knowledge base that has made the world a better place.

THE GROWING ENVIRONMENTAL HEALTH MOVEMENT

The experiences recounted here feed the growing environmental health movement, a movement that is profoundly concerned with both toxic exposures and environmental justice and that has deep roots in collaborative efforts between lay and expert, community and science. Social movements are not automatic responses, of course, and just because we need one does not mean we get one. Indeed, it is often hard to develop social movements in times of general social reaction because federal grant support is less available for progressive projects, academics may be more fearful of alternative approaches, and various levels of government regulators and legislators are less sympathetic. But in this case we have such a movement, one that is continually growing new offshoots. It would have been inconceivable even five years ago to predict where this movement would evolve, and much of what I have written in the book is testimony to that development. Even for scientists and academics who may not consider themselves formally part of a movement, the inspiration from this movement has altered their perspective.

From lay-expert collaboration, we have learned that research flourishes when it incorporates life science, social science, and natural science. Indeed, the breast cancer and environmental justice study I have mentioned so often is a good example of how we combine the biomonitoring in life science, the community forums and interviews in social science, and the air quality testing of natural science. Such integrated approaches can show scientists that value is created by transcending disciplines. The ethics, values, and advocacy concerns injected by community-based participatory research approaches can help science overall, especially in a period full of conflict-of-interest scandals. More broadly, citizen involvement in science is not just something for science. At present, it is one of the central manifestations of democratic practice *anywhere* in our society, so this burgeoning environmental health movement has effects far beyond its borders.

Academics have much work to do in building this new framework. In such efforts, we constantly need to show students and professionals how they can successfully build a career while also working with community partners for better science *and* for social betterment. We need to bring that message to our departments, other faculty on campus, and professional organizations. Many people are unaware of the nature of collaborations and can be shown their value; others already support such community-academic partnerships, but may not have a thorough appreciation of how to implement them. I put much effort into showing my university the value of such an approach in scientific, productive, financial, and moral or ethical terms. We must inculcate in our students, both undergraduate and graduate, these lessons and give them a new way to look at the world and a vision of what their role might be in bringing positive change to it.

We also need to invite community partners to talk on campuses so they can show professionals what they know and how they can lead. We must also show community partners that we are serious about egalitarian collaboration and that we are not simply using them to build our curricula vitae. We need to push our university research offices and Institutional Review Boards to provide both grant management and Institutional Review Board coverage for partner groups. This effort elicits more than just local benefit; it contributes to a national dialog on emerging forms of community-based and community-defined informed consent and human subjects protection.

Many hundreds of local organizations deal with toxics in neighborhoods, whether from oil refineries, textile mill production, factory farming, incinerators, oil waste injected into deep wells, landfill contamination, radioactive waste dumping, military toxics, and a host of other hazards. They work to remediate sites, prevent future contamination, seek financial settlements, and sometimes even relocation. National resource centers such as the Center for Health, Environment, and Justice and regional ones such as the Toxics Action Center in New England provide technical expertise and organizing assistance to these groups. For those groups working on environmental justice issues, further assistance comes from resource groups such as the Southeast Community Research Center in Atlanta, the Deep South Center for Environmental Justice in New Orleans, and the Southwest Network for Environmental and Economic Justice in Albuquerque.

Some of these groups grow and spread to become citywide organizations that take up larger sets of issues that impact their neighborhoods and the cities and areas around them: development and sprawl, transportation, housing policies and zoning, water resources, public safety, and sanitation. In this expansion of environmental justice to include everything about "where we work, live, and play," activists are integrating radical forms of sustainable development with innovative approaches to environmental justice.

Among the more well-resourced groups, Michael Lerner and his colleagues at Commonweal in northern California provide a unique synthesis of political advocacy, education, clinical training, social support for cancer victims, and spiritual practices. They were among the organizers who started Health Care Without Harm in 1996, a large coalition that works to remove toxics from health care workplaces. Commonweal started the Collaborative on Health and the Environment, now with many hundreds of organizations and even more individual members, which provides educational and research activities in environmental health that have become a national beacon. It is centrally involved with the Campaign for Safe Cosmetics, another national project, which seeks to remove potentially hazardous substances from personal care products. And its Biomonitoring Resource Center plays a central role in the growth of new "lay biomonitoring" activities.

Groups are becoming more forward looking, seeking a new model that can go beyond being reactive to individual problems. More recently, the Toward

Tomorrow Project, run by the Lowell Center for Sustainable Production (which has been a key player in precautionary principle development), began a national project that interviews elders of the environmental and related movements in order to develop a proactive environmental health agenda. In June 2006, with core energy from the Center for Health, Environment, and Justice, the first National Conference on Precaution took place in Baltimore, bringing hundreds of people together to move on new preventive, precautionary pathways. "Green" chemistry advocates and scholars who study clean production are finding ways to show businesses why it is in their best interest to act more safely in the manufacture, transport, and disposal of goods and their by-products. Transit-oriented development advocates are showing local and regional government bodies how to establish cleaner and more humane ways to live and work.

Although it is possible to view all these efforts as part of a growing environmental health movement, it is not always the case that all the players have that understanding of a collective endeavor. To make more linkages, it is necessary to get more groups talking to each other and participating in each other's activities. Otherwise, these linkages remain an unrealized potential. For instance, environmental justice groups increasingly are taking up a precautionary principle perspective because it rings true to them. But it is still necessary for some of the more well-resourced, more academic, and more scientific toxics reduction groups to seek out the environmental justice groups and to ask how they can collaborate. Activists of all types need to think seriously about what kinds of gatherings, rallies, and conferences can be held that will draw together the sometimes diverse groupings into productive conversation.

I am involved in one project where it has been wonderful to connect breast cancer activism and environmental justice activism, two arenas that have historically been very separate. Our partners Silent Spring Institute and Communities for a Better Environment, an environmental justice group, cooperate in a wide variety of research, education, and community organizing. Through active advisory boards in both the Boston and San Francisco Bay areas, this project—appropriately titled Linking Breast Cancer Activism and Environmental Justice—brings together many groups who might not otherwise be involved with each other and cements alliances across historical divides. Our project team worked hard to help WE ACT organize and staff its conference on breast cancer and women of color. Because the Massachusetts Breast Cancer Coalition and Silent Spring Institute are serious about trying to ally with environmental justice organizations and women of color health groups, they have gone to meetings, conferences, and health fairs run by those kinds of groups in order to set up a literature table and to speak with attendees.

We need to build these kinds of linkages if the environmental health movement is to expand in size and influence. Above all, we need to show up for other groups' events, even if there is no ongoing collaboration. If we can view ourselves as part of a larger collectivity, we can *make* that collectivity happen. It is truly an exciting time to be alive and be part of these developments. A clean, safe, democratic, artistic, and enjoyable world is out there waiting for us to create and expand it.

NOTES

PREFACE: TOXIC EXPOSURES AND THE CHALLENGE OF ENVIRONMENTAL HEALTH

1. Phil Brown and Edwin J. Mikkelsen, *No Safe Place: Toxic Waste, Leukemia, and Community Action,* rev. ed. (Berkeley: University of California Press, 1997).
2. Jonathan Harr, *A Civil Action* (New York: Vintage, 1995).
3. Kai Erikson, *Everything in Its Path: Destruction of Community in the Buffalo Creek Flood* (New York: Simon and Schuster, 1976).
4. Adeline Gordon Levine, *Love Canal: Science, Politics, and People* (Lexington, Mass.: Heath, 1982).
5. Michael Edelstein, *Contaminated Communities: The Social and Psychological Impacts of Residential Toxic Exposure* (Boulder, Colo.: Westview, 1988); Henry M. Vyner, *Invisible Trauma: The Psychosocial Effects of the Invisible Environmental Contaminants* (Lexington, Mass.: Lexington Books, 1988).
6. Lee Clarke, *Acceptable Risk? Making Decisions in a Toxic Environment* (Berkeley: University of California Press, 1989); Steve Kroll-Smith and Stephen R. Couch, *The Real Disaster Is above Ground: A Mine Fire and Social Conflict* (Lexington: University Press of Kentucky, 1990); Steven Picou, *Social Disruption and Psychological Stress in an Alaskan Fishing Community: The Impact of the Exxon Valdez Oil Spill* (Boulder: University of Colorado Natural Hazards Center, 1990).
7. Michael Reich, *Toxic Politics: Responding to Chemical Disasters* (Ithaca, N.Y.: Cornell University Press, 1991); Martha Balshem, *Cancer in the Community: Class and Medical Authority* (Washington, D.C.: Smithsonian Institution Press, 1993).
8. Sheldon Krimsky, *Hormonal Chaos: The Scientific and Social Origins of the Environmental Endocrine Hypothesis* (Baltimore: Johns Hopkins University Press, 2000).
9. Sabrina McCormick, Julia Brody, Phil Brown, and Ruth Polk, "Lay Involvement in Breast Cancer Research," *International Journal of Health Services* 34 (2004): 625–46.

1. CITIZEN-SCIENCE ALLIANCES AND HEALTH SOCIAL MOVEMENTS: CONTESTED ILLNESSES AND CHALLENGES TO THE DOMINANT EPIDEMIOLOGICAL PARADIGM

1. Baruch Fischhoff, Paul Slovic, and Sarah Lichtenstein, "Lay Foibles and Expert Fables in Judgments about Risk," Proceedings of the Sixth Symposium on Statistics and the Environment, *The American Statistician (Part 2)* 36 (3) (1982): 240–55.

2. Pew Environmental Health Commission, *America's Environmental Health Gap: Why the Country Needs a Nationwide Health Tracking Network* (Washington, D.C.: Pew Environmental Health Commission, 2000).
3. Allan Brandt, "The Cigarette, Risk, and American Culture," *Daedalus* 119 (4) (1990):155–76; Robert N. Proctor, *Cancer Wars: How Politics Shapes What We Know and Don't Know about Cancer* (New York: Basic Books, 1995); Stanton A. Glantz, *The Cigarette Papers* (Berkeley: University of California Press, 1996).
4. Gerald Markowitz and David Rosner, *Denial and Deceit: The Deadly Politics of Industrial Pollution* (Berkeley: University of California Press, 2002).
5. Michael Greenberg and Daniel Wartenberg, "Communicating to an Alarmed Community about Cancer Clusters: A Fifty State Survey," *Journal of Community Health* 16 (1991): 71–81.
6. Elliot Freidson, *Professional Dominance: The Social Structure of Medical Care* (Chicago: Aldine, 1970).
7. David Gee and Andrew Stirling, "Late Lessons from Early Warnings: Improving Science and Governance under Uncertainty and Ignorance," in *Precaution: Environmental Science and Preventive Public Policy*, ed. Joel Tickner, 195–214 (Washington, D.C.: Island Press, 2003).
8. Paul Brodeur, *Outrageous Misconduct: The Asbestos Industry on Trial* (New York: Pantheon, 1985).
9. Catherine Caufield, *Multiple Exposures: Chronicles of the Radiation Age* (Chicago: University of Chicago Press, 1990); Louise Kaplan, "Public Participation in Nuclear Facility Decisions," in *Science, Technology, and Democracy*, ed. Daniel Kleinman, 67–83 (Albany: State University of New York Press, 2000).
10. Ulrich Beck, *Risk Society: Towards a New Modernity* (London: Sage, 1992).
11. Suzanne M. Snedeker, "Pesticides and Breast Cancer Risk: A Review of DDT, DDE, and Dieldrin," *Environmental Health Perspectives* 109 (suppl. 1) (2001): 35–47.
12. Charles W. Hoge, Carl A. Castro, Stephen C. Messer, Dennis McGurk, Dave I. Cotting, and Robert L. Koffman, "Combat Duty in Iraq and Afghanistan, Mental Health Problems, and Barriers to Care," *New England Journal of Medicine* 351 (2004):13–22.
13. Peter Montague, "San Francisco Adopts the Precautionary Principle," *Rachel's Environment and Health News* 765 (March 20, 2003), available at http://www.rachel.org.
14. David N. Pellow, "Environmental Inequality Formation: Toward a Theory of Environmental Injustice," *American Behavioral Scientist* 43 (2000): 581–601.
15. Laura Pulido, *Environmentalism and Economic Justice: Two Chicano Struggles in the Southwest* (Tucson: University of Arizona Press, 1996), 4–5, 27.
16. Julian Agyeman, *Sustainable Communities and the Challenge of Environmental Justice* (New York: New York University Press, 2005), 11.
17. Sabrina McCormick, Phil Brown, and Stephen Zavestoski, "The Personal Is Scientific, the Scientific Is Political: The Public Paradigm of the Environmental Breast Cancer Movement," *Sociological Forum* 18 (2003): 545–76.
18. Ruthann A. Rudel, David E. Camann, John D. Spengler, Leo R. Korn, and Julia G. Brody, "Phthalates, Alkylphenols, Pesticides, Polybrominated Diphenyl Ethers, and Other Endocrine-Disrupting Compounds in Indoor Air and Dust," *Environmental Science and Technology* 37 (2003): 4543–553.
19. George E. Marcus, "Ethnography in/of the World System: The Emergence of Multi-sited Ethnography," *Annual Review of Anthropology* 24 (1) (1995): 95–118.
20. David Hess, personal communication, 2006; Thomas Kuhn, *The Structure of Scientific Revolution*, 2d ed. (Chicago: University of Chicago Press, 1970).

21. Freidson, *Professional Dominance*; Paul Starr, *The Social Transformation of American Medicine* (New York: Basic, 1988).
22. Rachel Morello-Frosch, Stephen Zavestoski, Phil Brown, Rebecca Gasior Altman, Sabrina McCormick, and Brian Mayer, "Embodied Health Movements: Responses to a 'Scientized' World," in *The New Political Sociology of Science: Institutions, Networks, and Power,* ed. Scott Frickel and Kelly Moore, 244–71 (Madison: University of Wisconsin Press, 2006).
23. Alvin Weinberg, "Science and Transcience," *Minerva* 10 (12) (1972): 209–22.
24. Morello-Frosch et al., "Embodied Health Movements."
25. Samuel S. Epstein, Nicholas A. Ashford, Barry Castleman, Edward Goldsmith, Anthony Mazzocchi, and Quentin D. Young, "The Crisis in U.S. and International Cancer Policy," *International Journal of Health Services* 32 (4) (2002): 669–707.
26. Krimsky, *Hormonal Chaos.*
27. Peter S. Houts, Paul D. Cleary, and Te-Wei Hu, *The Three Mile Island Crisis: Psychological, Social, and Economic Impacts on the Surrounding Population* (University Park: Pennsylvania State University, 1988).
28. Krimsky, *Hormonal Chaos.*
29. Phil Brown, Stephen Zavestoski , Sabrina McCormick, Brian Mayer, Rachel Morello-Frosch, and Rebecca Gasior, "Embodied Health Movements: Uncharted Territory in Social Movement Research," *Sociology of Health and Illness* 26 (2004): 1–31.
30. Anita Silvers, David Wasserman, and Mary Mahowald, *Disability, Difference, Discrimination: Perspectives on Justice in Bioethics and Public Policy* (Lanham, Md.: Rowman and Littlefield, 1998); Doris Fleischer and Frieda Zames, *The Disability Rights Movement: From Charity to Confrontation* (Philadelphia: Temple University Press, 2002); Sandra Morgen, *Into Our Own Hands: The Women's Health Movement in the United States, 1969–1990* (New Brunswick, N.J.: Rutgers University Press, 2002).
31. Brown et al., "Embodied Health Movements."
32. Ibid.
33. Steven Epstein, *Impure Science: AIDS, Activism, and the Politics of Knowledge* (Berkeley: University of California Press, 1996).
34. Ibid.
35. Peter Conrad, "The Experience of Illness: Recent and New Directions," *Research in the Sociology of Health Care* 6 (1987): 1–31; Kathy Charmaz, *Good Days, Bad Days: The Self in Chronic Illness* (New Brunswick, N.J.: Rutgers University Press, 1991); Michael Bury, "Chronic Illness as Biographical Disruption," *Sociology of Health and Illness* 4 (2) (1982): 167–82; Gareth Williams, "The Genesis of Chronic Illness: Narrative Re-construction," *Sociology of Health and Illness* 6 (1984): 176–200.
36. Erikson, *Everything in Its Path*; Balshem, *Cancer in the Community*; Anne Kasper and Susan Ferguson, eds., *Breast Cancer: Society Shapes an Epidemic* (New York: Palgrave, 2000).
37. Maren Klawiter, "Breast Cancer in Two Regimes: The Impact of Social Movements on Illness Experience," *Sociology of Health and Illness* 26 (2004): 347.
38. Brown et al., "Embodied Health Movements."
39. Mark Wolfson, *The Fight Against Big Tobacco: The Movement, the State, and the Public's Health* (Chicago: Aldine, 2001).
40. Susan Leigh Star and James R. Greisemer, "Institutional Ecology, 'Translations,' and Boundary Objects: Amateurs and Professionals in Berkeley's Museum of Vertebrate Zoology, 1907–39," *Social Studies of Science* 19 (1989): 387–420.
41. Rayka Ray, *Fields of Protest: A Comparison of Women's Movements in Two Indian Cities* (Minneapolis: University of Minnesota Press, 1998).

42. Krimsky, *Hormonal Chaos*.
43. Steven Epstein, "U.S. AIDS Activism and the Question of Epistemological Radicalism," paper presented at the Annual Meeting of the Society for the Social Study of Science, Cambridge, Mass., November 2001.
44. Phil Brown, Stephen Zavestoski, Sabrina McCormick, Joshua Mandelbaum, Theo Luebke, and Meadow Linder, "A Gulf of Difference: Disputes over Gulf War–Related Illnesses," *Journal of Health and Social Behavior* 42 (2001): 235–57.
45. U.S. Department of Health and Human Services, *Healthy People 2010: Understanding and Improving Health* (Washington, D.C.: U.S. Department of Health and Human Services, 2001).
46. Lesley Green, Mindy Fullilove, David Evans, and Peggy Shepard, "'Hey, Mom, Thanks!' Use of Focus Groups in the Development of Place-Specific Materials for a Community Environmental Action Campaign," *Environmental Health Perspectives* 110 (suppl. 2) (2002): 265–69.
47. Meredith Minkler and Nina Wallerstein, eds., *Community-Based Participatory Research for Health* (San Francisco: Jossey-Bass, 2003).
48. S. Zierler, personal communication, September 20, 2000.
49. McCormick et al., "Lay Involvement in Breast Cancer Research."
50. Bruce Link and Jo Phelan, "Social Conditions as Fundamental Causes of Disease," *Journal of Health and Social Behavior*, extra issue (1995): 80–94.
51. Steve Wing, "Limit of Epidemiology," *Medicine and Global Survival* 1 (1994): 74–86.
52. Krimsky, *Hormonal Chaos*, 2.
53. Scott Frickel, "Just Science? Organizing Scientist Activism in the US Environmental Justice Movement," *Science as Culture* 13 (2004): 449–69.
54. Kelly Moore, "Organizing Integrity: American Science and the Creation of Public Interest Science Organizations, 1955–1975," *American Journal of Sociology* 101 (1996): 1592–627; Kelly Moore, "Powered by the People: Varieties of Participatory Science as Challenges to Scientific Authority," in *The New Political Sociology of Science: Institutions, Networks, and Power*, ed. Scott Frickel and Kelly Moore, 299–323 (Madison: University of Wisconsin Press, 2006).
55. Marcus, "Ethnography in/of the World System."
56. Michael Burawoy, "Introduction: Reaching for the Global," in *Global Ethnography: Forces, Connections, and Imaginations in a Postmodern World*, ed. Michael Burawoy (Berkeley: University of California Press, 2000), 5.
57. Rayna Rapp, *Testing Women, Testing the Fetus: The Social Impact of Amniocentesis in America* (New York: Routledge, 1999).

2. BREAST CANCER: A POWERFUL MOVEMENT AND A STRUGGLE FOR SCIENCE

1. American Cancer Society (ACS), *Cancer Facts and Figures 2005* (Atlanta: ACS, 2005); S. H. Landis, T. Murray, S. Bolden, and P. A. Wingo, "Cancer Statistics, 1999," *CA: A Cancer Journal for Clinicians* 49 (1) (1999): 8–31.
2. ACS, *Cancer Facts and Figures 2005*.
3. L. A. G. Ries, M. P. Eisner, C. L. Kosary, B. F. Hankey, B. A. Miller, L. Clegg, and B. K. Edwards, eds., *SEER Cancer Statistics Review, 1973–1999* (Bethesda, Md.: National Cancer Institute, 2002).
4. Virginia L. Ernster, John Barclay, Karla Kerlikowske, Deborah Grady, and I. Craig Henderson, "Incidence of and Treatment for Ductal Carcinoma in Situ of the Breast," *Journal of the American Medical Association* 275 (1996): 913–18; Virginia L. Ernster, John Barclay, Karla Kerlikowske, H. Wilkie, and R. Ballard-Barbash, "Mortality

Among Women with Ductal Carcinoma in Situ of the Breast in the Population-Based Surveillance, Epidemiology, and End Results Program," *Archives of Internal Medicine* 160 (7) (2000): 953–58; ACS, *Cancer Facts and Figures 2005*.

5. J. R. Reiss and A. R. Martin, *Breast Cancer 2000: An Update on the Facts, Figures, and Issues* (San Francisco: Breast Cancer Fund, 2000).

6. Paula M. Lantz, Carol S. Weisman, and Zena Itani, "A Disease-Specific Medicaid Expansion for Women: The Breast and Cervical Cancer Prevention and Treatment Act of 2000," *Women's Health Issues* 13 (2003): 79–92.

7. Barbara Brenner, "The Breast Cancer Epidemic," paper presented at the American Public Health Association Conference, Boston, November 13, 2000. My terminology for this movement was discussed with movement actors. They generally agreed that the term *environmental breast cancer movement* is appropriate. However, as evidenced by Brenner's terminology, there is some differentiation in naming the movement, though the meaning and implications are similar.

8. Silent Spring Institute, *Grassroots Breast Cancer Advocacy and the Environment: A Report on Interviews with Grassroots Leaders* (Newton, Mass.: Silent Spring Institute, 2004).

9. A. K. Kant, A. Schatzkin, B. I. Graubard, and C. Schairer, "A Prospective Study of Diet Quality and Mortality in Women," *Journal of the American Medical Association* 283 (16) (2000): 2109–115; H. J. Thompson, "Effect of Amount and Type of Exercise on Experimentally Induced Breast Cancer," *Advances in Experimental Medicine and Biology* 322 (1992): 61–71.

10. Devra Lee Davis and H. Leon Bradlow, "Can Environmental Estrogens Cause Breast Cancer?" *Scientific American* 273 (4) (1995): 166–72.

11. I am grateful to Rachel Morello-Frosch for this formulation.

12. Julia Green Brody and Ruthann Rudel, "Environmental Pollutants and Breast Cancer," *Environmental Health Perspectives* 111 (2003): 1007–19.

13. Joan Fujimura, "Ecologies of Action: Recombining Genes, Molecularizing Cancer, and Transforming Biology," in *Ecologies of Knowledge: Work and Politics in Science and Technology*, ed. Susan Leigh Star, 302–46 (Albany: State University of New York Press, 1995).

14. Devra Davis, *When Smoke Ran Like Water: Tales of Environmental Deception and the Battle Against Pollution* (New York: Basic, 2002).

15. Emily S. Kolker, "Framing as a Cultural Resource in Health Social Movements: Funding Activism and the Breast Cancer Movement in the US 1990–1993," *Sociology of Health and Illness* 26 (2004):820–44.

16. Breast Cancer and the Environment Research Centers, *Early Environmental Exposures* (San Francisco: Breast Cancer and the Environment Research Centers, 2006), available at http://www.bcerc.org/research.htm, retrieved June 11, 2006.

17. Donna J. Haraway, "Situated Knowledges: The Science Question in Feminism and the Privilege of Partial Perspective," *Feminist Studies* 14 (1988): 575–99.

18. Nancy Hartsock, *The Feminist Standpoint Revisited and Other Essays* (Boulder, Colo.: Westview, 1998).

19. Ruth Hubbard, *The Politics of Women's Biology* (New Brunswick, N.J.: Rutgers University Press, 1990), 32.

20. Kay Dickersin and Lauren Schnaper, "Reinventing Medical Research," in *Man-Made Medicine: Women's Health, Public Policy, and Reform*, ed. K. L. Moss, 57–78 (Durham, N.C.: Duke University Press, 1996).

21. Harry Collins, "An Empirical Relativist Program in the Sociology of Scientific Knowledge," in *Science Observed*, ed. Karin Knorr-Cetina and Michael Mulkay, 85–114 (Beverly Hills, Calif.: Sage, 1983).

22. Theo Colborn, Dianne Dumanoski, and John Peterson Myers, *Our Stolen Future: How We Are Threatening Our Fertility, Intelligence, and Survival* (New York: Dutton, 1997).
23. Sandra Harding, *Is Science Multicultural?* (Bloomington: Indiana University Press, 1998).
24. Walter Adams and James W. Brock, *The Tobacco Wars* (Cincinnati: South-Western College Publications, 1999).
25. Grace E. Ziem and Barry I. Castleman, "Threshold Limit Values: Historical Perspectives and Current Practice," *Journal of Occupational and Environmental Medicine* 31 (1989): 910–18.
26. National Research Council, *Environmental Epidemiology: Public Health and Hazardous Wastes* (Washington, D.C.: National Academies Press, 1991).
27. National Research Council, *Use of the Gray Literature and Other Data in Environmental Epidemiology*, vol. 2 of *Environmental Epidemiology* (Washington, D.C.: National Academies Press, 1997).
28. U.S. Department of Energy, GrayLIT Network, 2003, available at http://www.osti.gov/graylit.
29. Brody and Rudel, "Environmental Pollutants and Breast Cancer," 1017.
30. Bruno Latour, *Science in Action: How to Follow Scientists and Engineers Through Society* (Cambridge, Mass.: Harvard University Press, 1987).
31. Krimsky, *Hormonal Chaos;* John Gaventa, "Science and Citizenship in a Global Context," paper presented at the Sussex Institute of Development Studies Conference, Sussex, England, December 12–13, 2002.
32. David Roe, William Pease, Karen Florin, and Ellen Silbergeld, *Toxic Ignorance: The Continuing Absence of Basic Health Testing for Top-Selling Chemicals in the United States* (New York: Environmental Defense Fund, 1997).
33. Joseph Thornton, *Pandora's Poison: Chlorine, Health, and a New Environmental Strategy* (Cambridge, Mass.: MIT Press, 2000).
34. Centers for Disease Control (CDC), *Third National Report on Human Exposures to Environmental Chemicals* (Atlanta: CDC, 2005), available at http://www.cdc.gov/exposurereport/3rd/.
35. Centers for Disease Control (CDC), *Third Report* conference call briefing, July 21, 2005, available through the CDC in Atlanta.
36. Carolyn Raffensperger and Joel Tickner, eds., *Protecting Public Health and the Environment: Implementing the Precautionary Principle* (Washington, D.C.: Island Press, 1999), 8.
37. Sandra Steingraber, *Living Downstream: An Ecologist Looks at Cancer and the Environment* (Reading, Mass.: Perseus Books, 1997).
38. Indeed, the "cancer establishment" (what critics consider the conservative mainstream of the NCI, the ACS, major cancer research centers, and other groups and agencies supporting them) historically characterized mammography as a key tool of "prevention." However, due to protests from many breast cancer activist groups, mammography is now more accurately billed as an early-detection technology. See Samuel S. Epstein, *The Politics of Cancer* (Garden City, N.Y.: Doubleday, 1998).
39. Public Broadcasting System (PBS), *Trade Secrets* (Boston: PBS, 2001), available at http://www.pbs.org/tradesecrets/transcript.html.
40. David Rosner and Gerald E. Markowitz, "From Dust to Dust: The Birth and Rebirth of National Concern About Silicosis," in *Illness and the Environment: A Reader in Contested Medicine,* ed. Stephen Kroll-Smith, Phil Brown, and Valerie Gunter, 167–74 (New York: New York University Press, 2000).

41. Krimsky, *Hormonal Chaos*, 2.
42. Joel Tickner, *European Chemicals Management Initiatives* (Lowell, Mass.: Lowell Center for Sustainable Production, 2000).
43. Colborn, Dumanoski, and Myers, *Our Stolen Future*.
44. Krimsky, *Hormonal Chaos*.
45. Ruthann A. Rudel, Julia Brody, John D. Spengler, Jose Vellarino, Paul W. Geno, and Alice Yau, "Identification of Selected Hormonally Active Agents and Animal Mammary Carcinogens in Commercial and Residential Air and Dust Samples," *Journal of the Air and Water Management Association* 51 (2001): 499–513.
46. Snedeker, "Pesticides and Breast Cancer Risk."
47. Ibid.
48. National Breast Cancer Coalition (NBCC), *Priority #6—Fact Sheet* (Washington, D.C.: NBCC, 2001), available at: http://www.natlbcc.org/, retrieved April 14, 2002.
49. Sylvia Noble Tesh, *Hidden Arguments: Political Ideology and Disease Prevention Policy* (New Brunswick, N.J.: Rutgers University Press, 1988).
50. Ibid.; S. Zierler and N. Krieger, "Reframing Women's Risk: Social Inequalities and HIV Infection," *Annual Review of Public Health* 18 (1997): 401–36.
51. N. Krieger and S. Zierler, "Accounting for the Health of Women," *Current Issues in Public Health* 1 (1995): 251–56.
52. Ibid.
53. Ibid.
54. Geoffrey Rose, "Sick Individuals and Sick Populations," *International Journal of Epidemiology* 14 (1985): 32–38.
55. Ibid.
56. Peter Conrad, "A Mirage of Genes," *Sociology of Health and Illness* 21 (1999): 228–41.
57. Davis and Bradlow, "Can Environmental Estrogens Cause Breast Cancer?"
58. Paul Lichtenstein, Niels Holm, Pia Verkasalo, Anastasia Iliadou, Jaakko Kaprio, Markku Koskenvuo, Eero Pukkala, Axel Skytthe, and Kari Hemminki, "Environmental and Heritable Factors in the Causation of Cancer—Analyses of Cohorts of Twins from Sweden, Denmark, and Finland," *New England Journal of Medicine* 343 (2000): 78–85.
59. Devra Lee Davis and Pamela S. Webster, "The Social Context of Science: Cancer and the Environment," *Annals of the American Academy of Political and Social Science* 584 (2002): 13–34.
60. S. D. Stellman and Q. S. Wang, "Cancer Mortality in Chinese Immigrants to New York City: Comparison with Chinese in Tianjin and with White Americans," *Cancer* 73 (1994): 1270–275.
61. David J. Hunter, Susan Hankinson, Francine Laden, Graham Colditz, JoAnne Manson, Walter Willett, Frank Speizer, and Mary S. Wolff, "Plasma Organochlorine Levels and the Risk of Breast Cancer," *New England Journal of Medicine* 337 (1997): 1253–258.
62. Julia Green Brody, Kirsten B. Moysich, Olivier Humblet, Kathleen R. Attfield, Gregory P. Beehler, and Ruthann Rudel, *Environmental Pollutants and Breast Cancer: Epidemiologic Studies,* report for the Susan G. Komen Breast Cancer Foundation (Newton, Mass.: Silent Spring Institute, 2005).
63. Mary S. Wolff, Paolo G. Toniolo, Eric W. Lee, Marilyn Rivera, and Neil Dublin, "Blood Levels of Organochlorine Residues and Risk of Breast Cancer," *Journal of the National Cancer Institute* 7 (1993): 579–83.
64. Hunter et al., "Plasma Organochlorine Levels and the Risk of Breast Cancer."
65. J. Griffith, R. C. Duncan, W. B. Riggan, and A. C. Pellon, "Cancer Mortality in U.S. Counties with Hazardous Waste Sites and Ground Water Pollution," *Archives of*

Environmental Health 44 (2) (1989): 69–74; G. R. Najem, D. B. Louria, M. A. Lavenhar, and M. Feurman, "Clusters of Cancer Mortality in New Jersey Municipalities with Special Reference to Chemical Toxic Waste Disposal Sites and per Capita Income," *International Journal of Epidemiology* 14 (1985): 528–37.

66. J. Dorgan, John Brock, Nathaniel Rothman, Larry Needham, and Rosetta Miller, "Serum Organochlorine Pesticides and PCBs and Breast Cancer Risk: Results from a Prospective Analysis," *Cancer Causes and Control* 10 (1999): 1–11.
67. A. Hoyer, A. M. Gerdes, T. Jorgensen, F. Rank, and H. Hartvig, "Organochlorines, P53 Mutations in Relations to Breast Cancer Risk and Survival: A Danish Cohort-Nested Case-Control Study," *Breast Cancer Research and Treatment* 71 (2002): 59–65.
68. Marcella Warner, Brenda Eskenazi, Paolo Mocarelli, Pier Mario Gerthoux, Steven Samuels, Larry Needham, Donald Patterson, and Paolo Brambilla, "Serum Dioxin Concentrations and Breast Cancer Risk in the Seveso Women's Health Study," *Environmental Health Perspectives* 110 (2002): 625–28.
69. Brody and Rudel, "Environmental Pollutants and Breast Cancer."
70. Snedeker, "Pesticides and Breast Cancer Risk."
71. Nancy Krieger, "Social Class and the Black/White Crossover in the Age-Specific Incidence of Breast Cancer: A Study Linking Census-Derived Data to Population-Based Registry Records," *American Journal of Epidemiology* 131 (1990): 804–14.
72. R. Millikan, E. DeVoto, E. Fuell, C. K. Tse, J. Beach, S. Edminston, S. Jackson, and B. Newman, "DDE, PCBs, and Breast Cancer Among African-American and White Residents of North Carolina," *Cancer Epidemiology Biomarkers and Prevention* 9 (11) (2000): 1233–240.
73. P. Olaya-Contreras, J. Rodriguez-Villami, H. J. Posso-Valencia, and J. E. Cortez, "Organochlorines and Breast Cancer Risk in Colombian Women," *Cadernos de Saude Publica* 14 (1998): 125–32; Isabelle Romieu, Mauricio Hernandez-Avila, Eduardo Lazcano-Ponce, Jean Phillippe Weber, and Eric Dewailly, "Breast Cancer, Lactation History, and Serum Organochlorines," *American Journal of Epidemiology* 152 (2000): 363–70; Gulnar A. S. Mendonça, José Eluf-Neto, Maria J. Andrada-Serpa, Pedro A. O. Carmo, Heloisa H. C. Barreto, Odete N. K. Inomata, and Tereza A. Kussumi, "Organochlorines and Breast Cancer: A Case-Control Study in Brazil," *International Journal of Cancer* 83 (1999): 596–600; L. Lopez-Carillo, A. Blair, M. Lopez-Cervantes, M. Cebrian, C. Rueda, R. Reyes, A. Mohar, and J. Bravo, "Dichlorodiphenyltrichloroethane Serum Levels and Breast Cancer Risk: A Case Study from Mexico," *Cancer Research* 57 (1997): 3728–732; A. Schecter, P. Toniolo, L. C. Dai, L. T. B. Thuy, and M. S. Wolff, "Blood Levels of DDT and Breast Cancer Risk Among Women Living in the North of Vietnam," *Archives of Environmental Contamination and Toxicology* 33 (1997): 453–56.
74. Ann Aschengrau, Sarah Rogers, and David Ozonoff, "Perchloroethylene-Contaminated Drinking Water and the Risk of Breast Cancer: Additional Results from Cape Cod, Massachusetts, USA," *Environmental Health Perspectives* 111 (2003): 167–73.
75. Y. M. Coyle, L. S. Hynan, D. M. Euhus, and A. T. Minhajuddin, "An Ecological Study of the Association of Environmental Chemicals on Breast Cancer Incidence in Texas," *Breast Cancer Research and Treatment* 92 (2005): 107–14.
76. B. MacMahon, "Pesticide Residues and Breast Cancer?" *Journal of the National Cancer Institute* 86 (1994): 572–73; Hunter et al., "Plasma Organochlorine Levels and the Risk of Breast Cancer"; Stephen H. Safe, "Xenoestrogens and Breast Cancer," *New England Journal of Medicine* 337 (1997): 1303–304.
77. Jonathan M. Samet and Thomas A. Burke, "Turning Science Into Junk: The Tobacco Industry and Passive Smoking," *American Journal of Public Health* 91 (2001): 1742–744;

E. K. Ong and S. Glantz, "Constructing 'Sound Science' and 'Good Epidemiology': Tobacco, Lawyers, and Public Relations Firms," *American Journal of Public Health* 91 (2001): 1749–757.
78. Hunter et al., "Plasma Organochlorine Levels and the Risk of Breast Cancer"; Safe, "Xenoestrogens and Breast Cancer."
79. Marilie D. Gammon, Regina M. Santella, Alfred I. Neugut, Sybil M. Eng, Susan L. Teitelbaum, Andrea Paykin, Bruce Levin, Mary Beth Terry, Tie Lan Young, Lian Wen Wang, Qiao Wang, Julie A. Britton, Mary S. Wolff, Steven D. Stellman, Maureen Hatch, Geoffrey C. Kabat, Ruby Senie, Gail Garbowski, Carla Maffeo, Pat Montalvan, Gertrud Berkowitz, Margaret Kemeny, Marc Citron, Freya Schnabel, Allan Schuss, Steven Hajdu, and Vincent Vinceguerra, "Environmental Toxins and Breast Cancer on Long Island. I. Polycyclic Aromatic Hydrocarbon DNA Adducts," *Cancer Epidemiology Biomarkers and Prevention* 11 (2002): 677–85; Marilie D. Gammon, Mary S. Wolff, Alfred I. Neugut, Sybil M. Eng, Susan L. Teitelbaum, Julie A. Britton, Mary Beth Terry, Bruce Levin, Steven D. Stellman, Geoffrey C. Kabat, Maureen Hatch, Ruby Senie, Gertrud Berkowitz, H. Leon Bradlow, Gail Garbowski, Carla Maffeo, Pat Montalvan, Margaret Kemeny, Marc Citron, Freya Schnabel, Allan Schuss, Steven Hajdu, Vincent Vinceguerra, Nancy Niguidula, Karen Ireland, and Regina M. Santella, "Environmental Toxins and Breast Cancer on Long Island. II. Organochlorine Compound Levels in Blood," *Cancer Epidemiology Biomarkers and Prevention* 11 (2002): 686–97.
80. F. Labreche and M. S. Goldberg, "Exposure to Organic Solvents and Breast Cancer in Women: A Hypothesis," *American Journal of Industrial Medicine* 32 (1997): 1–14.
81. Johnni Hansen, "Breast Cancer Risk among Relatively Young Women Employed in Solvent-Using Industries," *American Journal of Industrial Medicine* 36 (1999): 43–47.
82. Ibid.; John K. Thomas, Bibin Qin, Doris A. Howell, and Barbara A. Richardson, "Environmental Hazards and Rates of Female Breast Cancer Mortality in Texas," *Sociological Spectrum* 21 (2001): 359–75.
83. Brody et al., *Environmental Pollutants and Breast Cancer.*
84. Ibid.
85. Janice H. Platner, L. Michelle Bennet, Robert Millikan, and Mary D. G. Barker, "The Partnership Between Breast Cancer Advocates and Scientists," *Environmental and Molecular Mutagenesis* 39 (2002): 102–7.
86. Kay Dickersin, Lundy Braun, Margaret Mead, Robert Millikan, Anna M. Wu, Jennifer Pietenpol, Susan Troyan, Benjamin Anderson, and Frances Visco, "Development and Implementation of a Science Training Course for Breast Cancer Activists: Project LEAD (Leadership, Education, and Advocacy Development)," *Health Expectations* 4 (4) (2001): 213–20; Platner et al., "The Partnership Between Breast Cancer Advocates and Scientists."
87. Environmental Working Group (EWG), *Body Burden: Pollution in People* (Washington, D.C.: EWG, 2003); available at http://www.ewg.org/reports/bodyburden/.
88. Joseph Thornton, "Biomonitoring of Industrial Pollutants: Health and Policy Implications of the Chemical Body Burden," *Public Health Reports* 117 (2002):315–23.
89. EWG, *Body Burden: Pollution in People.*
90. Julia Green Brody, "No Smoking Gun or Magic Bullet," paper presented at the conference "Speak No Evil, Hear No Evil, See No Cure: Breast Cancer Truth and Consequences," sponsored by Hurricane Voices Breast Cancer Foundation and Massachusetts Breast Cancer Coalition, Waltham, Mass., March 25, 2004.
91. Rudel et al., "Identification of Selected Hormonally Active Agents."
92. Rudel et al., "Phthalates, Alkylphenols, Pesticides."

93. Wendy McKelvey, Julia Green Brody, Ann Aschengrau, and Christopher H. Schwartz, "Association Between Residence on Cape Cod, Massachusetts, and Breast Cancer," *Annals of Epidemiology* 14 (2004): 89–94.
94. Platner et al., "The Partnership Between Breast Cancer Advocates and Scientists"; Patricia Plummer, Susan Jackson, Jamie Konarski, Elizabeth Mahanna, Carolyn Dunsmore, Georgette Regan, Dianne Mattingly, Barbara Parker, Sara Williams, Catherine Andrews, Vani Vannappagari, Susan Hall, Sandra Deming, Elizabeth Hodgson, Patricia Moorman, Beth Newman, and Robert Millikan, "Making Epidemiological Studies Responsive to the Needs of Participants and Communities: The Carolina Breast Cancer Study Experience," *Environmental and Molecular Mutagenesis* 39 (2002): 96–101.
95. Maren Klawiter, "Racing for the Cure, Walking Women, and Toxic Touring: Mapping Cultures of Action within the Bay Area Terrain of Breast Cancer," *Social Problems* 46 (1999): 104–26.
96. Star and Griesemer, "Institutional Ecology, 'Translations,' and Boundary Objects."
97. Robert W. Clarke, Brent Coull, Ulrike Reinisch, Paul Catalano, Cheryl R. Killingsworth, Petros Koutrakis, Ilias Kavouras, Gopala Gazula Krishna Murthy, Joy Lawrence, Eric Lovett, J. Mikhail Wolfson, Richard L. Verrier, and John J. Godleski, "Inhaled Concentrated Ambient Particles Are Associated with Hematologic and Bronchoalveolar Lavage Changes in Canines," *Environmental Health Perspectives* 108 (2000): 1179–187.
98. Ann Aschengrau, David Ozonoff, Patricia Coogan, Richard Vezina, Timothy Heeren, and Yuqzng Zhang, "Cancer Risk and Residential Proximity to Cranberry Cultivation in Massachusetts," *American Journal of Public Health* 86 (1996): 1289–296; Dee West, Sally Glaser, and Angela Prehn, *Status of Breast Cancer Research in the San Francisco Bay Area* (Union City: Northern California Cancer Center, 1998).
99. Gammon, Santella, et al., "Environmental Toxins and Breast Cancer on Long Island. I"; Gammon, Wolff, et al., "Environmental Toxins and Breast Cancer on Long Island. II."
100. Deborah Winn, "The Long Island Breast Cancer Study Project," *Nature* 5 (2005): 986–94.
101. Silent Spring Institute, *The Cape Cod Breast Cancer and Environment Study: Results of the First Three Years of Study* (Newton, Mass.: Silent Spring Institute, 1998).
102. Rudel et al., "Phthalates, Alkylphenols, Pesticides."
103. Jennifer Fishman, "Assessing Breast Cancer: Risk, Science, and Environmental Activism in an 'at Risk' Community," in *Ideologies of Breast Cancer: Feminist Perspectives,* ed. Laura K. Potts, 181–204 (New York: St. Martin's, 2000).
104. Kasper and Ferguson, *Breast Cancer,* 5.
105. Jane S. Zones, "Profits from Pain: The Political Economy of Breast Cancer," in *Breast Cancer: Society Shapes an Epidemic,* ed. Anne S. Kasper and Susan J. Ferguson (New York: St. Martin's, 2000), 119.
106. Susan Orenstein, "The Selling of Breast Cancer: Is Corporate America's Love Affair with a Disease That Kills 40,000 Women a Year Good Marketing—or Bad Medicine?" *Business2.0* (February 2003), available at: http://www.business2.com/.
107. Samantha King, "An All-Consuming Cause: Breast Cancer, Corporate Philanthropy, and the Market for Generosity," *Social Text* 19 (4) (2001): 115–43.
108. Klawiter, "Racing for the Cure."
109. Jennifer R. Myhre, "Breast Cancer Activism: A True Political Movement?" *Breast Cancer Action Newsletter* no. 68 (November–December 2001), available at http://www.bcaction.org/Pages/GetInformed/Newsletters.html.
110. Zones, "Profits from Pain."

111. Carol S. Weisman, *Women's Health Care: Activist Traditions and Institutional Change* (Baltimore: Johns Hopkins University Press, 1998).
112. Sheryl Burt Ruzek, Virginia L. Olesen, and Adele E. Clarke, eds., *Women's Health: Complexities and Differences* (Columbus: Ohio State University Press, 1997); Morgen, *Into Our Own Hands*.
113. Epstein, *The Politics of Cancer*.
114. Nancy Evans, ed., *State of the Evidence: What Is the Connection Between the Environment and Breast Cancer?* 4th ed. (San Francisco: Breast Cancer Fund and Breast Cancer Action, 2004), 12, available at http://www.breastcancerfund.org/environment_evidence_main.htm.

3. ASTHMA, ENVIRONMENTAL FACTORS, AND ENVIRONMENTAL JUSTICE

1. Pew Environmental Health Commission, *America's Environmental Health Gap*.
2. David M. Mannino, David M. Homa, Carol A. Pertowski, Annette Ashizawa, Leah L. Nixon, Carol A. Johnson, Lauren B. Ball, Elizabeth Jack, and David S. Kang, "Surveillance for Asthma—United States 1960–1995," *Morbidity and Mortality Weekly Report* 47 (SS-1) (1998): 5.
3. American Lung Association (ALA), *Estimated Prevalence and Incidence of Lung Disease by Lung Association Territory* (Washington, D.C.: ALA Epidemiology and Statistic Unit Research and Program Services, May 2006), available at http://www.lungusa.org/atf/cf/{7A8D42C2-FCCA-4604-8ADE-7F5D5E762256}/ESTPREV06.PDF.
4. Centers for Disease Control and Prevention (CDC), "Adults Who Have Ever Been Told They Have Asthma," *BRFSS Prevalence Data* (2005), available at apps.nccd.cdc.gov/brfss/display.asp?cat = AS&yr = 2005&qkey = 4417&state = US, retrieved June 23, 2006.
5. American Lung Association (ALA), *Trends in Asthma Morbidity and Mortality* (Washington, D.C.: ALA Epidemiology and Statistic Unit Research and Program Services, May 2005), available at http://www.lungusa.org/atf/cf/{7A8D42C2-FCCA-4604-8ADE-7F5D5E762256}/ASTHMA1.PDF.
6. Ibid.
7. David M. Mannino, David M. Homa, Lara J. Akinbami, Jeanne E. Moorman, Charon Gwynn, and Stephen C. Redd, "Surveillance for Asthma—United States, 1980–1999," *Morbidity and Mortality Weekly Report* 51 (SS-1) (2002): 4; American Lung Association (ALA), "Trends in Asthma Morbidity and Mortality," April 2004, available at the American Lung Association Web site, http://www.lungusa.org.
8. ALA, *Trends in Asthma Mortality*.
9. Mannino et al., "Surveillance for Asthma—United States, 1980–1999."
10. Centers for Disease Control and Prevention (CDC), "Asthma Prevalence and Control Characteristics by Race/Ethnicity—United States, 2002," *Morbidity and Mortality Weekly Report* 53 (7) (2004): 145–48, available at http://www.cdc.gov/mmwr/preview/mmwrhtml/mm5307a1.htm.
11. Mannino et al., "Surveillance for Asthma—United States, 1960–1995."
12. ALA, *Trends in Asthma Mortality*; Pew Environmental Health Commission, *America's Environmental Health Gap*.
13. National Health Interview Survey, *NCHS Data Fact Sheet* (Hyattsville, Md.: National Center for Health Statistics, January 1997); Richard J. Jackson, "Habitat and Health: The Role of Environmental Factors in the Health of Urban Populations," *Journal of Urban Health* 75 (2) (1998): 258–62.

14. Child and Adolescent Health Measurement Initiative, *National Survey of Children's Health* (Portland, Ore.: Child and Adolescent Health Measurement Initiative, 2005), available at the Data Resource Center on Child and Adolescent Health Web site, http://www.nschdata.org, retrieved June 23, 2006.
15. Mannino et al., "Surveillance for Asthma—United States, 1960–1995."
16. National Health Interview Survey, *NCHS Data Fact Sheet*.
17. Pew Environmental Health Commission, *America's Environmental Health Gap*.
18. ALA, *Trends in Asthma Mortality*.
19. S. T. Weiss, "Environmental Risk Factors in Childhood Asthma," *Clinical and Experimental Allergy* 28 (suppl. 5) (1998): 29–34.
20. L. Claudio, L. Tulton, J. Doucette, and P. J. Landrigan, "Socioeconomic Factors and Asthma Hospitalization Rates in New York City," *Journal of Asthma* 36 (4) (1999): 343–50.
21. ALA, *Trends in Asthma Mortality*.
22. ALA, "Trends in Asthma Morbidity and Mortality."
23. U.S. Department of Health and Human Services, *Healthy People 2010*.
24. American Lung Association (ALA), *Breathless: Air Pollution and Hospital Admissions/Emergency Room Visits in 13 Cities* (Washington, D.C.: ALA, 1996).
25. Quoted in Charles F. Bostwick, "Child Asthma Rises in Los Angeles Area with Growing Exhaust Pollution," *Los Angeles Daily News*, February 23, 2004.
26. Laura Senier, Brian Mayer, and Phil Brown, "Report to Massachusetts Committee on Occupational Safety and Health and Boston Urban Asthma Coalition" (unpublished manuscript, 2005); Tolle Graham, Jean Zotter, and Marlene Camacho, *Who's Sick at School: Linking Poor School Conditions and Health Disparities for Boston's Children* (Boston: Massachusetts Coalition for Occupational Safety and Health and the Boston Urban Asthma Coalition, 2006).
27. Karen Hsu, "Boston Will Get $1.9 Million to Prevent Asthma," *Boston Globe*, February 24, 2000.
28. Davis, *When Smoke Ran Like Water*, 104–6, 120–22.
29. Pew Environmental Health Commission, *American's Environmental Health Gap*.
30. Mary Amdur, "Animal Toxicology," in *Particles in Our Air: Concentrations and Health Effects*, ed. Richard Wilson and John Spengler, 85–122 (Cambridge, Mass.: Harvard University Press, 1996).
31. Richard Wilson, "Introduction," in *Particles in Our Air: Concentrations and Health Effects*, ed. Richard Wilson and John Spengler, 1–14 (Cambridge, Mass.: Harvard University Press, 1996).
32. Francine Laden, Lucas M. Neas, Douglas W. Dockery, and Joel Schwartz, "Association of Fine Particulate Matter from Different Sources with Daily Mortality in Six U.S. Cities," *Environmental Health Perspectives* 108 (10) (2000): 941–47.
33. Douglas Dockery, Clyde Pope, X. Xu, Jack Spengler, John Ware, M. Ray, B. Ferris, and F. Speitzer, "An Association Between Air Pollution and Mortality in Six US Cities," *New England Journal of Medicine* 329 (1993): 1753–759.
34. Joel Schwartz, "Fine Particulate Air Pollution: Smoke and Mirrors of the '90s or Hazard of the New Millennium," paper presented at the Annual Meeting of the American Public Health Association, Boston, November 14, 2000.
35. Ibid.
36. Ibid.; Diane R. Gold, Augusto Litonjua, Joel Schwartz, Eric Lovett, Andrew Larson, Bruce Nearing, George Allen, Monique Verrier, Rebecca Cherry, and Richard Verrier, "Ambient Pollution and Heart Rate Variability," *Circulation* 101 (2000): 1267–273.

37. Scott L. Zeger, Francesca Dominici, and Jonathan Samet, "Harvesting-Resistant Estimates of Air Pollution Effects on Mortality," *Epidemiology* 10 (1999): 171–75.
38. C. Arden Pope, Richard T. Burnett, Michael J. Thun, Eugenia E. Calle, Daniel Krewski, Kazuhiko Ito, and George D. Thurston, "Lung Cancer, Cardiopulmonary Mortality, and Long-Term Exposure to Fine Particulate Air Pollution," *Journal of the American Medical Association* 287 (2002): 1132–141.
39. John Godleski, "Mechanisms of Particulate Air Pollution Health Effects," paper presented at the Annual Meeting of the American Public Health Association, Boston, November 14, 2000; Clark et al., "Inhaled Concentrated Ambient Particles."
40. Schwartz, "Fine Particulate Air Pollution."
41. Pope et al., "Lung Cancer."
42. C. Arden Pope, Richard T. Burnett, George Thurston, Michael J. Thun, Eugenia E. Calle, Daniel Krewski, and John J. Goldleski, "Cardiovascular Mortality and Long-Term Exposure to Particulate Air Pollution: Epidemiological Evidence of General Pathophysiological Pathways of Disease," *Circulation* 109 (2004): 71–77.
43. Michael Jerrett, Richard T. Burnett, Renjun Ma, C. Arden Pope, Daniel Krewski, Bruce K. Newbold, George Thurston, Yuanli Shi, Norm Finkelstein, Eugenia E. Calle, and Michael J. Thun, "Spatial Analysis of Air Pollution and Mortality in Los Angeles," *Epidemiology* 16 (6) (2005): 727–36.
44. G. Hoek, B. Brunekreef, S. Goldbohm, P. Fischer, and P. Van Den Brandt, "Association Between Mortality and Indicators of Traffic-Related Air Pollution in the Netherlands: A Cohort Study," *The Lancet* 360 (2002): 1203–209.
45. Jonathan M. Samet, Francesa Dominici, Frank C. Curriero, Ivan Coursac, and Scott L. Zeger, "Fine Particulate Air Pollution and Mortality in 20 U.S. Cities, 1987-1994," *New England Journal of Medicine* 343 (2000): 1724–729.
46. C. Arden Pope, "Respiratory Disease Associated with Community Air Pollution and a Steel Mill, Utah Valley," *American Journal of Public Health* 79 (1989): 623–28.
47. Michael S. Friedman, Kenneth E. Powell, Lori Hutwagner, LeRoy M. Graham, and W. Gerald Teague, "Impact of Changes in Transportation and Commuting Behaviors during the 1996 Summer Olympic Games in Atlanta on Air Quality and Childhood Asthma," *Journal of the American Medical Association* 285 (2001): 897–905.
48. Joel Schwartz, D. Slater, T. V. Larson, W. E. Pierson, and J. Q. Koenig, "Particulate Air Pollution and Hospital Emergency Visits for Asthma in Seattle," *American Review of Respiratory Disease* 147 (1993): 826–31.
49. Douglas Dockery and Arden Pope, "Epidemiology of Acute Health Effects: Summary of Times Series Studies," 123–36, and C. Arden Pope and Douglas Dockery, "Epidemiology of Chronic Health Effects: Cross-Sectional Studies," 136–48, both in *Particles in Our Air: Concentrations and Health Effects,* ed. Richard Wilson and John Spengler (Cambridge, Mass.: Harvard University Press, 1996).
50. Robert J. Pandya, Gina Solomon, Amy Kinner, and John R. Balme, "Diesel Exhaust and Asthma: Hypotheses and Molecular Mechanisms of Action," *Environmental Health Perspectives* 110 (suppl. 1) (2002): 103–12.
51. Shao Lin, Jean Pierre Munsie, Syni-An Hwand, Edward Fitzgerald, and Michael R. Cayo, "Childhood Asthma Hospitalization and Residential Exposure to State Route Traffic," *Environmental Research* 88 (2) (2002): 73–81.
52. Rob McConnell, Kiros Berhane, Frank Gilliland, Stephanie J. London, Talat Islam, W. James Gauderman, Edward Avol, Helene G. Margolis, and John M. Peters, "Asthma in Exercising Children Exposed to Ozone: A Cohort Study," *The Lancet* 359 (2002): 386–91.

53. Philip J. Landrigan, Clyde B. Schecter, Jeffrey M. Lipton, Marianne C. Fahs, and Joel Schwartz, "Environmental Pollutants and Disease in American Children: Estimates of Morbidity, Mortality, and Costs for Lead Poisoning, Asthma, Cancer, and Developmental Disabilities," *Environmental Health Perspectives* 110(7) (2002):721–28.
54. Conrad Schneider, *Death, Disease, and Dirty Power: Mortality and Health Damage due to Air Pollution from Power Plants* (Boston: Clean Air Task Force, 2000).
55. Clean Air Task Force, *Diesel and Health in America: The Lingering Threat* (Boston: Clean Air Task Force, 2005), available at http://www.catf.us/publications/view.php?id = 83, retrieved February 22, 2005.
56. Pope et al., "Cardiovascular Mortality."
57. Gerald J. Keeler, Timothy J. Dvonch, Fuyen Yip, Edith A. Parker, Barbara A. Israel, Frank J. Marsik, Masako Morishita, James A. Barres, Thomas G. Robins, Wilma Brakefield-Caldwell, and Mathew Sam, "Assessment of Personal and Community-Level Exposures to Particulate Matter (PM) among Children with Asthma in Detroit, Michigan, as Part of Community Action Against Asthma (CAAA)," *Environmental Health Perspectives* 110 (suppl. 2) (2002): 173–81.
58. Jason Corburn, "Combining Community-Based Research and Local Knowledge to Confront Asthma and Subsistence-Fishing Hazards in Greenpoint/Williamsburg, Brooklyn, New York," *Environmental Health Perspectives* 110 (suppl. 2) (2002): 241–48; Jason Corburn, *Street Science: Community Knowledge and Environmental Health Justice* (Cambridge, Mass.: MIT Press, 2005).
59. Ross Gelbspan, *The Heat Is On: The Climate Crisis, the Cover-Up, the Prescription* (Cambridge, Mass.: Perseus Books, 1997).
60. Davis, *When Smoke Ran Like Water*.
61. Daniel S. Greenbaum, John D. Bachmann, Daniel Krewski, Jonathan M. Samet, Ronald White, and Ronald E. Wyzga, "Particulate Air Pollution Standards and Morbidity and Mortality: Case Study," *American Journal of Epidemiology* 154 (2001): S78–S90.
62. L. Johannes, "Pollution Study Sparks Debate Over Secret Data," *Wall Street Journal*, April 2, 1997.
63. Greenbaum et al., "Particulate Air Pollution Standards."
64. Daniel Krewski, Richard T. Burnett, Mark S. Goldberg, B. Kristin Hoover, Jack Siemiatycki, Michael Jerrett, Michal Abrahamowicz, and Warren H. White, *Reanalysis of the Harvard Six Cities Study and the American Cancer Society Study of Particulate Air Pollution and Mortality* (Cambridge, Mass.: Health Effects Institute, 2000).
65. Greenbaum et al., "Particulate Air Pollution Standards."
66. Rebecca Klemm, Robert Mason Jr., Charles Heilig, Lucas Neas, and Douglas Dockery, "Is Daily Mortality Associated Specifically with Fine Particles? Data Reconstruction and Replication of Analysis," *Journal of the Air Waste Management Association* 50 (2000): 1215–222.
67. Greenbaum et al., "Particulate Air Pollution Standards."
68. *Whitman v. American Trucking Associations*, no. 99-1257, 175 F.3d 1027 and 195 F.3d 4 (D.C. Circuit 1999).
69. Carol Browner, Remarks Made to the U.S. Senate Environment and Public Works Subcommittee on Clean Air, May 20, 1999, available at http://www.epa.gov/ttn/oarpg/gen/cmbtest.html#remarks.
70. Greenbaum et al., "Particulate Air Pollution Standards."
71. Robert Weiss, "HHS Seeks Science Advice to Match Bush View," *Washington Post*, September 17, 2002.
72. Judy Pasternak, "Bush's Energy Plan Bares Industry Clout," *Los Angeles Times*, August 26, 2001.

73. Jeremy Symons, "How Bush and Co. Obscure the Science," *Washington Post,* July 13, 2003.
74. National Research Council, *Interim Report of the Committee on Changes in New Source Review Programs for Stationary Sources of Air Pollutants* (Washington, D.C.: National Academies Press, 2005), available at http://darwin.nap.edu/books/0309095786/html/R1.html.
75. Elizabeth Shogren, "EPA Drops Its Case Against Dozens of Alleged Polluters," *Los Angeles Times,* November 6, 2003, available at http://www.commondreams.org/headlines03/1106-01.htm.
76. Richard A. Oppel Jr. and Christopher Drew, "States Planning Own Lawsuits Over Pollution," *New York Times,* November 9, 2003.
77. National Research Council, *Interim Report.*
78. Greenwire, "Air Pollution: Supreme Court's Interest Expected to Shake Up NSR Debate," *New Source Review: An E&E Special Report* (May 16, 2006), available at http://www.eenews.net/Greenwire/Backissues/images/061402gwr1.pdf.
79. Union of Concerned Scientists, *Scientific Integrity in Policymaking* (Cambridge, Mass.: Union of Concerned Scientists, 2004), available at http://www.ucsusa.org/global_environment/rsi/page.cfm?pageID = 1363, retrieved June 7, 2005.
80. Peter Montague, "The Revolution, Part 3: Ultrafines," *Rachel's Environmental and Health News* 774 (July 24, 2003), available at http://www.rachel.org.
81. Dorceta E. Taylor, "The Rise of the Environmental Justice Paradigm: Injustice Framing and the Social Construction of Environmental Discourse," *American Behavioral Scientist* 43 (2000): 508–80.
82. Williams, "The Genesis of Chronic Illness."
83. National Institute of Environmental Health Sciences (NIEHS), *Environmental Justice—Archived Grantees* (Research Triangle Park, N.C.: NIEHS, 2005), available at http://www.niehs.nih.gov/translat/envjust/grantold.htm, retrieved June 7, 2005.
84. National Institute of Environmental Health Sciences (NIEHS), *Healthy Food, Healthy School, and Healthy Community* (Research Triangle Park, N.C.: NIEHS, 2005), available at http://www.niehs.nih.gov/translat/envjust/grantold.htm, retrieved June 8, 2005.
85. National Institute of Environmental Health Sciences (NIEHS), *Community Outreach for CTD Screening in High Risk Groups* (Research Triangle Park, N.C.: NIEHS, 2005), available at http://www.niehs.nih.gov/translat/envjust/projects/fraser.htm, retrieved June 11, 2005; Community-Based Participatory Research (CBPR) Institute for Biomedical Research, *Patricia Fraser, M.D.* (Boston: CBPR Institute for Biomedical Research, 2004), available at http://www.cbrinstitute.org/page.php?branch=pi&page=fraser&type=4, retrieved June 13, 2005.
86. Bob Weinhold, "Fuel for the Long Haul: Diesel in America," *Environmental Health Perspectives* 110 (2002): A458–A464.
87. West Harlem Environmental Action, Inc. (WE ACT), 2003. *WEACT History* (New York: WE ACT, 2003), available at http://www.weact.org/history.html, retrieved June 7, 2005.
88. Simon J. Williams, "Chronic Respiratory Illness and Disability: A Critical Review of the Psychosocial Literature," *Social Science and Medicine* 28 (1989): 791–903.
89. Jackie Joyner-Kersee, "Asthma and the Athlete's Challenge," *New York Times,* August 13, 2001.
90. Kirsten Rudestam, Phil Brown, Christina Zarcadoolas, and Catherine Mansell, "Children's Asthma Experience and the Importance of Place," *Health* 8 (4) (2004): 423–44.
91. Rebecca Center, "Poverty, Ethnicity, and Pediatric Asthma in Rhode Island," senior honors thesis, Center for Environmental Studies, Brown University, 2000.

92. Frederica P. Perera, Susan M. Illman, Patrick L. Kinney, Robin M. Whyatt, Elizabeth A. Kelvin, Peggy Shepard, David Evans, Mindy Fullilove, Jean Ford, Rachel L. Miller, Ilan H. Meyer, and Virginia A. Rauh, "The Challenge of Preventing Environmentally Related Disease in Young Children: Community-Based Research in New York City," *Environmental Health Perspectives* 110 (2002): 197–205.
93. Jonathan I. Levy, E. Andres Houseman, John D. Spengler, Penn Loh, and Louise Ryan, "Fine Particulate Matter and Polycyclic Hydrocarbon Concentration Patterns in Roxbury, Massachusetts: A Community-Based GIS Analysis," *Environmental Health Perspectives* 109 (2001): 341–47.
94. Peggy M. Shepard, Mary E. Northridge, Swati Prakash, and Gabriel Stover, "Preface: Advancing Environmental Justice Through Community-Based Participatory Action Research," *Environmental Health Perspectives* 110 (2002): 139–44; David Evans, Mindy Fullilove, Lesley Green, and Moshe Levison, "Awareness of Environmental Risks and Protective Actions Among Minority Women in Northern Manhattan," *Environmental Health Perspectives* 110 (suppl. 2) (2002): 271–75; Green et al., "'Hey, Mom, Thanks!'"; Penn Loh, Jodi Sugerman-Brozan, Standrick Wiggins, David Noiles, and Cecelia Archibald, "From Asthma to AirBeat: Community-Driven Monitoring of Fine Particles and Black Carbon in Roxbury, Massachusetts," *Environmental Health Perspectives* 110 (suppl. 2) (2002): 297–301.
95. ExxonMobil, "Clearing the Air on Asthma," advertisement, *New York Times,* November 15, 2001.
96. Asthma and Allergy Foundation of America, New England Chapter, *Asthma-Friendly Child Care: A Checklist for Parents and Providers* (Washington, D.C.: Asthma and Allergy Foundation of America, 2000).

4. GULF WAR–RELATED ILLNESSES AND THE HUNT FOR CAUSATION: THE "STRESS OF WAR" VERSUS THE "DIRTY BATTLEFIELD"

1. G. C. Gray, J. D. Knoke, S. W. Berg, F. S. Wignall, and E. Barrett-Connor, "Counterpoint: Responding to Suppositions and Misunderstandings," *American Journal of Epidemiology* 148 (1998):328–33; W. C. Reeves, K. Fukuda, R. Nisenbaum, and W. W. Thompson, "Letters: Chronic Multisystem Illness Among Gulf War Veterans," *Journal of the American Medical Association* 282 (1999): 327–29; National Institutes of Health (NIH), "The Persian Gulf Experience and Health," workshop statement, Technology Assessment Workshop, Washington, D.C., April 27–29, 1994; Institute of Medicine, *Health Consequences of Service during the Persian Gulf War: Recommendations for Research and Information Systems* (Washington, D.C.: National Academies Press, 1996); Lois Joellenback, Philip Russell, and Samuel B. Guze, eds., *Strategies to Protect the Health of Deployed U.S. Forces: Medical Surveillance, Record Keeping, and Risk Reduction* (Washington, D.C.: National Academies Press for the Institute of Medicine, 1999).
2. Carolyn E. Fulco, Catharyn T. Liverman, and Harold Sox, eds., *Depleted Uranium, Sarin, Pyridostigmine Bromide, and Vaccines,* vol. 1 of *Gulf War and Health* (Washington, D.C.: National Academies Press for the Institute of Medicine, 2000); D. J. Clauw, "The 'Gulf War Syndrome': Implications for Rheumatologists," *Journal of Clinical Rheumatology* 4 (4) (1998), 173–74.
3. Persian Gulf Veterans Coordinating Board, Research Working Group, *Research on Gulf War Veterans' Illnesses, Annual Report to Congress—1998* (Washington, D.C.: Persian Gulf Veterans Coordinating Board, 1999), available at http://www.va.gov/resdev/pgulf98/gwrpt98.htm, retrieved on October 1, 2000; U.S. General Accounting Office (GAO), Department of Veterans Affairs, *Report to the Chairman, Subcommittee on*

National Security, Emerging Threats, and International Relations, Committee on Government Reform, House of Representatives: Federal Gulf War Illnesses Research Strategy Needs Reassessment (Washington, D.C.: U.S. GAO, 2004).
4. Fulco, Liverman, and Sox, Depleted Uranium; Charles C. Engel, Xian Liu, Roy Clymer, Ronald F. Miller, Terry Sjoberg, and Jay R. Shapiro, "Rehabilitative Care of War-Related Health Concerns," Journal of Occupational and Environmental Medicine 42 (2000): 385–90.
5. Nortin M. Hadler, "If You Have to Prove You Are Ill, You Can't Get Well: The Object Lesson of Fibromyalgia," Spine 21 (20) (1996): 2397–400.
6. U.S. Department of Veterans Affairs, VA Creates Gulf War Advisory Committee (Washington, D.C.: U.S. Department of Veterans Affairs, January 23, 2002), available at http://www1.va.gov/rac-gwvi/docs/Pressrelease_ VA Creates Gulf War Advisory Committee_Jan2002.doc, retrieved July 21, 2005.
7. Philip Landrigan, "Illnesses in Gulf War Veterans: Causes and Consequences," Journal of the American Medical Association 277 (1997): 259–61.
8. Miriam Davis, "Gulf War Illnesses and Recognizing New Diseases," in Depleted Uranium, Sarin, Pyridostigmine Bromide, and Vaccines, vol. 1 of Gulf War and Health, ed. Carolyn E. Fulco, Catharyn T. Liverman, and Harold Sox, appendix D, 342–65 (Washington, D.C.: National Academies Press for the Institute of Medicine, 2000); NIH, "The Persian Gulf Experience and Health."
9. Institute of Medicine, Committee on Gulf War and Health, Fuels, Combustion Products, and Propellants, vol. 3 of Gulf War and Health (Washington, D.C.: National Academies Press, 2005).
10. Richard A Rettig, Military Use of Drugs Not Yet Approved by the FDA for CW/BW Defense (Santa Monica, Calif.: RAND, 1999).
11. U.S. Army Environmental Hygiene Agency, Final Report: Kuwait Oil Fire Health Risk Assessment, Report no. 39-26-L192-91, May 5–December 3, 1991 (Washington, D.C.: U.S. Department of Defense, 1994).
12. Jessica Wolfe, Darin J. Erickson, Erica J. Sharkansky, Daniel W. King, and Lynda A. King, "Course and Predictors of Posttraumatic Stress Disorder among Gulf War Veterans: A Prospective Analysis," Journal of Consulting and Clinical Psychology 67 (1999): 520–28.
13. NIH, "The Persian Gulf Experience and Health," 12.
14. Presidential Advisory Committee on Gulf War Veterans' Illnesses, Presidential Advisory Committee on Gulf War Veterans' Illnesses: Final Report (Washington, D.C.: U.S. Government Printing Office, December 1996).
15. Persian Gulf Veterans Coordinating Board, Research on Gulf War Veterans' Illnesses.
16. Robert Haley, Thomas Kurt, and Jim Hom, "Is There a Gulf War Syndrome? Searching for Syndromes by Factor Analysis of Symptoms," Journal of the American Medical Association 277 (1997): 215–22.
17. Han Kang and Tim Bullman, "Mortality Among U.S. Veterans of the Persian Gulf War," New England Journal of Medicine 335 (1996): 1498–504.
18. Gregory Gray, Bruce Coate, Christy Anderson, Han Kang, S. William Berg, Stephen Wignall, James Knoke, and Elizabeth Barret-Conner, "The Postwar Hospitalization Experience of U.S. Veterans of the Persian Gulf War," New England Journal of Medicine 335 (1996): 1505–513.
19. Philip Shenon, "Panel Disputes Studies on Gulf War Illnesses," New York Times, November 21, 1996.
20. Gina Kolata, "No Rise Found in Death Rates After Gulf War," New York Times, November 14, 1996.
21. Iowa Persian Gulf Study Group, "Self-reported Illness and Health Status Among Gulf War Veterans," Journal of the American Medical Association 277 (1997): 238–45; K.

Fukuda, R. Nisenbaum, G. Stewart, W. W. Thompson, L. Robin, R. M. Washko, D. L. Noah, D. H. Barrett, B. Randall, B. L. Herwaldt, A. C. Mawle, and W. C. Reeves, "Chronic Multisymptom Illness Affecting Air Force Veterans of the Gulf War," *Journal of the American Medical Association* 280 (1998): 981–88; S. P. Proctor, T. Heeren, R. F. White, J. Wolfe, M. S. Borgos, J. D. Davis, L. Pepper, R. Clapp, P. B. Sutker, J. J. Vasterling, and D. Ozonoff, "Health Status of Persian Gulf War Veterans: Self-reported Symptoms, Environmental Exposures, and the Effect of Stress," *International Journal of Epidemiology* 27 (1998): 1000–1010.

22. Proctor et al., "Health Status of Persian Gulf War Veterans."
23. Kimberly Sullivan, Maxine Krengel, Susan P. Proctor, Sherral Devine, Timothy Heeren, and Roberta F. White, "Cognitive Functioning in Treatment-Seeking Gulf War Veterans: Pyridostigmine Bromide Use and PTSD," *Journal of Psychopathology and Behavioral Assessment* 25 (2) (2003): 95–103.
24. Rogene F. Henderson, Edward B. Barr, Walter B. Blackwell, Connie R. Clark, Carole A. Conn, Roma Kalra, Thomas H. March, Mohan L. Sopori, Yohannes Tesfaigzi, Margaret G. Ménache, and Deborah C. Mash, "Response of Rats to Low Levels of Sarin," *Journal of Toxicology and Applied Pharmacology* 184 (2002): 67–76.; A. W. Abu-Qare and M. B. Abou-Donia, "Sarin: Health Effects, Metabolism, and Methods of Analysis," *Food and Chemical Toxicology* 40 (2002): 1327–333; K. Husain, R. Vijayaraghavan, S. C. Pant, S. K. Raza, and K. S. Pandey, "Delayed Neurotoxic Effect of Sarin in Mice After Repeated Inhalation Exposure," *Journal of Applied Toxicology* 13 (1993): 143–45.
25. Bharti Mackness, Paul N. Durrington, and Michael I. Mackness, "Low Paraoxonase in Persian Gulf War Veterans Self-Reporting Gulf War Syndrome," *Biochemical and Biophysical Research Communications* 276 (2000): 729–33; Robert Haley, Scott Billecke, and Bert La Du, "Association of Low PON1 Type Q (Type A) Arylesterase Activity with Neurological Symptom Complexes in Gulf War Veterans," *Toxicology and Applied Pharmacology* 157 (1999): 227–33; Haley, Kurt, and Hom, "Is There a Gulf War Syndrome?"
26. Robert Haley, James Fleckenstein, W. Wesley Marshall, George McDonald, Gerald Kramer, and Frederick Petty, "Effect of Basal Ganglia Injury on Central Dopamine Activity in Gulf War Syndrome: Correlation of Proton Magnetic Resonance Spectroscopy and Plasma Homovanillic Acid," *Archives of Neurology* 57 (2000): 1280–285; Robert Haley, W. Wesley Marshall, George McDonald, Mark Daugherty, Frederick Petty, and James Fleckenstein, "Brain Abnormalities in Gulf War Syndrome: Evaluation by H Magnetic Resonance Spectroscopy," *Radiology* 215 (2000): 807–17; D. J. Meyerhoff, J. Lindgren, D. Hardin, J. M. Griffis, and M. W. Weiner, "Reduced N-Acetylaspartate in the Right Basal Ganglia of Ill Gulf War Veterans by Magnetic Resonance Spectroscopy," *Proceedings of the International Society of Magnetic Resonance Medicine* 9 (2001): 994.
27. James Meikle, "U.S. Scientist Challenges UK on Gulf War Illness," *The Guardian*, August 4, 2004.
28. Maria Rosario G. Araneta, Karen M. Schlangen, Larry D. Edmonds, Daniel A. Destiche, Ruth D. Merz, Charlotte A. Hobbs, Timothy J. Flood, John A. Harris, Diane Krishnamurti, and Gregory C. Gray, "Prevalence of Birth Defects Among Infants of Gulf War Veterans in Arkansas, Arizona, California, Georgia, Hawaii, and Iowa, 1989–1993," *Birth Defects Research (Part A)* 67 (2002): 246–60.
29. Susan P. Proctor, Sucharita Gopal, Asuka Imai, Jessica Wolfe, David Ozonoff, and Roberta F. White, "Spatial Analysis of 1991 Gulf War Troop Locations in Relationship with Postwar Health Symptom Reports Using GIS Techniques," *Transactions in GIS* 9 (3) (2005): 386.

30. Simon Wessely, "A Controlled Epidemiological and Clinical Study Into the Effect of Gulf War Service on Servicemen and Women of the United Kingdom Armed Forces," 2000, available at http://www.csa.com.
31. Gulf War Illness Advisory Committee, Department of National Defence, *Health Study of Canadian Forces Personnel Involved in the 1991 Conflict in the Persian Gulf*, vol. 1 (Ottawa, Ontario: Goss Gilroy, 1998).
32. Simon Wessely, "Ten Years On: What Do We Know About the Gulf War Syndrome?" *Clinical Medicine* 1 (2001): 1–10.
33. R. Salamon, C. Verret, M. A. Jutand, M. Begassat, F. Laoudj, F. Conso, and P. Brochard, "Health Consequences of the First Persian Gulf War on French Troops," *International Journal of Epidemiology* 35 (2) (2006): 479–87.
34. Gregory Gray, Gary Gackstetter, Han Kang, John Graham, and Ken Scott, "After More Than 10 Years of Gulf War Veteran Medical Evaluations, What Have We Learned?" *American Journal of Preventive Medicine* 26 (2004): 443–52.
35. Institute of Medicine, *Health Effects of Serving in the Gulf War*, vol. 4 of *Gulf War and Health* (Washington, D.C.: National Academies Press, 2006).
36. Roberta White, "Service in the Gulf War and Significant Health Problems: Focus on the Central Nervous System," *Journal of Psychopathology and Behavioral Assessment* 25 (2) (2003): 77–83.
37. U.S. General Accounting Office (GAO), *Gulf War Illnesses: Federal Research Efforts Have Waned, and Research Findings Have Not Been Reassessed*, GAO-04-815T (Washington, D.C.: U.S. GAO, June 1, 2004); U.S. General Accounting Office (GAO), *Department of Veterans Affairs: Federal Gulf War Illnesses Research Needs Reassessment*, GAO-04-767 (Washington, D.C.: U.S. GAO, June 1, 2004).
38. Kenneth Hyams, Stephen Wignall, and Robert Roswell, "War Syndromes and Their Evaluation: From the U.S. Civil War to the Persian Gulf War," *Annals of Internal Medicine* 125 (1996): 298–305.
39. NIH, "The Persian Gulf Experience and Health."
40. Joellenback, Russell, and Guze, *Strategies to Protect the Health of Deployed U.S. Forces*.
41. See ibid. for the Institute of Medicine report; the information here comes from an interview with a researcher rather than from consulting either the DOD report or the Millennium Cohort Study.
42. Philip Shenon, "Defense Secretary Vows Thorough Inquiry on Gulf War Illnesses," *New York Times*, March 6, 1997.
43. James J. Tuite III, "Testimony to the Presidential Advisory Committee on Gulf War Veterans' Illnesses," Salt Lake City, Utah, March 18, 1997, available at http://www.chronic illnet.org/PGWS/Tuite/SLC.html, retrieved October 1, 2000; NIH, "The Persian Gulf Experience and Health."
44. U.S. General Accounting Office (GAO), *Gulf War Illnesses: DOD's Conclusions About U.S. Troops' Exposure Cannot Be Adequately Supported*, GAO-04-821T (Washington, D.C.: U.S. GAO, June 1, 2004).
45. Tim A. Bullman, Clare M. Mahan, Han K. Jang, and William F. Page, "Mortality in US Army Gulf War Veterans Exposed to 1991 Khamisiyah Chemical Munitions Destruction," *American Journal of Public Health* 95 (2005): 1382–388.
46. NIH, "The Persian Gulf Experience and Health," 13.
47. Caufield, *Multiple Exposures*.
48. Institute of Medicine, *Veterans and Agent Orange: Health Effects of Herbicides Used in Vietnam* (Washington, D.C.: Academy Press, 1994).
49. Charles Aldinger, "Study Links Agent Orange, Diabetes," *New York Times*, March 29, 2000.

50. Quoted in Kolata, "No Rise Found in Death Rates After Gulf War."
51. Quoted in Philip Shenon, "Advisers Condemn Pentagon Review of Gulf Ailments," *New York Times,* November 8, 1996.
52. Philip Shenon, "Gulf War Panel Reviews Researcher's Ouster," *New York Times,* December 23, 1996.
53. Shenon, "Panel Disputes Studies on Gulf War Illnesses."
54. Philip Shenon, "Oversight Suggested for Study of Gulf War Ills," *New York Times,* November 14, 1996.
55. Philip Shenon, "House Committee Assails Pentagon on Gulf War Ills," *New York Times,* October 26, 1997.
56. National Gulf War Resource Center, *Uncounted Casualties: America's Ailing Veterans—A Special Report* (Silver Spring, Md.: National Gulf War Resource Center, January 17, 2001), 4.
57. Sheryl Gay Stolberg, "U.S. Reports Disease Link to Gulf War," *New York Times,* December 11, 2001.
58. K. L. Ng and D. M. Hamby, "Fundamentals for Establishing a Risk Communication Program," *Health Physics* 73 (3) (1997): 465–72; Gerald Middendorf and Lawrence Busch, "Inquiry for the Public Good: Democratic Participation in Agricultural Research," *Agriculture and Human Values* 14 (1) (1997): 45–57.
59. Hyams, Wignall, and Roswell, "War Syndromes and Their Evaluation."
60. Quoted in Eric Schmitt, "Panel Criticizes Pentagon Inquiry on Gulf Illnesses," *New York Times,* January 8, 1997.
61. Wilbur Scott, "Competing Paradigms in the Assessment of Latent Disorders: The Case of Agent Orange," *Social Problems* 35 (1988): 145–61.
62. NIH, "The Persian Gulf Experience and Health."
63. Hyams, Wignall, and Roswell, "War Syndromes and Their Evaluation."
64. Ali Abdel-Rahman, Ashok K. Shetty, and Mohamed B. Abou-Donia, "Disruption of the Blood-Brain Barrier and Neuronal Cell Death in Cingulate Cortex, Dentate Gyrus, Thalamus, and Hypothalamus in a Rat Model of Gulf-War Syndrome," *Neurobiology of Disease* 10 (2002): 306–26; Rogene Henderson, Edward Barr, Walter Blackwell, Connie Clark, Carole Conn, Roma Kalra, Thomas March, Mohan Sopori, Yohannes Tesfaigzi, Margaret Ménache, Deborah Mash, Karol Doklandny, Wieslaw Kozak, Anna Kozak, Maceij Wachulec, Karin Rudolph, Matthew Kluger, Shashi Singh, Seddigheh Razani-Boroujerdi, and Raymond Langley, "Response of F344 Rats to Inhalation of Subclinical Levels of Sarin: Exploring Potential Causes of Gulf War Illness," *Journal of Toxicology and Industrial Health* 17 (5–10) (2001): 294–97.
65. Engel et al., "Rehabilitative Care of War-Related Health Concerns."
66. Michael Balint, *The Doctor, His Patient, and the Illness* (New York: International Universities Press, 1957).
67. Robert Aronowitz, *Making Sense of Illness: Science, Society, and Disease* (Cambridge, England: Cambridge University Press, 1998); David B. Morris, *Illness and Culture in the Postmodern Age* (Berkeley: University of California Press, 1998).
68. Charles C. Engel and Wayne J. Katon, "Population and Need-Based Prevention of Unexplained Symptoms in the Community," in *Strategies to Protect the Health of Deployed U.S. Forces: Medical Surveillance, Record Keeping, and Risk Reduction,* ed. Lois Joellenback, Philip Russell, and Samuel B. Guze, 173–212 (Washington, D.C.: National Academies Press for the Institute of Medicine, 1999).
69. Charles C. Engel, Joyce A. Adkins, and David Cowan, "Caring for Medically Unexplained Physical Symptoms After Toxic Environmental Exposures: Effects of Contested Causation," *Environmental Health Perspectives* 110 (suppl. 4) (2002): 641–47.

70. Simon Wessely, Chaichana Nimnuan, and Michael Sharpe, "Functional Somatic Syndromes: One or Many?" *The Lancet* 354 (1999): 936–39.
71. Dedra Buchwald and Deborah Garrity, "Comparison of Patients with Chronic Fatigue Syndrome, Fibromyalgia, and Multiple Chemical Sensitivities," *Archives of Internal Medicine* 154 (18) (1994): 2049–53.
72. Howard M. Kipen, William Hallman, Han Kang, Nancy Fiedler, and Benjamin H. Natelson, "Prevalence of Chronic Fatigue and Chemical Sensitivities in Gulf Registry Veterans," *Archives of Environmental Health* 54 (1999): 313–18.
73. Seth A. Eisen, Han K. Kang, Frances M. Murphy, Melvin S. Blanchant, Domenic J. Reda, William G. Henderson, Rosemary Toomey, Leila W. Jackson, Renee Alpem, Becky J. Parks, Nancy Klimas, Coleen Hall, Hon S. Pak, Joyce Hunter, Joel Karlinsky, Michael J. Battistone, Michael J. Lyons, and the Gulf War Study Participating Investigators, "Gulf War Veterans' Health: Medical Evaluation of a U.S. Cohort," *Annals of Internal Medicine* 142 (2005): 881–90.
74. Wessely, "Ten Years On."
75. Engel, Adkins, and Cowan, "Caring for Medically Unexplained Physical Symptoms."
76. Ibid.
77. Susie Kilshaw, "Friendly Fire: The Construction of Gulf War Syndrome Narratives," *Anthropology and Medicine* 11 (2) (2004): 149–60.
78. Quoted in Wessely, "Ten Years On," 2.
79. Thomas E. Shriver, "Environmental Hazards and Veterans' Framing of Gulf War Illnesses," *Sociological Inquiry* 71 (4) (2001): 403–20; Thomas E. Shriver, Amy L. Chasteen, and Brent D. Adams, "Cultural and Political Constraints in the Gulf War Illness Social Movement," *Sociological Focus* 35(2) (2002):123–43; Thomas E. Shriver and Sherry Cable, "Fault Lines and Frictions: Intramovement Conflicts in the Gulf War Illness Movement," paper presented at the Annual Meeting of the Southern Sociological Society, New Orleans, March 26–29, 2003.
80. Thomas E. Shriver, Gary R. Webb, and Brent Adams, "Environmental Exposures, Contested Illnesses, and Collective Action: The Controversy Over Gulf War Illness," *Humboldt Journal of Social Relations* 27 (1) (2002): 73–105.
81. Thomas E. Shriver, Amy Chasteen Miller, and Sherry Cable, "Women's Work: Women's Involvement in the Gulf War Illness Movement," *Sociological Quarterly* 44 (4) (2003): 639–58.
82. Wessely, "Ten Years On."
83. J. A Lipscomb, K. P. Satin, and R. R. Neutra, "Reported Symptom Prevalence Rates from Comparison Populations in Community-Based Environmental Studies," *Archives of Environmental Health* 47 (1992): 263–69; Lewis H. Roht, Sally W. Vernon, Francis W. Weir, Stanley M. Pier, Peggy Sullivan, and Lindsay J. Reed, "Community Exposure to Hazardous Waste Disposal Sites: Assessing Reporting Bias," *American Journal of Epidemiology* 122 (3) (1985): 418–33; D. Shusterman, J. Lipscomb, R. Neutra, and K. Satin, "Symptom Prevalence and Odor-Worry Interaction Near Hazardous Waste Sites," *Environmental Health Perspectives* 94 (1991): 25–30.
84. J. A. Muir Gray, "Postmodern Medicine," *The Lancet* 354 (1999): 1550–553.
85. Steven Kroll-Smith and Hugh H. Floyd, *Bodies in Protest: Environmental Illness and the Struggle over Medical Knowledge* (New York: New York University Press, 1997).
86. Jonathan Rest, "The Chronic Fatigue Syndrome," *Annals of Internal Medicine* 123 (1) (1995): 74–76.
87. Morris, *Illness and Culture in the Postmodern Age*; Wessely, "Ten Years On."
88. Gray, "Postmodern Medicine."

89. Michael Sharpe and Alan Carson, "'Unexplained' Somatic Symptoms, Functional Syndromes, and Somatization: Do We Need a Paradigm Shift?" *Annals of Internal Medicine* 134 (9) (2001): 926–30.
90. Wessely, Nimnuan, and Sharpe, "Functional Somatic Syndromes."
91. Richard Mayou and Michael Sharpe, "Treating Medically Unexplained Physical Symptoms," *British Medical Journal* 315 (1997): 561–62.
92. Thomas Ricks, "Anthrax Shots Cause Military Exodus," *Washington Post,* October 11, 2000.
93. James Vicini, "Judge Bars Mandatory Anthrax Shots for Troops," *Boston Globe,* October 28, 2004.
94. Pauline Jelinek, "Judge OK's Voluntary Anthrax Vaccination," *Boston Globe,* April 8, 2005.
95. Eileen Kelley, "GIs Link Ills to Malaria Drug: Two Fort Carson Soldiers Suspect the Medication Given Them in Iraq Is the Cause of Their Mental Problems," *Denver Post,* June 11, 2004.
96. Research Advisory Committee on Gulf War Veterans' Illnesses, *Scientific Progress in Understanding Gulf War Veterans' Illnesses: Report and Recommendations* (Washington, D.C.: U.S. Government Printing Office, 2004).
97. Jonathan B. Perlin, M.D., Ph.D, Deputy Undersecretary for Health, Department of Veterans Affairs, "Testimony on Pre- and Post Deployment Health Assessments," 2003, in *Federal Document Clearing House Congressional Testimony* (Bethesda, Md.: Congressional Information Service, July 9, 2004), available from *LexisNexis™ Congressional* (online service).

5. SIMILARITIES AND DIFFERENCES AMONG ASTHMA, BREAST CANCER, AND GULF WAR ILLNESSES

1. Pew Environmental Health Commission, *Attack Asthma: Why America Needs a Public Health Defense System to Battle Environmental Threats* (Washington, D.C.: Pew Environmental Health Commission, May 16, 2000).
2. Evans, ed., *State of the Evidence.*
3. National Breast Cancer Coalition, *Priority #6—Fact Sheet.*
4. Davis Baltz, "Endocrine Regulation Comes into Regulatory Focus," *New Solutions* 9 (1999): 29–35.
5. CDC, *Third National Report on Human Exposure to Environmental Chemicals.*
6. Fulco, Liverman, and Sox, *Depleted Uranium.*
7. Stolberg, "U.S. Reports Disease Link to Gulf War."

6. THE NEW PRECAUTIONARY APPROACH: A PUBLIC PARADIGM IN PROGRESS

1. Quoted in Raffensperger and Tickner, eds., *Protecting Public Health and the Environment,* 350.
2. Gee and Stirling, "Late Lessons from Early Warnings."
3. Brian Mayer, Phil Brown, and Meadow Linder, "Moving Further Upstream: From Toxics Reduction to the Precautionary Principle," *Public Health Reports* 117 (2002): 574–86.
4. Romeo Quijano, "Elements of the Precautionary Principle," in *Precaution: Environmental Science and Preventive Public Policy,* ed. Joel Tickner, 21–27 (Washington, D.C.: Island Press, 2003).

5. Joel Tickner, "A Map Toward Precautionary Decision Making," in *Protecting Public Health and the Environment: Implementing the Precautionary Principle,* ed. Carolyn Raffensperger and Joel Tickner, 162–86 (Washington, D.C.: Island Press, 1999).
6. Healthy Building Network, *The Louisville Charter for Safer Chemicals—A Platform for Creating a Safe and Healthy Environment Through Innovation* (Washington, D.C.: Healthy Building Network, 2005), available at http://www.healthybuilding.net/environmental_justice/louisville_charter.pdf.
7. Joel Tickner and Nancy Myers, "Current Status and Implementation of the Precautionary Principle," *Science and Environmental Health Network* (2000), available at http://www.sehn.org/ppcurrenstatus.html, retrieved June 6, 2005.
8. Lowell Center for Sustainable Production, *Integrated Chemicals Policy: Seeking New Directions in Chemical Management* (Lowell, Mass.: Lowell Center for Sustainable Production, 2003).
9. Frank Ackerman and Rachel Massey, *Prospering with Precaution* (Medford, Mass.: Tufts University Global Development and Environment Institute, 2002).
10. Maria Cone, "Flame Retardants to Be Extinguished," *Los Angeles Times,* November 4, 2003.
11. Paul Meller, "Europe Proposes Overhaul of Chemical Industry," *New York Times,* October 30, 2003; Sandra Steingraber, "Report from Europe: Precaution Ascending," *Rachel's Environment and Health News* 786 (March 4, 2004), available at http://www.rachel.org.
12. Steingraber, "Report from Europe: Precaution Ascending."
13. Ackerman and Massey, *Prospering with Precaution.*
14. European Industrial Relations Observatory (EIRO) On-Line, *Skanska Managers Found Guilty of Work Environment Crime* (Dublin: EIRO, 2002), available at http://www.eiro.eurofound.eu.int/2002/02/feature/se0202106f.html.
15. Joel Tickner and Ken Geiser, "The Precautionary Principle Stimulus for Solutions- and Alternatives-Based Environmental Policy," *Environmental Impact Assessment Review* 34 (2004): 801–24.
16. Terry Collins, "Toward Sustainable Chemistry," in *Precaution: Environmental Science and Preventive Public Policy,* ed. Joel Tickner, 297–301 (Washington, D.C.: Island Press, 2003); Kenneth Geiser, *Materials Matter: Toward a Sustainable Materials Policy* (Cambridge, Mass.: MIT Press, 2001).
17. Rudel et al., "Phthalates, Alkylphenols, Pesticides."
18. Richard Levins, "Whose Scientific Method: Scientific Methods for a Complex World," in *Precaution: Environmental Science and Preventive Public Policy,* ed. Joel Tickner, 355–68 (Washington, D.C.: Island Press, 2003).
19. Mayer, Brown, and Linder, "Moving Further Upstream."
20. Tickner and Geiser, "The Precautionary Principle Stimulus for Solutions."
21. Mayer, Brown, and Linder, "Moving Further Upstream."
22. Toxics Action Center and Maine Environmental Policy Institute, *Overkill: Why Pesticide Spraying for West Nile Virus May Cause More Harm Than Good,* pamphlet (Boston: Toxics Action Center, July 2001).
23. Environmental Advocates, New York Public Interest Research Group, and New York Coalition for Alternatives to Pesticides, *Toward Safer Mosquito Control in New York,* pamphlet (Albany: New York Environmental Advocates, New York Public Interest Research Group, and New York Coalition for Alternatives to Pesticides, January 2000).
24. Mark Weiner, "CDC: Back Off on West Nile Spraying," *Syracuse Herald Journal,* April 5, 2001.

25. Marilyn Massey-Stokes, "Foreword: Environmental Issues in the Health of Children," *Family and Community Health* 24 (4) (2002): viii–ix.
26. Pew Environmental Health Commission, *America's Environmental Health Gap.*
27. Breast Cancer Action, *Pill for "Prevention" vs. the Precautionary Principle,* Breast Cancer Action Fact Sheet (San Francisco: Breast Cancer Action, 2001), n.p.
28. Women's Community Cancer Project, *Fact Sheet on the Precautionary Principle* (Cambridge, Mass.: Women's Community Cancer Project, 2001), n.p.
29. Ibid.
30. Rachel Morello-Frosch, Manuel Pastor Jr., and James Sadd, "Integrating Environmental Justice and the Precautionary Principle in Research and Policy-Making: The Case of Ambient Air Toxics Exposures and Health Risks Among School Children in Los Angeles," *Annals of the American Academy of Political and Social Science* 584 (2002): 47–68.
31. Ecology Center, *Community Water Rights Project* (Berkeley, Calif.: Ecology Center, 2005), available at http://www.ecologycenter.org/cwrp/index.html, retrieved June 9, 2005.
32. Ibid.
33. Martha Matsuoka, ed., *Building Healthy Communities from the Ground Up: Environmental Justice in California* (Oakland, Calif.: Asian Pacific Environmental Network, Communities for a Better Environment, Environmental Health Coalition, People Organizing to Demand Environmental and Economic Rights, Silicon Valley Toxics Coalition Health, and Environmental Justice Project, 2003), 14.
34. Markowitz and Rosner, *Denial and Deceit.*
35. Center for Health, Environment, and Justice, *U.S. Facing Waste Crisis from Disposal of 70 Billion Pounds of PVC in Next Decade* (Falls Church, Va.: Center for Health, Environment, and Justice, 2004), available at http://www.chej.org.
36. Grassroots Recycling Network, "RE. [greenyes] Great News on PVC Campaign" (2005), available at http://greenyes.grrn.org/2005/02/msg00141.html, retrieved June 7, 2005; Steve Rosenberg, "Limiting Burning of PVC Is Urged," *Boston Globe,* December 9, 2004, available at http://www.besafenet.com/pvcnewspage1.htm, retrieved June 8, 2005.
37. Jane Houlihan, Charlotte Brody, and Bryony Schwan, *Not Too Pretty: Phthalates, Beauty Products, and the FDA* (Washington, D.C.: Environmental Working Group, 2002); Shanna H. Swan, Katharina M. Main, Fan Liu, Sara L. Stewart, Robin L. Kruse, Antonia M. Calafat, Catherine S. Mao, J. Bruce Redmon, Christine L. Ternand, Shannon Sullivan, J. Lynn Teague, and the Study for Future Families Research Team, "Decrease in Anogenital Distance among Male Infants with Prenatal Phthalate Exposure," *Environmental Health Perspectives* 113 (2005): 1056–61.
38. Joseph DiGangi and Helena Norrin, *Pretty Nasty: Phthalates in European Cosmetics Products* (Uppsala, Sweden: Health Care Without Harm, 2002).
39. Campaign for Safe Cosmetics, Web site, 2005, at http://www.safecosmetics.org.
40. Momo Chang, "Youth Raise Awareness of Cosmetic Chemicals," *Inside Bay Area,* August 7, 2005.
41. Montague, "San Francisco Adopts the Precautionary Principle."
42. Bay Area Working Group on the Precautionary Principle, "The Precautionary Principle in Action: Bay Area Working Group Action Locally and Regionally," 2005, available at http://www.takingprecaution.org.
43. Neha Patel, "Portland, Oregon, City and County Adopt Precautionary Approach," *Everyone's Backyard* 22 (4) (2004): 3, 10.
44. Maria Pellerano, "Precautionary Mister Rogers, Part 2," *Rachel's Environment and Health News* 803 (October 28, 2004), available at http://www.rachel.org.

45. Daniel L. Kleinman, ed., *Science, Technology, and Democracy* (Albany: State University of New York Press, 2000).
46. Toxic Use Reduction Institute (TURI), *Outreach Successes and Challenges* (Lowell, Mass.: TURI, 2004), available at http://www.turi.org/content/content/view/full/2125/, retrieved June 6, 2005.
47. Richard Sclove, "Better Approaches to Science Policy," *Science* 279 (1998): 1283.
48. See also Levins, "Whose Scientific Method."

7. IMPLICATIONS OF THE CONTESTED ILLNESSES PERSPECTIVE

1. Sara Shostak, "Locating Gene-Environment Interaction: At the Intersections of Genetics and Public Health," *Social Science and Medicine* 56 (2003): 2327–342.
2. Wolfson, *The Fight Against Big Tobacco*.
3. C. Wright Mills, *The Sociological Imagination* (New York: Oxford University Press, 1959).
4. Institute of Medicine, *Addressing the Physician Shortage in Occupational and Environmental Medicine: Report of a Study* (Washington, D.C.: National Academy Press, 1991). See also Joseph Castorina and Linda Rosenstock, "Physician Shortage in Occupational and Environmental Medicine," *Annals of Internal Medicine* 113 (1990):983–86; L. M. Frazier, J. W. Cromer, K. M. Andolsek, G. N. Greenberg, W. R. Thomann, and W. Stopford, "Teaching Occupational and Environmental Medicine in Primary Care Residency Training Programs: Experience Using Three Approaches during 1984–1991," *American Journal of the Medical Sciences* 302 (1) (1991): 42–45.
5. Brown and Mikkelsen, *No Safe Place*; Levine, *Love Canal*; Edelstein, *Contaminated Communities*; Stella Capek, "Toxic Hazards in Arkansas: Emerging Coalitions," paper presented at the Annual Meeting of the Society for the Study of Social Problems, Chicago, August 16, 1987; Lin Nelson, "Women's Lives Against the Industrial Chemical Landscape: Environmental Health and the Health of the Environment," in *Healing Technologies: Feminist Perspectives,* ed. Kathryn Strother Ratcliff, 347–69 (Ann Arbor: University of Michigan Press, 1989).
6. Phil Brown and Judith Kelley, "Physicians' Knowledge of and Actions Concerning Environmental Health Hazards: Analysis of Survey of Massachusetts Physicians," *Industrial and Environmental Crisis Quarterly* 9 (1996): 512–42.
7. Doug Brugge, Janelle Bagley, and James Hyde, "Environmental Management of Asthma at Top-Ranked U.S. Managed Care Organizations," *Journal of Asthma* 40 (2003): 605–14.
8. Randall M. Packard, Peter J. Brown, Ruth L. Berkelman, and Howard Frumkin, "Emerging Illnesses as Social Process," in *Emerging Illnesses and Society: Negotiating the Public Health Agenda,* ed. Randall M. Packard, Peter J. Brown, Ruth L. Berkelman, and Howard Frumkin (Baltimore: Johns Hopkins University Press, 2004), 2.
9. Arlene Kaplan Daniels, "The Captive Professional: Bureaucratic Limitations in the Practice of Military Psychiatry," *Journal of Health and Social Behavior* 10 (4) (1969): 255–65.
10. Ackerman and Massey, *Prospering with Precaution*.
11. John Warner, "Green Chemistry," lecture at Brown University, Providence, R.I., September 30, 2005.
12. Thaddeus Herrick, "Cosmetics Companies Shun Contentious Chemical," *Wall Street Journal,* January 14, 2005.
13. Dara O'Rourke, "Outsourcing Regulation: Analyzing Nongovernmental Systems of Labor Standards and Monitoring," *Policy Studies Journal* 31 (2003): 1–29; Dara O'Rourke,

Community-Driven Regulation: Balancing Development and the Environment in Vietnam (Cambridge, Mass.: MIT Press, 2004).
14. Baltz, "Endocrine Regulation Comes Into Regulatory Focus."
15. Thornton, *Pandora's Poison*.
16. Gee and Stirling, "Late Lessons from Early Warnings."
17. Valerie Beral, Emily Banks, and Gillian Reeves, "Evidence from Randomised Trials on the Long-Term Effects of Hormone Replacement Therapy," *The Lancet* 360 (2002): 942–44; Writing Group for the Women's Health Initiative Investigators, "Risks and Benefits of Estrogen Plus Progestorone in Healthy Post-menopausal women," *Journal of the American Medical Association* 288 (2002): 321–33; Esteve Fernandez, Silvano Gallus, Cristina Bosetti, Silvia Franceschi, Eva Negri, and Carlo La Vecchia, "Hormone Replacement Therapy and Cancer Risk: A Systematic Analysis from a Network of Case-Control Studies," *International Journal of Cancer* 105 (2003): 408–12; National Women's Health Network, *The Truth about Hormone Replacement Therapy: How to Break Free from the Medical Myths of Menopause* (Roseville, Calif.: Prima, 2002).
18. Brody and Rudell, "Environmental Pollutants and Breast Cancer."
19. Centers for Disease Control and Prevention (CDC), *Second National Report on Human Exposure to Environmental Chemicals* (Atlanta: CDC, 2003), available at http://www.cdc.gov/exposurereport.
20. CDC, *Third National Report on Human Exposures to Environmental Chemicals*.
21. CDC, *Third Report* conference call briefing.
22. Douglas Fischer, "CDC Results Are In: We're Full of Contaminants," *Oakland Tribune*, July 22, 2005.
23. National Children's Study, *What Is the National Children's Study?* (Rockville, Md.: National Children's Study, 2005), available at http://nationalchildrensstudy.gov/about/mission/overview.cfm, retrieved June 22, 2005.
24. Ibid.
25. National Children's Study, *What Makes This Study Different from Other U.S. Health Studies?* (Rockville, Md.: National Children's Study, 2005), available at http://nationalchildrensstudy.gov/about/mission/unique.cfm, retrieved June 22, 2005; Jeff Nesmith, "Children's Health Study in Need of Money," *Atlanta Journal-Constitution*, May 11, 2005.
26. Nesmith, "Children's Health Study in Need of Money."
27. Environmental Protection Agency (EPA), *America's Children and the Environment: Measures of Contaminants, Body Burdens, and Illnesses* (Washington, D.C.: EPA, 2003); see also Environmental Protection Agency (EPA), *America's Children and the Environment—Project Background* (Washington, D.C.: EPA, 2004), available at http://www.epa.gov/envirohealth/children/background/index.htm, retrieved June 22, 2005.
28. Trust for America's Health, *Health Tracking Network* (Washington, D.C.: Trust for America's Health, 2005), available at http://healthyamericans.org/topics/index.php?TopicID = 28, retrieved June 21, 2005.
29. Trust for America's Health, *Nationwide Health Tracking: Investigating Life-Saving Discoveries* (Washington, D.C.: Trust for America's Health, 2005), available at http://healthyamericans.org/reports/files/HealthTrackingBackgrounder.pdf, retrieved June 21, 2005.
30. Ibid.
31. William Winkenwerder Jr., M.D., MBA, Assistant Secretary of Defense for Health Affairs, "Testimony on Pre- and Post-deployment Health Assessments," 2003, in *Federal Document Clearing House Congressional Testimony* (Bethesda, Md.: Congressional Information Service, July 9, 2004), available from *LexisNexis*™ *Congressional* (on-line service).

32. Joellenbeck, Russell, and Guze, *Strategies to Protect the Health of Deployed U.S. Forces*.
33. Christine Lehmann, "Military Boosts Monitoring of Soldiers' Mental Health," *Psychiatric News* 40 (5) (2005): 8.
34. Evans, *State of the Evidence*.
35. Breast Cancer Action, *Cancer Research and Policy* (San Francisco: Breast Cancer Action, 2005), available at http://www.bcaction.org/Pages/LearnAboutUs/CancerResearch&Policy.html, retrieved June 21, 2005.
36. Liam R. O'Fallon and Allen Dearry, "Community-Based Participatory Research as a Tool to Advance Environmental Health Sciences," *Environmental Health Perspectives* 110 (2002): 155–59; Shepard et al., "Preface."
37. Lorraine Halinka Malcoe, Robert A. Lynch, Michelle Crozier Kegler, and Valerie J. Skaggs, "Lead Sources, Behaviors, and Socioeconomic Factors in Relation to Blood Lead of Native American and White Children: A Community-Based Assessment of a Former Mining Area," *Environmental Health Perspectives* 110 (suppl. 2) (2002): 221–31.
38. Loh et al., "From Asthma to AirBeat."
39. Alternatives for Community and Environment (ACE), *Facts About the Boston University Proposal to Build a National Biocontainment Laboratory at Boston University Medical Center* (Roxbury, Mass.: ACE, 2003), available at http://www.ace-ej.org/BiolabWeb/Biolabdocs/BiolabfactrptDec012003.pdf, retrieved July 18, 2005.
40. Angela Ledford, "Foreword: The Dirty Secret Behind Dirty Air: Dirty Power," in *Dirty Air Dirty Power: Mortality and Health Damage Due to Pollution from Power Plants*, ed. Conrad Schneider, 1–3 (Boston: Clean Air Task Force, 2004).
41. David Michaels, Eula Bingham, Les Boden, Richard Clapp, Lynn R. Goldman, Polly Hoppin, Sheldon Krimsky, Celeste Monforton, David Ozonoff, and Anthony Robbins, "Advice Without Dissent," *Science* 298 (2002): 703.
42. Chris Mooney, *The Republican War on Science* (Cambridge, Mass.: Basic Books, 2005).
43. Michaels et al., "Advice Without Dissent."
44. Robert Steinbrook, "Science, Politics, and Federal Advisory Committees," *New England Journal of Medicine* 350 (2004): 1454–460.
45. Dan Ferber, "Overhaul of CDC Panel Revives Lead Safety Debate," *Science* 298 (2002): 732.
46. Donald Kennedy, "An Epidemic of Politics," *Science* 299 (2003): 625.
47. Environmental Protection Agency (EPA), "Draft Report on the Environment," 2003, available at http://www.epa.gov/indicators/roe/html/roeAirGlo.htm.
48. Andrew Revkin and Katharine Seelye, "Report by EPA Leaves Out Data on Climate Change," *New York Times*, June 19, 2003.
49. Quoted in ibid.
50. Miguel Llanos, "EPA Official Quits, Rips White House Regulatory Chief, Cites Push 'to Weaken the Rules,'" *MSNBC*, February 28, 2002.
51. National Environmental Trust, *The Bush Administration's Anti-environmental Actions* (Washington, D.C.: National Environmental Trust, 2005), available at http://www.net.org/reports/rollbacks.vtml, retrieved June 22, 2005.
52. National Resources Defense Council, "Court Strikes Down Bush Rollbacks of Clean Air Act, Upholds Others," 2005, available at the National Resources Defense Council Press Archive, http://nrdc.org/media/pressReleases/050624.asp; National Environmental Trust, *The Bush Administration's Anti-environmental Actions*.
53. Bush Greenwatch, "Administration Misleads on Alaska Drilling Claims," April 26, 2005, available at http://www.bushgreenwatch.org/mt_archives/000260.php.
54. Earthjustice, "Judge: Bush Administration Pesticide Rules for Endangered Species Are Illegal," August 24, 2006, available at http://www.earthjustice.org/.

55. Bush Greenwatch, "EPA Allows Dumping in Chesapeake and Shenandoah River," September 6, 2006, available at http://www.bushgreenwatch.org/mt_archives/000317.php.
56. Bush Greenwatch, "Bush Administration Announces Weak New Fuel Economy Standards," March 30, 2006, available at http://www.bushgreenwatch.org/mt_archives/000307.php.
57. Project on Scientific Knowledge and Public Policy, *Daubert: The Most Influential Supreme Court Ruling You've Never Heard Of* (Boston: Tellus Institute, 2003), available at http://www.defendingscience.org, retrieved June 22, 2005.
58. Peter Waldman, "EPA Bans Staff from Discussing Issue of Perchlorate Pollution," *Wall Street Journal*, April 28, 2003.
59. Engel et al., "Rehabilitative Care of War-Related Health Concerns."
60. Perlin, "Testimony on Pre- and Post-deployment Health Assessments."
61. Weinhold, "Fuel for the Long Haul."
62. Pandya et al., "Diesel Exhaust and Asthma."
63. John Snow, *On the Mode of Communication of Cholera* (London: John Churchill, 1855), 85.
64. Patricia H. Hynes, *The Recurring Silent Spring* (Elmsford, N.Y.: Pergamon, 1989).
65. Ibid.
66. Davis, *When Smoke Ran Like Water*, 72–80.
67. Sheldon Krimsky, *Science in the Private Interest: Has the Lure of Profits Corrupted Biomedical Research?* (Lanham, Md.: Rowman and Littlefield, 2003), 190–91.
68. Steve Wing, "Social Responsibility and Research Ethics in Community-Driven Studies of Industrialized Hog Production," *Environmental Health Perspectives* 110 (2002): 437–44.
69. Steingraber, *Living Downstream*, 182.
70. David Hess, "The Problem of Undone Science: Values, Interests, and the Selection of Research Programs," paper presented at the Annual Meeting of the Society for the Social Study of Science, Halifax, Nova Scotia, October 30–November 2, 1999.
71. Richard Clapp, Genevieve Howe, and Molly Jacobs Lefebvre, *Environmental and Occupational Causes of Cancer: A Review of Recent Scientific Literature* (Lowell: Lowell Center for Sustainable Production at the University of Massachusetts–Lowell, 2005).
72. Sandra Steingraber, *Having Faith: An Ecologist's Journey to Motherhood* (New York: Perseus Books, 2001).
73. Barbara L. Allen, *Uneasy Alchemy: Citizens and Experts in Louisiana's Chemical Corridor Disputes* (Cambridge, Mass.: MIT Press, 2003).
74. EWG, *Body Burden: Pollution in People*.
75. Ibid.
76. Barbara A. Israel, Amy J. Schulz, Edith A. Parker, and Adam B. Becker, "Review of Community-Based Research: Assessing Partnership Approaches to Improve Public Health," *Annual Review of Public Health* 19 (1998): 173–202; Minkler and Wallerstein, *Community-Based Participatory Research for Health*.
77. Akwesasne Research Advisory Committee, *Protocol for Review of Environmental and Scientific Research Proposals* (Hogansburg, N.Y.: Akwesasne Task Force on the Environment Research Advisory Committee, 1996).
78. Steingraber, *Having Faith*; Maria Boswell-Penc, *Tainted Milk: Breastmilk, Feminisms, and the Politics of Environmental Degradation* (Albany: State University of New York Press, 2006).
79. Environmental Working Group (EWG), *Study Finds Record High Levels of Toxic Fire Retardants in Breast Milk from American Mothers* (Washington, D.C.: EWG, 2003), available at http://www.ewg.org/reports/mothersmilk/es.php.

80. Environmental Working Group (EWG), *Body Burden—The Pollution in Newborns: A Benchmark Investigation of Industrial Chemicals, Pollutants, and Pesticides in Umbilical Cord Blood. Executive Summary* (Washington, D.C.: EWG, 2005), available at http://www.ewg.org/reports/bodyburden2/execsumm.php, retrieved July 18, 2005.
81. Jack Spengler, "Persistent Organic Pollutants," presentation given at the Silent Spring Institute, Newton, Mass., November 30, 2004.

BIBLIOGRAPHY

Abdel-Rahman, Ali, Ashok K. Shetty, and Mohamed B. Abou-Donia. "Disruption of the Blood-Brain Barrier and Neuronal Cell Death in Cingulate Cortex, Dentate Gyrus, Thalamus, and Hypothalamus in a Rat Model of Gulf-War Syndrome." *Neurobiology of Disease* 10 (2002): 306–26.
Abou-Donia, Mohamed, Kenneth Wilmarth, Karl Jensen, Fredrick Oeheme, and Thomas Kurt. "Neurotoxicity Resulting from Coexposure to Pyrodostigmine Bromide, DEET, and Permethrin: Implications of Gulf War Chemical Exposures." *Journal of Toxicology and Environmental Health* 48 (1996): 35–56.
Abu-Qare, A. W. and M. B. Abou-Donia. "Sarin: Health Effects, Metabolism, and Methods of Analysis." *Food and Chemical Toxicology* 40 (2002): 1327–333.
Ackerman, Frank and Rachel Massey. *Prospering with Precaution.* Medford, Mass.: Tufts University Global Development and Environment Institute, 2002.
Adams, Walter and James W. Brock. *The Tobacco Wars.* Cincinnati: South-Western College Publications, 1999.
Agyeman, Julian. *Sustainable Communities and the Challenge of Environmental Justice.* New York: New York University Press, 2005.
Akwesasne Research Advisory Committee. *Protocol for Review of Environmental and Scientific Research Proposals.* Hogansburg, N.Y.: Akwesasne Task Force on the Environment Research Advisory Committee, 1996.
Aldinger, Charles. "Study Links Agent Orange, Diabetes." *New York Times,* March 29, 2000.
Allen, Barbara L. *Uneasy Alchemy: Citizens and Experts in Louisiana's Chemical Corridor Disputes.* Cambridge, Mass.: MIT Press, 2003.
Alternatives for Community and Environment (ACE). *Facts About the Boston University Proposal to Build a National Biocontainment Laboratory at Boston University Medical Center.* Roxbury, Mass.: ACE, 2003. Available at http://www.ace-ej.org/BiolabWeb/Biolabdocs/BUlabfactrptDec012003.pdf. Retrieved July 18, 2005.
Amdur, Mary. "Animal Toxicology." In *Particles in Our Air: Concentrations and Health Effects,* ed. Richard Wilson and John Spengler, 85–122. Cambridge, Mass.: Harvard University Press, 1996.
American Cancer Society (ACS). *Cancer Facts and Figures 2001.* Atlanta: ACS, 2001.
——. *Cancer Facts and Figures 2005.* Atlanta: ACS, 2005.

American Lung Association (ALA). *Breathless: Air Pollution and Hospital Admissions/Emergency Room Visits in 13 Cities.* Washington, D.C.: ALA, 1996.

———. *Estimated Prevalence and Incidence of Lung Disease by Lung Association Territory.* Washington, D.C.: ALA Epidemiology and Statistic Unit Research and Program Services, May 2006. Available at http://www.lungusa.org/atf/cf/{7A8D42C2-FCCA-4604-8ADE-7F5D5E762256}/ESTPREV06.PDF.

———. "Trends in Asthma Morbidity and Mortality." April 2004. Available at http://www.lung usa.org.

———. *Trends in Asthma Morbidity and Mortality.* Washington, D.C.: ALA Epidemiology and Statistic Unit Research and Program Services, May 2005. Available at http://www.lungusa.org/atf/cf/{7A8D42C2-FCCA-4604-8ADE-7F5D5E762256}/ASTHMA1.PDF.

American Trucking Association, Inc., et al., v. United States Environmental Protection Agency. 175 F.3d 1027 (D.C. Circuit 1999).

Araneta, Maria Rosario G., Karen M. Schlangen, Larry D. Edmonds, Daniel A. Destiche, Ruth D. Merz, Charlotte A. Hobbs, Timothy J. Flood, John A. Harris, Diane Krishnamurti, and Gregory C. Gray. "Prevalence of Birth Defects Among Infants of Gulf War Veterans in Arkansas, Arizona, California, Georgia, Hawaii, and Iowa, 1989–1993." *Birth Defects Research (Part A)* 67 (2002): 246–60.

Aronowitz, Robert. *Making Sense of Illness: Science, Society, and Disease.* Cambridge, England: Cambridge University Press, 1998.

Aschengrau, Ann, David Ozonoff, Patricia Coogan, Richard Vezina, Timothy Heeren, and Yuqzng Zhang. "Cancer Risk and Residential Proximity to Cranberry Cultivation in Massachusetts." *American Journal of Public Health* 86 (1996): 1289–296.

Aschengrau, Ann, Sarah Rogers, and David Ozonoff. "Perchloroethylene-Contaminated Drinking Water and the Risk of Breast Cancer: Additional Results from Cape Cod, Massachusetts, USA." *Environmental Health Perspectives* 111 (2003): 167–73.

Ashford, Nicholas and Charles Caldart. *Technology Law and the Working Environment.* Washington, D.C.: Island Press, 1996.

Asthma and Allergy Foundation of America, New England Chapter. *Asthma-Friendly Child Care: A Checklist for Parents and Providers.* Washington, D.C.: Asthma and Allergy Foundation of America, 2000.

Balint, Michael. *The Doctor, His Patient, and the Illness.* New York: International Universities Press, 1957.

Balshem, Martha. *Cancer in the Community: Class and Medical Authority.* Washington, D.C.: Smithsonian Institution Press, 1993.

Baltz, Davis. "Endocrine Regulation Comes into Regulatory Focus." *New Solutions* 9 (1999): 29–35.

Bay Area Working Group on the Precautionary Principle. "The Precautionary Principle in Action: Bay Area Working Group Action Locally and Regionally." 2005. Available at: http://www.takingprecaution.org.

Beck, Ulrich. *Ecological Enlightenment.* Translated by M. A. Ritter. Atlantic Highlands, N.J.: Humanities Press, 1995.

———. *Risk Society: Towards a New Modernity.* London: Sage, 1992.

Beral, Valerie, Emily Banks, and Gillian Reeves. "Evidence from Randomised Trials on the Long-Term Effects of Hormone Replacement Therapy." *The Lancet* 360 (2002): 942–44.

"Boston Public Schools Green Cleaners Project: Pilot Program Assessment." Unpublished report, Department of Sociology, Brown University, 2005.

Bostwick, Charles F. "Child Asthma Rises in Los Angeles Area with Growing Exhaust Pollution." *Los Angeles Daily News,* February 23, 2004.

Boswell-Penc, Maria. *Tainted Milk: Breastmilk, Feminisms, and the Politics of Environmental Degradation*. Albany: State University of New York Press, 2006.

Brandt, Allan. "The Cigarette, Risk, and American Culture." *Daedalus* 119 (4) (1990): 155–76.

Breast Cancer Action. *Cancer Research and Policy*. San Francisco: Breast Cancer Action, 2005. Available at http://www.bcaction.org/Pages/LearnAboutUs/CancerResearch&Policy.html. Retrieved June 21, 2005.

———. *Pill for "Prevention" vs. the Precautionary Principle*. Breast Cancer Action Fact Sheet. San Francisco: Breast Cancer Action, 2001.

Breast Cancer and the Environment Research Centers. *Early Environmental Exposures*. San Francisco: Breast Cancer and the Environment Research Centers, 2006. Available at http://www.bcerc.org/research.htm. Retrieved June 11, 2006.

Brenner, Barbara. "The Breast Cancer Epidemic." Paper presented at the American Public Health Association Conference, Boston, November 13, 2000.

Brodeur, Paul. *Outrageous Misconduct: The Asbestos Industry on Trial*. New York: Pantheon, 1985.

Brody, Julia Green. "No Smoking Gun or Magic Bullet." Paper presented at the conference "Speak No Evil, Hear No Evil, See No Cure: Breast Cancer Truth and Consequences," sponsored by Hurricane Voices Breast Cancer Foundation and Massachusetts Breast Cancer Coalition, Waltham, Mass., March 25, 2004.

Brody, Julia Green, Kirsten B. Moysich, Olivier Humblet, Kathleen R. Attfield, Gregory P. Beehler, and Ruthann Rudel. *Environmental Pollutants and Breast Cancer: Epidemiologic Studies*. Report for the Susan G. Komen Breast Cancer Foundation. Newton, Mass.: Silent Spring Institute, 2005.

Brody, Julia Green and Ruthann Rudel. "Environmental Pollutants and Breast Cancer." *Environmental Health Perspectives* 111 (2003): 1007–19.

Brown, Phil and Judith Kelley. "Physicians' Knowledge of and Actions Concerning Environmental Health Hazards: Analysis of Survey of Massachusetts Physicians." *Industrial and Environmental Crisis Quarterly* 9 (1996): 512–42.

Brown, Phil, J. Stephen Kroll-Smith, and Valerie Gunter. "Knowledge, Citizens, and Organizations: An Overview of Environments, Diseases, and Social Conflict." In *Illness and the Environment: A Reader in Contested Medicine*, ed. J. Stephen Kroll-Smith, Phil Brown, and Valerie Gunter, 9–25. New York: New York University Press, 2000.

Brown, Phil, Sabrina McCormick, Brian Mayer, Stephen Zavestoski, Rachel Morello-Frosch, Rebecca Gasior Altman, and Laura Senier. "'A Lab of Our Own': Environmental Causation of Breast Cancer and Challenges to the Dominant Epidemiological Paradigm." *Science, Technology, and Human Values* 31 (2006): 499–536.

Brown, Phil and Edwin J. Mikkelsen. *No Safe Place: Toxic Waste, Leukemia, and Community Action*. Rev. ed. Berkeley: University of California Press, 1997.

Brown, Phil, Stephen Zavestoski, Sabrina McCormick, Joshua Mandelbaum, Theo Luebke, and Meadow Linder. "A Gulf of Difference: Disputes Over Gulf War–Related Illnesses." *Journal of Health and Social Behavior* 42 (2001): 235–57.

Brown, Phil, Stephen Zavestoski, Sabrina McCormick, Brian Mayer, Rachel Morello-Frosch, and Rebecca Gasior. "Embodied Health Movements: Uncharted Territory in Social Movement Research." *Sociology of Health and Illness* 26 (2004): 1–31.

Browner, Carol. Remarks made to the U.S. Senate Environment and Public Works Subcommittee on Clean Air, May 20, 1999. Available at http://www.epa.gov/ttn/oarpg/gen/cmbtest.html#remarks.

Brugge, Doug, Janelle Bagley, and James Hyde. "Environmental Management of Asthma at Top-Ranked U.S. Managed Care Organizations." *Journal of Asthma* 40 (2003): 605–14.

Buchwald, Dedra and Deborah Garrity. "Comparison of Patients with Chronic Fatigue Syndrome, Fibromyalgia, and Multiple Chemical Sensitivities." *Archives of Internal Medicine* 154 (18) (1994): 2049-53.

Bullman, Tim A., Clare M. Mahan, Han K. Jang, and William F. Page. "Mortality in US Army Gulf War Veterans Exposed to 1991 Khamisiyah Chemical Munitions Destruction." *American Journal of Public Health* 95 (2005): 1382-388.

Burawoy, Michael. "Introduction: Reaching for the Global." In *Global Ethnography: Forces, Connections, and Imaginations in a Postmodern World,* ed. Michael Burawoy, 1-59. Berkeley: University of California Press, 2000.

Bury, Michael. "Chronic Illness as Biographical Disruption." *Sociology of Health and Illness* 4 (2) (1982): 167-82.

Bush Greenwatch. "Administration Misleads on Alaska Drilling Claims." April 26, 2005. Available at http://www.bushgreenwatch.org/mt_archives/000260.php.

———. "Bush Administration Announces Weak New Fuel Economy Standards." March 30, 2006. Available at http://www.bushgreenwatch.org/mt_archives/000307.php.

———. "EPA Allows Dumping in Chesapeake and Shenandoah River." September 6, 2006. Available at http://www.bushgreenwatch.org/mt_archives/000317.php.

Campaign for Safe Cosmetics. Web site, 2005, at http://www.safecosmetics.org.

Capek, Stella. "Toxic Hazards in Arkansas: Emerging Coalitions." Paper presented at the Annual Meeting of the Society for the Study of Social Problems, Chicago, August 16, 1987.

Castorina, Joseph and Linda Rosenstock. "Physician Shortage in Occupational and Environmental Medicine." *Annals of Internal Medicine* 113 (1990): 983-86.

Caufield, Catherine. *Multiple Exposures: Chronicles of the Radiation Age.* Chicago: University of Chicago Press, 1990.

Center, Rebecca. "Poverty, Ethnicity, and Pediatric Asthma in Rhode Island." Senior honors thesis, Center for Environmental Studies, Brown University, 2000.

Center for Health, Environment, and Justice. *U.S. Facing Waste Crisis from Disposal of 70 Billion Pounds of PVC in Next Decade.* Falls Church, Va.: Center for Health, Environment, and Justice, 2004. Available at http://www.chej.org.

Centers for Disease Control and Prevention (CDC). "Adults Who Have Ever Been Told They Have Asthma." *BRFSS Prevalence Data* (2005). Available at http://apps.nccd.cdc.gov/brfss/display.asp?cat = AS&yr = 2005&qkey = 4417&state = US. Retrieved June 23, 2006.

———. "Asthma Prevalence and Control Characteristics by Race/Ethnicity—United States, 2002." *Morbidity and Mortality Weekly Report* 53 (2004): 145-48. Available at http://www.cdc.gov/mmwr/preview/mmwrhtml/mm5307a1.htm.

———. *Second National Report on Human Exposure to Environmental Chemicals.* Atlanta: CDC, 2003. Available at http://www.cdc.gov/exposurereport.

———. "Self-reported Asthma Prevalence and Control Among Adults—United States, 2001." *Morbidity and Mortality Weekly Report* 52 (2003): 381-84. Available at http://www.cdc.gov/mmwr/preview/mmwrhtml/mm5217a2.htm.

———. *Third Report* conference call briefing, July 21, 2005. Available from the CDC in Atlanta.

———. *Third National Report on Human Exposure to Environmental Chemicals.* Atlanta: CDC, 2005. Available at http://www.cdc.gov/exposurereport/3rd/.

Chang, Momo. "Youth Raise Awareness of Cosmetic Chemicals." *Inside Bay Area,* August 7, 2005.

Charmaz, Kathy. *Good Days, Bad Days: The Self in Chronic Illness.* New Brunswick, N.J.: Rutgers University Press, 1991.

Child and Adolescent Health Measurement Initiative. *National Survey of Children's Health.* Portland, Ore.: Child and Adolescent Health Measurement Initiative, 2005. Available at the Data Resource Center on Child and Adolescent Health Web site, http://www.nschdata.org. Retrieved June 23, 2006.

Clapp, Richard, Genevieve Howe, and Molly Jacobs Lefebvre. *Environmental and Occupational Causes of Cancer: A Review of Recent Scientific Literature*. Lowell: Lowell Center for Sustainable Production at the University of Massachusetts–Lowell, 2005.

Clapp, Richard and David Ozonoff. "Environment and Health: Vital Intersection or Contested Territory?" *American Journal of Law and Medicine* 30 (2004): 189–215.

Clarke, Lee. *Acceptable Risk? Making Decisions in a Toxic Environment*. Berkeley: University of California Press, 1989.

Clarke, Robert W., Brent Coull, Ulrike Reinisch, Paul Catalano, Cheryl R. Killingsworth, Petros Koutrakis, Ilias Kavouras, Gopala Gazula Krishna Murthy, Joy Lawrence, Eric Lovett, J. Mikhail Wolfson, Richard L. Verrier, and John J. Godleski. "Inhaled Concentrated Ambient Particles Are Associated with Hematologic and Bronchoalveolar Lavage Changes in Canines." *Environmental Health Perspectives* 108 (2000): 1179–187.

Claudio L., L. Tulton, J. Doucette, and P. J. Landrigan. "Socioeconomic Factors and Asthma Hospitalization Rates in New York City." *Journal of Asthma* 36 (4) (1999): 343–50.

Clauw, D. J. "The 'Gulf War Syndrome': Implications for Rheumatologists." *Journal of Clinical Rheumatology* 4 (4) (1998): 173–74.

Clauw, Daniel, Charles Engel, Robert Aronowitz, Edgar Jones, Howard Kipen, Jurt Jroenke, Scott Ratzan, Michael Sharpe, and Wessely Simon. "Unexplained Symptoms After Terrorism and War: An Expert Consensus Statement." *Journal of Occupational and Environmental Medicine* 45 (10) (2003): 1040–48.

Clean Air Task Force. *Diesel and Health in America: The Lingering Threat*. Boston: Clean Air Task Force, 2005. Available at http://www.catf.us/publications/view.php?id = 83. Retrieved February 22, 2005.

Colborn, Theo, Dianne Dumanoski, and John Peterson Myers. *Our Stolen Future: How We Are Threatening Our Fertility, Intelligence, and Survival*. New York: Dutton, 1997.

Collins, Harry. "An Empirical Relativist Program in the Sociology of Scientific Knowledge." In *Science Observed*, ed. Karin Knorr-Cetina and Michael Mulkay, 85–114. Beverly Hills, Calif.: Sage, 1983.

Collins, Terry. "Toward Sustainable Chemistry." In *Precaution: Environmental Science and Preventive Public Policy*, ed. Joel Tickner, 297–301. Washington, D.C.: Island Press, 2003.

Community-Based Participatory Research (CBPR) Institute for Biomedical Research. *Patricia Fraser, M.D.* Boston: CBPR Institute for Biomedical Research, 2004. Available at http://www.cbrinstitute.org/page.php?branch = pi&page = fraser&type = 4. Retrieved June 13, 2005.

Cone, Maria. "Flame Retardants to Be Extinguished." *Los Angeles Times*, November 4, 2003.

Conrad, Peter. "The Experience of Illness: Recent and New Directions." *Research in the Sociology of Health Care* 6 (1987): 1–31.

——. "A Mirage of Genes." *Sociology of Health and Illness* 21 (1999): 228–41.

Corburn, Jason. "Combining Community-Based Research and Local Knowledge to Confront Asthma and Subsistence-Fishing Hazards in Greenpoint/Williamsburg, Brooklyn, New York." *Environmental Health Perspectives* 110 (suppl. 2) (2002): 241–48.

——. *Street Science: Community Knowledge and Environmental Health Justice*. Cambridge, Mass.: MIT Press, 2005.

Couch, Steven R. and Steve Kroll-Smith. "Environmental Movements and Expert Knowledge: Evidence for a New Populism." *International Journal of Contemporary Sociology* 34 (1997): 185–210.

Coyle, Y. M., L. S. Hynan, D. M. Euhus, and A. T. Minhajuddin. "An Ecological Study of the Association of Environmental Chemicals on Breast Cancer Incidence in Texas." *Breast Cancer Research and Treatment* 92 (2005): 107–14.

Daniels, Arlene Kaplan. "The Captive Professional: Bureaucratic Limitations in the Practice of Military Psychiatry." *Journal of Health and Social Behavior* 10 (4) (1969): 255–65.

Daubert v. Merrell Dow Pharmaceuticals, Inc., 1993. 113 S.Ct. 2786.

Davis, Devra. *When Smoke Ran Like Water: Tales of Environmental Deception and the Battle Against Pollution.* New York: Basic Books, 2002.

Davis, Devra Lee and H. Leon Bradlow. "Can Environmental Estrogens Cause Breast Cancer?" *Scientific American* 273 (4) (1995): 166–72.

Davis, Devra Lee and Pamela S. Webster. "The Social Context of Science: Cancer and the Environment." *Annals of the American Academy of Political and Social Science* 584 (2002): 13–34.

Davis, Miriam. "Gulf War Illnesses and Recognizing New Diseases." In *Depleted Uranium, Sarin, Pyridostigmine Bromide, and Vaccines,* vol. 1 of *Gulf War and Health,* ed. Carolyn E. Fulco, Catharyn T. Liverman, and Harold Sox, appendix D, 342–65. Washington, D.C.: National Academies Press for the Institute of Medicine, 2000.

Dickersin, Kay, Lundy Braun, Margaret Mead, Robert Millikan, Anna M. Wu, Jennifer Pietenpol, Susan Troyan, Benjamin Anderson, and Frances Visco. "Development and Implementation of a Science Training Course for Breast Cancer Activists: Project LEAD (Leadership, Education, and Advocacy Development)." *Health Expectations* 4 (4) (2001): 213–20.

Dickersin, Kay and Lauren Schnaper. "Reinventing Medical Research." In *Man-Made Medicine: Women's Health, Public Policy, and Reform,* ed. K. L. Moss, 57–78. Durham, N.C.: Duke University Press, 1996.

DiGangi, Joseph, and Helena Norrin. *Pretty Nasty: Phthalates in European Cosmetics Products.* Uppsala, Sweden: Health Care Without Harm, 2002.

Dockery, Douglas and Arden Pope. "Epidemiology of Acute Health Effects: Summary of Times Series Studies." In *Particles in Our Air: Concentrations and Health Effects,* ed. Richard Wilson and John Spengler, 123–36. Cambridge, Mass.: Harvard University Press, 1996.

Dockery, Douglas, Clyde Pope, X. Xu, Jack Spengler, John Ware, M. Ray, B. Ferris, and F. Speitzer. "An Association Between Air Pollution and Mortality in Six US Cities." *New England Journal of Medicine* 329 (1993): 1753–1759.

Dorgan, J., John Brock, Nathaniel Rothman, Larry Needham, and Rosetta Miller. "Serum Organochlorine Pesticides and PCBs and Breast Cancer Risk: Results from a Prospective Analysis." *Cancer Causes and Control* 10 (1999): 1–11.

Earthjustice. "Judge: Bush Administration Pesticide Rules for Endangered Species Are Illegal." August 24, 2006. Available at http://www.earthjustice.org/.

Ecology Center. *Community Water Rights Project.* Berkeley, Calif.: Ecology Center, 2005. Available at http://www.ecologycenter.org/cwrp/index.html. Retrieved June 9, 2005.

Edelstein, Michael. *Contaminated Communities: The Social and Psychological Impacts of Residential Toxic Exposure.* Boulder, Colo.: Westview, 1988.

Eisen, Seth A., Han K. Kang, Frances M. Murphy, Melvin S. Blanchant, Domenic J. Reda, William G. Henderson, Rosemary Toomey, Leila W. Jackson, Renee Alpem, Becky J. Parks, Nancy Klimas, Coleen Hall, Hon S. Pak, Joyce Hunter, Joel Karlinsky, Michael J. Battistone, Michael J. Lyons, and the Gulf War Study Participating Investigators. "Gulf War Veterans' Health: Medical Evaluation of a U.S. Cohort." *Annals of Internal Medicine* 142 (2005): 881–90.

Engel, Charles C., Joyce A. Adkins, and David Cowan. "Caring for Medically Unexplained Physical Symptoms After Toxic Environmental Exposures: Effects of Contested Causation." *Environmental Health Perspectives* 110 (suppl. 4) (2002): 641–47.

Engel, Charles C. and Wayne J. Katon. "Population and Need-Based Prevention of Unexplained Symptoms in the Community." In *Strategies to Protect the Health of Deployed U.S. Forces:*

Medical Surveillance, Record Keeping, and Risk Reduction, ed. Lois Joellenback, Philip Russell, and Samuel B. Guze, 173–212. Washington, D.C.: National Academies Press for the Institute of Medicine, 1999.

Engel, Charles C., Xian Liu, Roy Clymer, Ronald F. Miller, Terry Sjoberg, and Jay R. Shapiro. "Rehabilitative Care of War-Related Health Concerns." *Journal of Occupational and Environmental Medicine* 42 (2000): 385–90.

Environmental Advocates, New York Public Interest Research Group, and New York Coalition for Alternatives to Pesticides. *Toward Safer Mosquito Control in New York.* Pamphlet. Albany: Environmental Advocates, New York Public Interest Group, and New York Coalition for Alternatives to Pesticides, January 2000.

Environmental Protection Agency (EPA). *America's Children and the Environment: Measures of Contaminants, Body Burdens, and Illnesses.* Washington, D.C.: EPA, 2003.

——. *America's Children and the Environment—Project Background.* Washington, D.C.: EPA, 2004. Available at http://www.epa.gov/envirohealth/children/background/index.htm. Retrieved June 22, 2005.

——. "Draft Report on the Environment." 2003. Available at http://www.epa.gov/indicators/roe/html/roeAirGlo.htm.

Environmental Working Group (EWG). *Body Burden: Pollution in People.* Washington, D.C.: EWG, 2003. Available at http://www.ewg.org/reports/bodyburden/.

——. *Body Burden—The Pollution in Newborns: A Benchmark Investigation of Industrial Chemicals, Pollutants, and Pesticides in Umbilical Cord Blood. Executive Summary.* Washington, D.C.: EWG, 2005. Available at http://www.ewg.org/reports/bodyburden2/execsumm.php. Retrieved July 18, 2005.

——. *Study Finds Record High Levels of Toxic Fire Retardants in Breast Milk from American Mothers.* Washington, D.C.: EWG, 2003. Available at http://www.ewg.org/reports/mothersmilk/es.php.

Epstein, Samuel S. *The Politics of Cancer.* Garden City, N.Y.: Doubleday, 1998.

Epstein, Samuel S., Nicholas A. Ashford, Barry Castleman, Edward Goldsmith, Anthony Mazzocchi, and Quentin D. Young. "The Crisis in U.S. and International Cancer Policy." *International Journal of Health Services* 32 (4) (2002): 669–707.

Epstein, Steven. *Impure Science: AIDS, Activism, and the Politics of Knowledge.* Berkeley: University of California Press, 1996.

——. "U.S. AIDS Activism and the Question of Epistemological Radicalism." Paper presented at the Annual Meeting of the Society for the Social Study of Science, Cambridge, Mass., November 2001.

Erikson, Kai. *Everything in Its Path: Destruction of Community in the Buffalo Creek Flood.* New York: Simon and Schuster, 1976.

Ernster, Virginia L., John Barclay, Karla Kerlikowske, Deborah Grady, and I. Craig Henderson. "Incidence of and Treatment for Ductal Carcinoma in Situ of the Breast." *Journal of the American Medical Association* 275 (1996): 913–18.

Ernster, Virginia L., John Barclay, Karla Kerlikowske, H. Wilkie, and R. Ballard-Barbash. "Mortality Among Women with Ductal Carcinoma in Situ of the Breast in the Population-Based Surveillance, Epidemiology, and End Results Program." *Archives of Internal Medicine* 160 (7) (2000): 953–58.

European Industrial Relations Observatory (EIRO) On-Line. *Skanska Managers Found Guilty of Work Environment Crime.* Dublin: EIRO, 2002. Available at http://www.eiro.eurofound.eu.int/2002/02/feature/se0202106f.html.

Evans, David, Mindy Fullilove, Lesley Green, and Moshe Levison. "Awareness of Environmental Risks and Protective Actions Among Minority Women in Northern Manhattan." *Environmental Health Perspectives* 110 (suppl. 2) (2002): 271–75.

Evans, David, Mindy Fullilove, and Peggy Shepard. "Healthy Home, Healthy Child Campaign: A Community Intervention by the Columbia Center for Children's Environmental Health." Paper presented at the Annual Meeting of the American Public Health Association, Boston, November 15, 2000.

Evans, Nancy, ed. *State of the Evidence: What Is the Connection Between the Environment and Breast Cancer?* 3rd ed. San Francisco: Breast Cancer Fund and Breast Cancer Action, 2004.

——. *State of the Evidence: What Is the Connection Between the Environment and Breast Cancer?* 4th ed. San Francisco: Breast Cancer Fund and Breast Cancer Action, 2006. Available at http://www.breastcancerfund.org/environment_evidence_main.htm.

ExxonMobil. "Clearing the Air on Asthma." Advertisement, *New York Times,* November 15, 2001.

Ferber, Dan. "Overhaul of CDC Panel Revives Lead Safety Debate." *Science* 298 (2002): 732.

Fernandez, Esteve, Silvano Gallus, Cristina Bosetti, Silvia Franceschi, Eva Negri, and Carlo La Vecchia. "Hormone Replacement Therapy and Cancer Risk: A Systematic Analysis from a Network of Case-Control Studies." *International Journal of Cancer* 105 (2003): 408–12.

Fischer, Douglas. "CDC Results Are In: We're Full of Contaminants." *Oakland Tribute,* July 22, 2005.

Fischer, Frank. *Citizens, Experts, and the Environment: The Politics of Local Knowledge.* Durham, N.C.: Duke University Press, 2000.

Fischhoff, Baruch, Paul Slovic, and Sarah Lichtenstein. "Lay Foibles and Expert Fables in Judgments About Risk." Proceedings of the Sixth Symposium on Statistics and the Environment, *The American Statistician (Part 2)* 36 (3) (1982): 240–55.

Fishman, Jennifer. "Assessing Breast Cancer: Risk, Science, and Environmental Activism in an 'at Risk' Community." In *Ideologies of Breast Cancer: Feminist Perspectives,* ed. Laura K. Potts, 181–204. New York: St. Martin's, 2000.

Fleischer, Doris and Frieda Zames. *The Disability Rights Movement: From Charity to Confrontation.* Philadelphia: Temple University Press, 2002.

Frazier, L. M., J. W. Cromer, K. M. Andolsek, G. N. Greenberg, W. R. Thomann, and W. Stopford. "Teaching Occupational and Environmental Medicine in Primary Care Residency Training Programs: Experience Using Three Approaches During 1984–1991." *American Journal of the Medical Sciences* 302 (1) (1991): 42–45.

Freidson, Elliot. *Professional Dominance: The Social Structure of Medical Care.* Chicago: Aldine, 1970.

Frickel, Scott. "Just Science? Organizing Scientist Activism in the US Environmental Justice Movement." *Science as Culture* 13 (2004): 449–69.

Friedman, Michael S., Kenneth E. Powell, Lori Hutwagner, LeRoy M. Graham, and W. Gerald Teague. "Impact of Changes in Transportation and Commuting Behaviors during the 1996 Summer Olympic Games in Atlanta on Air Quality and Childhood Asthma." *Journal of the American Medical Association* 285 (2001): 897–905.

Fujimura. Joan. "Ecologies of Action: Recombining Genes, Molecularizing Cancer, and Transforming Biology." In *Ecologies of Knowledge: Work and Politics in Science and Technology,* ed. Susan Leigh Star, 302–46. Albany: State University of New York Press, 1995.

Fukuda, K., R. Nisenbaum, G. Stewart, W. W. Thompson, L. Robin, R. M. Washko, D. L. Noah, D. H. Barrett, B. Randall, B. L. Herwaldt, A. C. Mawle, and W. C. Reeves. "Chronic Multisymptom Illness Affecting Air Force Veterans of the Gulf War." *Journal of the American Medical Association* 280 (1998): 981–88.

Fulco, Carolyn E., Catharyn T. Liverman, and Harold Sox, eds. *Depleted Uranium, Sarin, Pyridostigmine Bromide, and Vaccines.* Vol. 1 of *Gulf War and Health.* Washington, D.C.: National Academies Press for the Institute of Medicine, 2000.

Gammon, Marilie D., Regina M. Santella, Alfred I. Neugut, Sybil M. Eng, Susan L. Teitelbaum, Andrea Paykin, Bruce Levin, Mary Beth Terry, Tie Lan Young, Lian Wen Wang, Qiao Wang, Julie A. Britton, Mary S. Wolff, Steven D. Stellman, Maureen Hatch, Geoffrey C. Kabat, Ruby Senie, Gail Garbowski, Carla Maffeo, Pat Montalvan, Gertrud Berkowitz, Margaret Kemeny, Marc Citron, Freya Schnabel, Allan Schuss, Steven Hajdu, and Vincent Vinceguerra. "Environmental Toxins and Breast Cancer on Long Island. I. Polycyclic Aromatic Hydrocarbon DNA Adducts." *Cancer Epidemiology Biomarkers and Prevention* 11 (2002): 677–85.

Gammon, Marilie D., Mary S. Wolff, Alfred I. Neugut, Sybil M. Eng, Susan L. Teitelbaum, Julie A. Britton, Mary Beth Terry, Bruce Levin, Steven D. Stellman, Geoffrey C. Kabat, Maureen Hatch, Ruby Senie, Gertrud Berkowitz, H. Leon Bradlow, Gail Garbowski, Carla Maffeo, Pat Montalvan, Margaret Kemeny, Marc Citron, Freya Schnabel, Allan Schuss, Steven Hajdu, Vincent Vinceguerra, Nancy Niguidula, Karen Ireland, and Regina M. Santella. "Environmental Toxins and Breast Cancer on Long Island. II. Organochlorine Compound Levels in Blood." *Cancer Epidemiology Biomarkers and Prevention* 11 (2002): 686–97.

Gamson, William, Bruce Fireman, and Steven Rytina. *Encounters with Unjust Authority.* Homewood, Ill.: Dorsey, 1982.

Gaventa, John. "Science and Citizenship in a Global Context." Paper presented at the Sussex Institute of Development Studies conference, Sussex, England, December 12–13, 2002.

Gee, David and Andrew Stirling. "Late Lessons from Early Warnings: Improving Science and Governance under Uncertainty and Ignorance." In *Precaution: Environmental Science and Preventive Public Policy,* ed. Joel Tickner, 195–214. Washington, D.C.: Island Press, 2003.

Geiser, Kenneth. *Materials Matter: Toward a Sustainable Materials Policy.* Cambridge, Mass.: MIT Press, 2001.

Gelbspan, Ross. *The Heat Is On: The Climate Crisis, the Cover-up, the Prescription.* Cambridge, Mass.: Perseus Books, 1997.

Glantz, Stanton A. *The Cigarette Papers.* Berkeley: University of California Press, 1996.

Godleski, John. "Mechanisms of Particulate Air Pollution Health Effects." Paper presented at the Annual Meeting of the American Public Health Association, Boston, November 14, 2000.

Gold, Diane R., Augusto Litonjua, Joel Schwartz, Eric Lovett, Andrew Larson, Bruce Nearing, George Allen, Monique Verrier, Rebecca Cherry, and Richard Verrier. "Ambient Pollution and Heart Rate Variability." *Circulation* 101 (2000): 1267–273.

Graham, Tolle, Jean Zotter, and Marlene Camacho. *Who's Sick at School: Linking Poor School Conditions and Health Disparities for Boston's Children.* Boston: Massachusetts Coalition for Occupational Safety and Health and the Boston Urban Asthma Coalition, 2006.

Grassroots Recycling Network. "RE. [greenyes] Great News on PVC Campaign." 2005. Available at http://greenyes.grrn.org/2005/02/msg00141.html. Retrieved June 7, 2005.

Gray, G. C., J. D. Knoke, S. W. Berg, F. S. Wignall, and E. Barrett-Connor. "Counterpoint: Responding to Suppositions and Misunderstandings." *American Journal of Epidemiology* 148 (1998): 328–33.

Gray, Gregory, Bruce Coate, Christy Anderson, Han Kang, S. William Berg, Stephen Wignall, James Knoke, and Elizabeth Barret-Conner. "The Postwar Hospitalization Experience of U.S. Veterans of the Persian Gulf War." *New England Journal of Medicine* 335 (1996): 1505–513.

Gray, Gregory, Gary Gackstetter, Han Kang, John Graham, and Ken Scott. "After More Than 10 Years of Gulf War Veteran Medical Evaluations, What Have We Learned?" *American Journal of Preventive Medicine* 26 (2004): 443–52.

Gray, J. A. Muir. "Postmodern Medicine." *The Lancet* 354 (1999): 1550–553.

Green, Lesley, Mindy Fullilove, David Evans, and Peggy Shepard. "'Hey, Mom, Thanks!' Use of Focus Groups in the Development of Place-Specific Materials for a Community Environmental Action Campaign." *Environmental Health Perspectives* 110 (suppl. 2) (2002): 265–69.

Greenbaum, Dan. "Interface of Science with Policy." Paper presented at the Annual Meeting of the American Public Health Association, Boston, November 14, 2000.

Greenbaum, Daniel S., John D. Bachmann, Daniel Krewski, Jonathan M. Samet, Ronald White, and Ronald E. Wyzga. "Particulate Air Pollution Standards and Morbidity and Mortality: Case Study." *American Journal of Epidemiology* 154 (2001): S78–S90.

Greenberg, Michael and Daniel Wartenberg. "Communicating to an Alarmed Community About Cancer Clusters: A Fifty State Survey." *Journal of Community Health* 16 (1991): 71–81.

Greenwire. "Air Pollution: Supreme Court's Interest Expected to Shake Up NSR Debate." *New Source Review: An E&E Special Report*, May 16, 2006. Available at http://www.eenews.net/Greenwire/Backissues/images/061402gwr1.pdf.

Griffith, J., R. C. Duncan, W. B. Riggan, and A. C. Pellon. "Cancer Mortality in U.S. Counties with Hazardous Waste Sites and Ground Water Pollution." *Archives of Environmental Health* 44 (2) (1989): 69–74.

Gulf War Illness Advisory Committee, Department of National Defence. *Health Study of Canadian Forces Personnel Involved in the 1991 Conflict in the Persian Gulf*. Vol. 1. Ottawa, Ontario: Goss Gilroy, Inc., 1998.

Guttes, S., K. Failing, K. Neumann, J. Kleinstein, S. Georgii, and H. Brunn. "Chlororganic Pesticides and Polychlorinated Biphenyls in Breast Tissue of Women with Benign and Malignant Breast Disease." *Archives of Environmental Contamination and Toxicology* 35 (1998): 140–47.

Hadler, Nortin M. "If You Have to Prove You Are Ill, You Can't Get Well: The Object Lesson of Fibromyalgia." *Spine* 21 (20) (1996): 2397–400.

Haley, Robert, Scott Billecke, and Bert La Du. "Association of Low PON1 Type Q (Type A) Arylesterase Activity with Neurological Symptom Complexes in Gulf War Veterans." *Toxicology and Applied Pharmacology* 157 (1999): 227–33.

Haley, Robert, James Fleckenstein, W. Wesley Marshall, George McDonald, Gerald Kramer, and Frederick Petty. "Effect of Basal Ganglia Injury on Central Dopamine Activity in Gulf War Syndrome: Correlation of Proton Magnetic Resonance Spectroscopy and Plasma Homovanillic Acid." *Archives of Neurology* 57 (2000): 1280–285.

Haley, Robert, Thomas Kurt, and Jim Hom. "Is There a Gulf War Syndrome? Searching for Syndromes by Factor Analysis of Symptoms." *Journal of the American Medical Association* 277 (1997): 215–22.

Haley, Robert, W. Wesley Marshall, George McDonald, Mark Daugherty, Frederick Petty, and James Fleckenstein. "Brain Abnormalities in Gulf War Syndrome: Evaluation by H Magnetic Resonance Spectroscopy." *Radiology* 215 (2000): 807–17.

Hansen, Johnni. "Breast Cancer Risk Among Relatively Young Women Employed in Solvent-Using Industries." *American Journal of Industrial Medicine* 36 (1999): 43–47.

Haraway, Donna J. "Situated Knowledges: The Science Question in Feminism and the Privilege of Partial Perspective." *Feminist Studies* 14 (1988): 575–99.

Harding, Sandra. *Is Science Multicultural?* Bloomington: Indiana University Press, 1998.

Harr, Jonathan. *A Civil Action*. New York: Vintage, 1995.

Hartsock, Nancy. *The Feminist Standpoint Revisited and Other Essays*. Boulder, Colo.: Westview, 1998.

Healthy Building Network. *The Louisville Charter for Safer Chemicals—A Platform for Creating a Safe and Healthy Environment Through Innovation*. Washington, D.C.: Healthy Building Network, 2005. Available at http://www.healthybuilding.net/environmental_justice/louisville_charter.pdf.

Henderson, Rogene F., Edward B. Barr, Walter B. Blackwell, Connie R. Clark, Carole A. Conn, Roma Kalra, Thomas H. March, Mohan L. Sopori, Yohannes Tesfaigzi, Margaret G. Ménache, and Deborah C. Mash. "Response of Rats to Low Levels of Sarin." *Journal of Toxicology and Applied Pharmacology* 184 (2002): 67–76.

Henderson, Rogene, Edward Barr, Walter Blackwell, Connie Clark, Carole Conn, Roma Kalra, Thomas March, Mohan Sopori, Yohannes Tesfaigzi, Margaret Ménache, Deborah Mash, Karol Doklandny, Wieslaw Kozak, Anna Kozak, Maceij Wachulec, Karin Rudolph, Matthew Kluger, Shashi Singh, Seddigheh Razani-Boroujerdi, and Raymond Langley. "Response of F344 Rats to Inhalation of Subclinical Levels of Sarin: Exploring Potential Causes of Gulf War Illness." *Journal of Toxicology and Industrial Health* 17 (5–10) (2001): 294–97.

Herrick, Thaddeus. "Cosmetics Companies Shun Contentious Chemical." *Wall Street Journal*, January 14, 2005.

Hess, David. "The Problem of Undone Science: Values, Interests, and the Selection of Research Programs." Paper presented at the Annual Meeting of the Society for the Social Study of Science, Halifax, Nova Scotia, October 30–November 2, 1999.

Hoek, G., B. Brunekreef, S. Goldbohm, P. Fischer, and P. Van Den Brandt. "Association Between Mortality and Indicators of Traffic-Related Air Pollution in the Netherlands: A Cohort Study." *The Lancet* 360 (2002): 1203–209.

Hoge, Charles W., Carl A. Castro, Stephen C. Messer, Dennis McGurk, Dave I. Cotting, and Robert L. Koffman. "Combat Duty in Iraq and Afghanistan, Mental Health Problems, and Barriers to Care." *New England Journal of Medicine* 351 (2004): 13–22.

Houlihan, Jane, Charlotte Brody, and Bryony Schwan. *Not Too Pretty: Phthalates, Beauty Products, and the FDA*. Washington, D.C.: Environmental Working Group, 2002.

Houts, Peter S., Paul D. Cleary, and Te-Wei Hu. *The Three Mile Island Crisis: Psychological, Social, and Economic Impacts on the Surrounding Population*. University Park: Pennsylvania State University, 1988.

Hoyer, A., A. M. Gerdes, T. Jorgensen, F. Rank, and H. Hartvig. "Organochlorines, P53 Mutations in Relation to Breast Cancer Risk and Survival: A Danish Cohort-Nested Case-Control Study." *Breast Cancer Research and Treatment* 71 (2002): 59–65.

Hoyer, A. P., T. Jorgensen, J. W. Brock, and P. Grandjean. "Organochlorine Exposure and Breast Cancer Survival." *Journal of Clinical Epidemiology* 53 (3) (2000): 323–30.

Hoyer, A., Torben Jorgensen, Philippe Grandjean, and Helle Boggild Hartvig. "Repeated Measures of Organochlorine Exposure and Breast Cancer Risk." *Cancer Causes and Control* 11 (1999): 177–84.

Hsu, Karen. "Boston Will Get $1.9 Million to Prevent Asthma." *Boston Globe*, February 24, 2000.

Hubbard, Ruth. *The Politics of Women's Biology*. New Brunswick, N.J.: Rutgers University Press, 1990.

Hunter, David J., Susan Hankinson, Francine Laden, Graham Colditz, JoAnne Manson, Walter Willett, Frank Speizer, and Mary S. Wolff. "Plasma Organochlorine Levels and the Risk of Breast Cancer." *New England Journal of Medicine* 337 (1997): 1253–258.

Hunter, David and Karl T. Kelsey. "Pesticide Residues and Breast Cancer: The Harvest of the Silent Spring?" *Journal of the National Cancer Institute* 85 (8) (1993): 598–99.

Husain, K., R. Vijayaraghavan, S. C. Pant, S. K. Raza, and K. S. Pandey. "Delayed Neurotoxic Effect of Sarin in Mice After Repeated Inhalation Exposure." *Journal of Applied Toxicology* 13 (1993): 143–45.

Hyams, Kenneth, Stephen Wignall, and Robert Roswell. "War Syndromes and Their Evaluation: From the U.S. Civil War to the Persian Gulf War." *Annals of Internal Medicine* 125 (1996): 298–305.

Hynes, Patricia H. *The Recurring Silent Spring*. Elmsford, N.Y.: Pergamon, 1989.

Institute of Medicine. *Addressing the Physician Shortage in Occupational and Environmental Medicine: Report of a Study*. Washington, D.C.: National Academies Press, 1991.
———. *Health Consequences of Service During the Persian Gulf War: Recommendations for Research and Information Systems*. Washington, D.C.: National Academies Press, 1996.
———. *Health Effects of Serving in the Gulf War*. Vol. 4 of *Gulf War and Health*. Washington, D.C.: National Academies Press, 2006.
———. *Veterans and Agent Orange: Health Effects of Herbicides Used in Vietnam*. Washington, D.C.: National Academies Press, 1994.
Institute of Medicine, Committee on Gulf War and Health. *Fuels, Combustion Products, and Propellants*. Vol. 3 of *Gulf War and Health*. Washington, D.C.: National Academies Press, 2005.
Iowa Persian Gulf Study Group. "Self-Reported Illness and Health Status Among Gulf War Veterans." *Journal of the American Medical Association* 277 (1997): 238–45.
Irwin, Alan. *Citizen Science: A Study of People, Expertise, and Sustainable Development*. London: Routledge, 1995.
Israel, Barbara A., Amy J. Schulz, Edith A. Parker, and Adam B. Becker. "Review of Community-Based Research: Assessing Partnership Approaches to Improve Public Health." *Annual Review of Public Health* 19 (1998): 173–202.
Jackson, Richard J. "Habitat and Health: The Role of Environmental Factors in the Health of Urban Populations." *Journal of Urban Health* 75 (2) (1998): 258–62.
Jelinek, Pauline. "Judge OK's Voluntary Anthrax Vaccination." *Boston Globe*, April 8, 2005.
Jerrett, Michael, Richard T. Burnett, Renjun Ma, C. Arden Pope, Daniel Krewski, Bruce K. Newbold, George Thurston, Yuanli Shi, Norm Finkelstein, Eugenia E. Calle, and Michael J. Thun. "Spatial Analysis of Air Pollution and Mortality in Los Angeles." *Epidemiology* 16 (6) (2005): 727–36.
Joellenback, Lois, Philip Russell, and Samuel B. Guze, eds. 1999. *Strategies to Protect the Health of Deployed U.S. Forces: Medical Surveillance, Record Keeping, and Risk Reduction*. Washington, D.C.: National Academies Press for the Institute of Medicine, 1999.
Johannes, L. "Pollution Study Sparks Debate over Secret Data." *Wall Street Journal*, April 2, 1997.
Joyner-Kersee, Jackie. "Asthma and the Athlete's Challenge." *New York Times*, August 13, 2001.
Kang, Han and Tim Bullman. "Mortality Among U.S. Veterans of the Persian Gulf War." *New England Journal of Medicine* 335 (1996): 1498–504.
Kant, A. K., A. Schatzkin, B. I. Graubard, and C. Schairer. "A Prospective Study of Diet Quality and Mortality in Women." *Journal of the American Medical Association* 283 (2000): 2109–115.
Kaplan, Louise. "Public Participation in Nuclear Facility Decisions." In *Science, Technology, and Democracy*, ed. Daniel Kleinman, 67–83. Albany: State University of New York Press, 2000.
Kasper, Anne and Susan Ferguson, eds. *Breast Cancer: Society Shapes an Epidemic*. New York: Palgrave, 2000.
Keeler, Gerald J., Timothy J. Dvonch, Fuyen Yip, Edith A. Parker, Barbara A. Israel, Frank J. Marsik, Masako Morishita, James A. Barres, Thomas G. Robins, Wilma Brakefield-Caldwell, and Mathew Sam. "Assessment of Personal and Community-Level Exposures to Particulate Matter (PM) among Children with Asthma in Detroit, Michigan, as Part of Community Action Against Asthma (CAAA)." *Environmental Health Perspectives* 110 (suppl. 2) (2002): 173–81.
Kelley, Eileen. "GIs Link Ills to Malaria Drug: Two Fort Carson Soldiers Suspect the Medication Given Them in Iraq Is the Cause of Their Mental Problems." *Denver Post*, June 11, 2004.
Kennedy, Donald. "An Epidemic of Politics." *Science* 299 (2003): 625.

Kilshaw, Susie. "Friendly Fire: The Construction of Gulf War Syndrome Narratives." *Anthropology and Medicine* 11 (2) (2004): 149–60.

King, Samantha. "An All-Consuming Cause: Breast Cancer, Corporate Philanthropy, and the Market for Generosity." *Social Text* 19 (4) (2001): 115–43.

Kipen, Howard M., William Hallman, Han Kang, Nancy Fiedler, and Benjamin H. Natelson. "Prevalence of Chronic Fatigue and Chemical Sensitivities in Gulf Registry Veterans." *Archives of Environmental Health* 54 (1999): 313–18.

Klawiter, Maren. "Breast Cancer in Two Regimes: The Impact of Social Movements on Illness Experience." *Sociology of Health and Illness* 26 (2004): 845–74.

——. "Racing for the Cure, Walking Women, and Toxic Touring: Mapping Cultures of Action within the Bay Area Terrain of Breast Cancer." *Social Problems* 46 (1999): 104–26.

Kleinman, Daniel. "Democratization of Science and Technology." In *Science, Technology, and Democracy*, ed. Daniel Kleinman, 139–63. Albany: State University of New York Press, 2000.

Kleinman, Daniel L., ed. *Science, Technology, and Democracy*. Albany: State University of New York Press, 2000.

Klemm, Rebecca, Robert Mason Jr., Charles Heilig, Lucas Neas, and Douglas Dockery. "Is Daily Mortality Associated Specifically with Fine Particles? Data Reconstruction and Replication of Analysis." *Journal of the Air Waste Management Association* 50 (2000): 1215–222.

Kolata, Gina. "No Rise Found in Death Rates after Gulf War." *New York Times*, November 14, 1996.

Kolker, Emily S. "Framing as a Cultural Resource in Health Social Movements: Funding Activism and the Breast Cancer Movement in the US 1990–1993." *Sociology of Health and Illness* 26 (2004): 820–44.

Krewski, Daniel, Richard T. Burnett, Mark S. Goldberg, B. Kristin Hoover, Jack Siemiatycki, Michael Jerrett, Michal Abrahamowicz, and Warren H. White. *Reanalysis of the Harvard Six Cities Study and the American Cancer Society Study of Particulate Air Pollution and Mortality*. Cambridge, Mass.: Health Effects Institute, 2000.

Krieger, Nancy. "Social Class and the Black/White Crossover in the Age-Specific Incidence of Breast Cancer: A Study Linking Census-Derived Data to Population-Based Registry Records." *American Journal of Epidemiology* 131 (1990): 804–14.

Krieger, N. and S. Zierler. "Accounting for the Health of Women." *Current Issues in Public Health* 1 (1995): 251–56.

Krimsky, Sheldon. *Hormonal Chaos: The Scientific and Social Origins of the Environmental Endocrine Hypothesis*. Baltimore: Johns Hopkins University Press, 2000.

——. *Science in the Private Interest: Has the Lure of Profits Corrupted Biomedical Research?* Lanham, Md.: Rowman and Littlefield, 2003.

Kroll-Smith, Steven and Stephen R. Couch. *The Real Disaster Is Above Ground: A Mine Fire and Social Conflict*. Lexington: University Press of Kentucky, 1990.

Kroll-Smith, Steven and Hugh H. Floyd. *Bodies in Protest: Environmental Illness and the Struggle Over Medical Knowledge*. New York: New York University Press, 1997.

Kuhn, Thomas. *The Structure of Scientific Revolution*. 2d ed. Chicago: University of Chicago Press, 1970.

Labreche, F. and M. S. Goldberg. "Exposure to Organic Solvents and Breast Cancer in Women: A Hypothesis." *American Journal of Industrial Medicine* 32 (1997): 1–14.

Laden, Francine, Lucas M. Neas, Douglas W. Dockery, and Joel Schwartz. "Association of Fine Particulate Matter from Different Sources with Daily Mortality in Six U.S. Cities." *Environmental Health Perspectives* 108 (2000): 941–47. Available at http://www.ehponline.org/members/2000/108p941-947laden/laden.pdf. Retrieved June 23, 2006.

Landis, S. H., T. Murray, S. Bolden, and P. A. Wingo. "Cancer Statistics, 1999." *CA: A Cancer Journal for Clinicians* 49 (1) (1999): 8–31.

Landrigan, Philip. "Illnesses in Gulf War Veterans: Causes and Consequences." *Journal of the American Medical Association* 277 (1997): 259–61.

Landrigan, Philip J., Clyde B. Schecter, Jeffrey M. Lipton, Marianne C. Fahs, and Joel Schwartz. "Environmental Pollutants and Disease in American Children: Estimates of Morbidity, Mortality, and Costs for Lead Poisoning, Asthma, Cancer, and Developmental Disabilities." *Environmental Health Perspectives* 110 (2002): 721–28.

Lantz, Paula M., Carol S. Weisman, and Zena Itani. "A Disease-Specific Medicaid Expansion for Women: The Breast and Cervical Cancer Prevention and Treatment Act of 2000." *Women's Health Issues* 13 (2003): 79–92.

Latour, Bruno. *Science in Action: How to Follow Scientists and Engineers Through Society.* Cambridge, Mass.: Harvard University Press, 1987.

Ledford, Angela. "Foreword: The Dirty Secret Behind Dirty Air: Dirty Power." In *Dirty Air, Dirty Power: Mortality and Health Damage due to Pollution from Power Plants*, by Conrad Schneider, 1–3. Boston: Clean Air Task Force, 2004.

Legator, Marvin and Sabrina Strawn. *Chemical Alert: A Community Action Handbook.* Austin: University of Texas Press, 1993.

Lehmann, Christine. "Military Boosts Monitoring of Soldiers' Mental Health." *Psychiatric News* 40 (5) (2005): 8.

Levine, Adeline Gordon. *Love Canal: Science, Politics, and People.* Lexington, Mass.: Heath, 1982.

Levins, Richard. "Whose Scientific Method: Scientific Methods for a Complex World." In *Precaution: Environmental Science and Preventive Public Policy*, ed. Joel Tickner, 355–68. Washington, D.C.: Island Press, 2003.

Levy, Jonathan I., E. Andres Houseman, John D. Spengler, Penn Loh, and Louise Ryan. "Fine Particulate Matter and Polycyclic Hydrocarbon Concentration Patterns in Roxbury, Massachusetts: A Community-Based GIS Analysis." *Environmental Health Perspectives* 109 (2001): 341–47.

Lichtenstein, Paul, Niels Holm, Pia Verkasalo, Anastasia Iliadou, Jaakko Kaprio, Markku Koskenvuo, Eero Pukkala, Axel Skytthe, and Kari Hemminki. "Environmental and Heritable Factors in the Causation of Cancer: Analyses of Cohorts of Twins from Sweden, Denmark, and Finland." *New England Journal of Medicine* 343 (2000): 78–85.

Lin, Shao, Jean Pierre Munsie, Syni-An Hwand, Edward Fitzgerald, and Michael R. Cayo. "Childhood Asthma Hospitalization and Residential Exposure to State Route Traffic." *Environmental Research* 88 (2) (2002): 73–81.

Link, Bruce and Jo Phelan. "Social Conditions as Fundamental Causes of Disease." *Journal of Health and Social Behavior* extra issue (1995): 80–94.

Lipscomb, J. A., K. P. Satin, and R. R. Neutra. "Reported Symptom Prevalence Rates from Comparison Populations in Community-Based Environmental Studies." *Archives of Environmental Health* 47 (1992): 263–69.

Llanos, Miguel. "EPA Official Quits, Rips White House Regulatory Chief, Cites Push 'to Weaken the Rules.'" *MSNBC*, February 28, 2002.

Loh, Penn and Jodi Sugerman-Brozan. "Environmental Justice Organizing for Environmental Health: Case Study on Asthma and Diesel Exhaust in Roxbury, Massachusetts." *Annals of the American Academy of Political and Social Science* 584 (2002): 110–24.

Loh, Penn, Jodi Sugerman-Brozan, Standrick Wiggins, David Noiles, and Cecelia Archibald. "From Asthma to AirBeat: Community-Driven Monitoring of Fine Particles and Black Carbon in Roxbury, Massachusetts." *Environmental Health Perspectives* 110 (suppl. 2) (2002): 297–301.

Lopez-Carillo, L., A. Blair, M. Lopez-Cervantes, M. Cebrian, C. Rueda, R. Reyes, A. Mohar, and J. Bravo. "Dichlorodiphenyltrichloroethane Serum Levels and Breast Cancer Risk: A Case Study from Mexico." *Cancer Research* 57 (1997): 3728–732.

Lowell Center for Sustainable Production. *Integrated Chemicals Policy: Seeking New Directions in Chemical Management*. Lowell, Mass.: Lowell Center for Sustainable Production, 2003.

Mackness, Bharti, Paul N. Durrington, and Michael I. Mackness. "Low Paraoxonase in Persian Gulf War Veterans Self-Reporting Gulf War Syndrome." *Biochemical and Biophysical Research Communications* 276 (2000): 729–33.

MacMahon, B. "Pesticide Residues and Breast Cancer?" *Journal of the National Cancer Institute* 86 (1994): 572–73.

Madigan, M. Patricia, Regina Ziegler, Jacques Benichou, Celia Byrne, and Robert Hoover. "Proportion of Breast Cancer Cases in the United States Explained by Well-Established Risk Factors." *Journal of the National Cancer Institute* 87 (1995): 1681–685.

Malcoe, Lorraine Halinka, Robert A. Lynch, Michelle Crozier Kegler, and Valerie J. Skaggs. "Lead Sources, Behaviors, and Socioeconomic Factors in Relation to Blood Lead of Native American and White Children: A Community-Based Assessment of a Former Mining Area." *Environmental Health Perspectives* 110 (suppl. 2) (2002): 221–31.

Mannino, David M., David M. Homa, Lara J. Akinbami, Jeanne E. Moorman, Charon Gwynn, and Stephen C. Redd. "Surveillance for Asthma–United States, 1980–1999." *Morbidity and Mortality Weekly Report* 51 (SS-1) (2002): 1–13.

Mannino, David M., David M. Homa, Carol A. Pertowski, Annette Ashizawa, Leah L. Nixon, Carol A. Johnson, Lauren B. Ball, Elizabeth Jack, and David S. Kang. "Surveillance for Asthma—United States 1960–1995." *Morbidity and Mortality Weekly Report* 47 (SS-1) (1998): 1–27.

Marcus, George E. "Ethnography in/of the World System: The Emergence of Multi-Sited Ethnography." *Annual Review of Anthropology* 24 (1) (1995): 95–118.

Markowitz, Gerald and David Rosner. *Denial and Deceit: The Deadly Politics of Industrial Pollution*. Berkeley: University of California Press, 2002.

Massey-Stokes, Marilyn. "Foreword: Environmental Issues in the Health of Children." *Family and Community Health* 24 (4) (2002): viii–ix.

Matsuoka, Martha, ed. *Building Healthy Communities from the Ground Up: Environmental Justice in California*. Oakland, Calif.: Asian Pacific Environmental Network, Communities for a Better Environment, Environmental Health Coalition, People Organizing to Demand Environmental and Economic Rights, Silicon Valley Toxics Coalition Health, and Environmental Justice Project, 2003.

Mayer, Brian, Phil Brown, and Meadow Linder. "Moving Further Upstream: From Toxics Reduction to the Precautionary Principle." *Public Health Reports* 117 (2002): 574–86.

Mayou, Richard and Michael Sharpe. "Treating Medically Unexplained Physical Symptoms." *British Medical Journal* 315 (1997): 561–62.

McAdam, Doug, John D. McCarthy, and Mayer N. Zald. "Social Movements." In *Handbook of Sociology*, ed. Neil Smelser, 695–737. Beverly Hills, Calif.: Sage, 1989.

McConnell, Rob, Kiros Berhane, Frank Gilliland, Stephanie J. London, Talat Islam, W. James Gauderman, Edward Avol, Helene G. Margolis, and John M. Peters. "Asthma in Exercising Children Exposed to Ozone: A Cohort Study." *The Lancet* 359 (2002): 386–91.

McCormick, Sabrina, Julia Brody, Phil Brown, and Ruth Polk. "Lay Involvement in Breast Cancer Research." *International Journal of Health Services* 34 (2004): 625–46.

McCormick, Sabrina, Phil Brown, and Stephen Zavestoski. "The Personal Is Scientific, the Scientific Is Political: The Public Paradigm of the Environmental Breast Cancer Movement." *Sociological Forum* 18 (2003): 545–76.

McKelvey, Wendy, Julia Green Brody, Ann Aschengrau, and Christopher H. Schwartz. "Association Between Residence on Cape Cod, Massachusetts, and Breast Cancer." *Annals of Epidemiology* 14 (2004): 89–94.

Meikle, James. "U.S. Scientist Challenges UK on Gulf War Illness." *The Guardian*, August 4, 2004.

Meller, Paul. "Europe Proposes Overhaul of Chemical Industry." *New York Times*, October 30, 2003.

Mendonça, Gulnar A. S., José Eluf-Neto, Maria J. Andrada-Serpa, Pedro A. O. Carmo, Heloisa H. C. Barreto, Odete N. K. Inomata, and Tereza A. Kussumi. "Organochlorines and Breast Cancer: A Case-Control Study in Brazil." *International Journal of Cancer* 83 (1999): 596–600.

Meyerhoff, D. J., J. Lindgren, D. Hardin, J. M. Griffis, and M. W. Weiner. "Reduced N-Acetylaspartate in the Right Basal Ganglia of Ill Gulf War Veterans by Magnetic Resonance Spectroscopy." *Proceedings of the International Society of Magnetic Resonance Medicine* 9 (2001): 994.

Michaels, David, Eula Bingham, Les Boden, Richard Clapp, Lynn R. Goldman, Polly Hoppin, Sheldon Krimsky, Celeste Monforton, David Ozonoff, and Anthony Robbins. "Advice Without Dissent." *Science* 298 (2002): 703.

Middendorf, Gerald and Lawrence Busch. "Inquiry for the Public Good: Democratic Participation in Agricultural Research." *Agriculture and Human Values* 14(1) (1997): 45–57.

Millikan, R., E. DeVoto, E. Fuell, C. K. Tse, J. Beach, S. Edminston, S. Jackson, and B. Newman. "DDE, PCBs, and Breast Cancer Among African-American and White Residents of North Carolina." *Cancer Epidemiology Biomarkers and Prevention* 9 (11) (2000): 1233–240.

Mills, C. Wright. *The Sociological Imagination*. New York: Oxford University Press, 1959.

Minkler, Meredith and Nina Wallerstein, eds. *Community-Based Participatory Research for Health*. San Francisco: Jossey-Bass, 2003.

Montague, Peter. "The Revolution, Part 3: Ultrafines." *Rachel's Environmental and Health News* 774 (July 24, 2003). Available at http://www.rachel.org.

———. "San Francisco Adopts the Precautionary Principle." *Rachel's Environment and Health News* 765 (March 20, 2003). Available at http://www.rachel.org.

Mooney, Chris. *The Republican War on Science*. Cambridge, Mass.: Basic Books, 2005.

Moore, Kelly. "Organizing Integrity: American Science and the Creation of Public Interest Science Organizations, 1955–1975." *American Journal of Sociology* 101 (1996): 1592–627.

———. "Powered by the People: Varieties of Participatory Science as Challenges to Scientific Authority." In *The New Political Sociology of Science: Institutions, Networks, and Power*, ed. Scott Frickel and Kelly Moore, 299–323. Madison: University of Wisconsin Press, 2006.

Morello-Frosch, Rachel, Manuel Pastor Jr., and James Sadd. "Integrating Environmental Justice and the Precautionary Principle in Research and Policy-Making: The Case of Ambient Air Toxics Exposures and Health Risks Among School Children in Los Angeles." *Annals of the American Academy of Political and Social Science* 584 (2002): 47–68.

Morello-Frosch, Rachel, Stephen Zavestoski, Phil Brown, Rebecca Gasior Altman, Sabrina McCormick, and Brian Mayer. "Embodied Health Movements: Responses to a 'Scientized' World." In *The New Political Sociology of Science: Institutions, Networks, and Power*, ed. Scott Frickel and Kelly Moore, 244–71. Madison: University of Wisconsin Press, 2006.

Morgen, Sandra. *Into Our Own Hands: The Women's Health Movement in the United States, 1969–1990*. New Brunswick, N.J.: Rutgers University Press, 2002.

Morris, David B. *Illness and Culture in the Postmodern Age*. Berkeley: University of California Press, 1998.

Moysich, Kirsten. "Thoughts on Recent Findings Regarding Organochlorines and Breast Cancer Risk." *The Ribbon* 6 (3) (2001): 5–6.

Myers, Nancy. "Debating the Precautionary Principle." *Science and Environmental Health Network* (March 2000). Available at http://www.sehn.org/ppdebate.html. Retrieved June 6, 2005.

Myhre, Jennifer R. "Breast Cancer Activism: A True Political Movement?" *Breast Cancer Action Newsletter* 68 (November–December 2001). Available at http://www.bcaction.org/Pages/GetInformed/Newsletters.html.

Najem, G. R., D. B. Louria, M. A. Lavenhar, and M. Feurman. "Clusters of Cancer Mortality in New Jersey Municipalities with Special Reference to Chemical Toxic Waste Disposal Sites and per Capita Income." *International Journal of Epidemiology* 14 (1985): 528–37.

National Breast Cancer Coalition (NBCC). *Priority #6—Fact Sheet*. Washington, D.C.: NBCC, 2001. Available at http://www.natlbcc.org/. Retrieved April 14, 2002.

National Children's Study. *What Is the National Children's Study?* Rockville, Md.: National Children's Study, 2005. Available at http://nationalchildrensstudy.gov/about/mission/overview.cfm. Retrieved June 22, 2005.

———. *What Makes This Study Different from Other U.S. Health Studies?* Rockville, Md.: National Children's Study, 2005. Available at http://nationalchildrensstudy.gov/about/mission/unique.cfm. Retrieved June 22, 2005.

National Environmental Trust. *The Bush Administration's Anti-environmental Actions*. Washington, D.C.: National Environmental Trust, 2005. Available at http://www.net.org/reports/rollbacks.vtml. Retrieved June 22, 2005.

National Gulf War Resource Center, *Uncounted Casualties: America's Ailing Veterans—A Special Report* (Silver Spring, Md.: National Gulf War Resource Center, January 17, 2001), 4.

National Health Interview Survey. *NCHS Data Fact Sheet*. Hyattsville, Md.: National Center for Health Statistics, January 1997.

National Institute of Environmental Health Sciences (NIEHS). *Community Outreach for CTD Screening in High Risk Groups*. Research Triangle Park, N.C.: NIEHS, 2005. Available at http://www.niehs.nih.gov/translat/envjust/projects/fraser.htm. Retrieved June 11, 2005.

———. *Environmental Justice—Archived Grantees*. Research Triangle Park, N.C.: NIEHS, 2005. Available at http://www.niehs.nih.gov/translat/envjust/grantold.htm. Retrieved June 7, 2005.

———. *Healthy Food, Healthy School, and Healthy Community*. Research Triangle Park, N.C.: NIEHS, 2005. Available at: http://www.niehs.nih.gov/translat/envjust/grantold.htm. Retrieved June 8, 2005.

National Institutes of Health (NIH). "The Persian Gulf Experience and Health." Workshop statement, Technology Assessment Workshop, Washington D.C., April 27–29, 1994.

National Research Council. *Environmental Epidemiology: Public Health and Hazardous Wastes*. Washington, D.C.: National Academies Press, 1991.

———. *Interim Report of the Committee on Changes in New Source Review Programs for Stationary Sources of Air Pollutants*. Washington, D.C.: National Academies Press, 2005. Available at http://darwin.nap.edu/books/0309095786/html/R1.html.

———. *Use of the Gray Literature and Other Data in Environmental Epidemiology*. Vol. 2 of *Environmental Epidemiology*. Washington, D.C.: National Academies Press, 1997.

National Resources Defense Council. "Court Strikes Down Bush Rollbacks of Clean Air Act, Upholds Others." 2005. Available at the National Resources Defense Council Press Archive, http://nrdc.org/media/pressReleases/050624.asp.

National Women's Health Network. *The Truth about Hormone Replacement Therapy: How to Break Free from the Medical Myths of Menopause*. Roseville, Calif.: Prima, 2002.

Nelson, Lin. "Women's Lives Against the Industrial Chemical Landscape: Environmental Health and the Health of the Environment." In *Healing Technologies: Feminist Perspectives*, ed. Kathryn Strother Ratcliff, 347–69. Ann Arbor: University of Michigan Press, 1989.

Nesmith, Jeff. "Children's Health Study in Need of Money." *Atlanta Journal-Constitution,* May 11, 2005.

Ng, K. L. and D. M. Hamby. "Fundamentals for Establishing a Risk Communication Program." *Health Physics* 73 (3) (1997): 465–72.

O'Fallon, Liam R. and Allen Dearry. "Community-Based Participatory Research as a Tool to Advance Environmental Health Sciences." *Environmental Health Perspectives* 110 (2002): 155–59.

Olaya-Contreras, P., J. Rodriguez-Villami, H. J. Posso-Valencia, and J. E. Cortez. "Organochlorines and Breast Cancer Risk in Colombian Women." *Cadernos de Saude Publica* 14 (1998):125–32.

Ong, E. K. and S. Glantz. "Constructing 'Sound Science' and 'Good Epidemiology': Tobacco, Lawyers, and Public Relations Firms." *American Journal of Public Health* 91 (2001): 1749–757.

Oppel, Richard A., Jr., and Christopher Drew. "States Planning Own Lawsuits Over Pollution." *New York Times,* November 9, 2003.

Orenstein, Susan. "The Selling of Breast Cancer: Is Corporate America's Love Affair with a Disease That Kills 40,000 Women a Year Good Marketing—or Bad Medicine?" *Business2.0* (February 2003). Available at http://www.business2.com/.

O'Rourke, Dara. *Community-Driven Regulation: Balancing Development and the Environment in Vietnam.* Cambridge, Mass.: MIT Press, 2004.

———. "Outsourcing Regulation: Analyzing Nongovernmental Systems of Labor Standards and Monitoring." *Policy Studies Journal* 31 (2003): 1–29.

Packard, Randall M., Peter J. Brown, Ruth L. Berkelman, and Howard Frumkin. "Emerging Illnesses as Social Process." In *Emerging Illnesses and Society: Negotiating the Public Health Agenda,* ed. Randall M. Packard, Peter J. Brown, Ruth L. Berkelman, and Howard Frumkin, 1–31. Baltimore: Johns Hopkins University Press, 2004.

Pandya, Robert J., Gina Solomon, Amy Kinner, and John R. Balme. "Diesel Exhaust and Asthma: Hypotheses and Molecular Mechanisms of Action." *Environmental Health Perspectives* 110 (suppl. 1) (2002): 103–12.

Pasternak, Judy. "Bush's Energy Plan Bares Industry Clout." *Los Angeles Times,* August 26, 2001.

Patel, Neha. "Portland, Oregon, City and County Adopt Precautionary Approach." *Everyone's Backyard* 22 (4) (2004): 3, 10.

Pearlin, Leonard and Carol Aneshensel. "Coping and Social Supports: Their Functions and Applications." In *Applications of Social Science to Clinical Medicine and Health Policy,* ed. Linda Aiken and David Mechanic, 417–37. New Brunswick, N.J.: Rutgers University Press, 1986.

Pellerano, Maria. "Precautionary Mister Rogers, Part 2." *Rachel's Environment and Health News* 803 (October 28, 2004). Available at http://www.rachel.org.

Pellow, David N. "Environmental Inequality Formation: Toward a Theory of Environmental Injustice." *American Behavioral Scientist* 43 (2000): 581–601.

Perera, Frederica P., Susan M. Illman, Patrick L. Kinney, Robin M. Whyatt, Elizabeth A. Kelvin, Peggy Shepard, David Evans, Mindy Fullilove, Jean Ford, Rachel L. Miller, Ilan H. Meyer, and Virginia A. Rauh. "The Challenge of Preventing Environmentally Related Disease in Young Children: Community-Based Research in New York City." *Environmental Health Perspectives* 110 (2002): 197–205.

Perlin, Jonathan B., M.D., Ph.D, Deputy Undersecretary for Health, Department of Veterans Affairs. "Testimony on Pre- and Post Deployment Health Assessments," 2003. In *Federal Document Clearing House Congressional Testimony.* Bethesda, MD: Congressional

Information Service, July 9, 2004. Available from *LexisNexis™ Congressional* (on-line service).

Persian Gulf Veterans Coordinating Board, Research Working Group. *Research on Gulf War Veterans' Illnesses, Annual Report to Congress—1998*. Washington, D.C.: Persian Gulf Veterans Coordinating Board, 1999. Available at http://www.va.gov/resdev/pgulf98/gwrpt98.htm. Retrieved October 1, 2000.

Pew Environmental Health Commission. *America's Environmental Health Gap: Why the Country Needs a Nationwide Health Tracking Network*. Washington, D.C.: Pew Environmental Health Commission, 2000.

———. *Attack Asthma: Why America Needs a Public Health Defense System to Battle Environmental Threats*. Washington, D.C.: Pew Environmental Health Commission, May 16, 2000.

Picou, Steven. *Social Disruption and Psychological Stress in an Alaskan Fishing Community: The Impact of the Exxon Valdez Oil Spill*. Boulder: University of Colorado Natural Hazards Center, 1990.

Platner, Janice H., L. Michelle Bennet, Robert Millikan, and Mary D. G. Barker. "The Partnership Between Breast Cancer Advocates and Scientists." *Environmental and Molecular Mutagenesis* 39 (2002): 102–7.

Plummer, Patricia, Susan Jackson, Jamie Konarski, Elizabeth Mahanna, Carolyn Dunsmore, Georgette Regan, Dianne Mattingly, Barbara Parker, Sara Williams, Catherine Andrews, Vani Vannappagari, Susan Hall, Sandra Deming, Elizabeth Hodgson, Patricia Moorman, Beth Newman, and Robert Millikan. "Making Epidemiological Studies Responsive to the Needs of Participants and Communities: The Carolina Breast Cancer Study Experience." *Environmental and Molecular Mutagenesis* 39 (2002): 96–101.

Pope, C. Arden. "Respiratory Disease Associated with Community Air Pollution and a Steel Mill, Utah Valley." *American Journal of Public Health* 79 (1989): 623–28.

Pope, C. Arden, Richard T. Burnett, Michael J. Thun, Eugenia E. Calle, Daniel Krewski, Kazuhiko Ito, and George D. Thurston. "Lung Cancer, Cardiopulmonary Mortality, and Long-Term Exposure to Fine Particulate Air Pollution." *Journal of the American Medical Association* 287 (2002): 1132–141.

Pope, C. Arden, Richard T. Burnett, George Thurston, Michael J. Thun, Eugenia E. Calle, Daniel Krewski, and John J. Goldleski. "Cardiovascular Mortality and Long-Term Exposure to Particulate Air Pollution: Epidemiological Evidence of General Pathophysiological Pathways of Disease." *Circulation* 109 (2004): 71–77.

Pope, C. Arden and Douglas Dockery. "Epidemiology of Chronic Health Effects: Cross-Sectional Studies." In *Particles in Our Air: Concentrations and Health Effects*, ed. Richard Wilson and John Spengler, 136–48. Cambridge, Mass.: Harvard University Press, 1996.

Presidential Advisory Committee on Gulf War Veterans' Illnesses. *Presidential Advisory Committee on Gulf War Veterans' Illnesses: Final Report*. Washington, D.C.: U.S. Government Printing Office, December 1996.

Princeton Research Associates, for Health-Track. *National Survey of Public Perceptions of Environmental Health Risks*. Princeton, N.J.: Princeton Research Associates, June 2000.

Proctor, Robert N. *Cancer Wars: How Politics Shapes What We Know and Don't Know About Cancer*. New York: Basic Books, 1995.

Proctor, Susan P., Sucharita Gopal, Asuka Imai, Jessica Wolfe, David Ozonoff, and Roberta F. White. "Spatial Analysis of 1991 Gulf War Troop Locations in Relationship with Postwar Health Symptom Reports Using GIS Techniques." *Transactions in GIS* 9 (3) (2005): 381–96.

Proctor, S. P., T. Heeren, R. F. White, J. Wolfe, M. S. Borgos, J. D. Davis, L. Pepper, R. Clapp, P. B. Sutker, J. J. Vasterling, and D. Ozonoff. "Health Status of Persian Gulf War Veterans:

Self-Reported Symptoms, Environmental Exposures, and the Effect of Stress." *International Journal of Epidemiology* 27 (1998): 1000–1010.

Project on Scientific Knowledge and Public Policy. *Daubert: The Most Influential Supreme Court Ruling You've Never Heard Of.* Boston: Tellus Institute, 2003. Available at http://www.defendingscience.org. Retrieved June 22, 2005.

Public Broadcasting System (PBS). *Trade Secrets.* Boston: PBS, 2001. Available at http://www.pbs.org/tradesecrets/transcript.html.

Pulido, Laura. *Environmentalism and Economic Justice: Two Chicano Struggles in the Southwest.* Tucson: University of Arizona Press, 1996.

Quijano, Romeo. "Elements of the Precautionary Principle." In *Precaution: Environmental Science and Preventive Public Policy,* ed. Joel Tickner, 21–27. Washington, D.C.: Island Press, 2003.

Raffensperger, Carolyn and Joel Tickner, eds. *Protecting Public Health and the Environment: Implementing the Precautionary Principle.* Washington, D.C.: Island Press, 1999.

Rapp, Rayna. *Testing Women, Testing the Fetus: The Social Impact of Amniocentesis in America.* New York: Routledge, 1999.

Ray, Rayka. *Fields of Protest: A Comparison of Women's Movements in Two Indian Cities.* Minneapolis: University of Minnesota Press, 1998.

Reeves, W. C., K. Fukuda, R. Nisenbaum, and W. W. Thompson. "Letters: Chronic Multisystem Illness among Gulf War Veterans." *Journal of the American Medical Association* 282 (1999): 327–29.

Reich, Michael. *Toxic Politics: Responding to Chemical Disasters.* Ithaca, N.Y.: Cornell University Press, 1991.

Reiss, J. R. and A. R. Martin. *Breast Cancer 2000: An Update on the Facts, Figures, and Issues.* San Francisco: Breast Cancer Fund, 2000.

Research Advisory Committee on Gulf War Veterans' Illnesses. *Scientific Progress in Understanding Gulf War Veterans' Illnesses: Report and Recommendations.* Washington, D.C.: U.S. Government Printing Office, 2004.

Rest, Jonathan. "The Chronic Fatigue Syndrome." *Annals of Internal Medicine* 123 (1) (1995): 74–76.

Rettig, Richard A. *Military Use of Drugs Not Yet Approved by the FDA for CW/BW Defense.* Santa Monica, Calif.: RAND, 1999.

Revkin, Andrew and Katharine Seelye. "Report by EPA Leaves Out Data on Climate Change." *New York Times,* June 19, 2003.

Ricks, Thomas. "Anthrax Shots Cause Military Exodus." *Washington Post,* October 11, 2000.

Ries, L. A. G., M. P. Eisner, C. L. Kosary, B. F. Hankey, B. A. Miller, L. Clegg, and B. K. Edwards, eds. *SEER Cancer Statistics Review, 1973–1999.* Bethesda, Md.: National Cancer Institute, 2002.

Robbins, Anthony S., Sonia Brescianini, and Jennifer Kelsey. "Regional Differences in Known Risk Factors and the Higher Incidence of Breast Cancer in San Francisco." *Journal of the National Cancer Institute* 89 (1997): 960–65.

Roe, David, William Pease, Karen Florin, and Ellen Silbergeld. *Toxic Ignorance: The Continuing Absence of Basic Health Testing for Top-Selling Chemicals in the United States.* New York: Environmental Defense Fund, 1997.

Roht, Lewis H., Sally W. Vernon, Francis W. Weir, Stanley M. Pier, Peggy Sullivan, and Lindsay J. Reed. "Community Exposure to Hazardous Waste Disposal Sites: Assessing Reporting Bias." *American Journal of Epidemiology* 122 (1985): 418–33.

Romieu, Isabelle, Mauricio Hernandez-Avila, Eduardo Lazcano-Ponce, Jean Phillippe Weber, and Eric Dewailly. "Breast Cancer, Lactation History, and Serum Organochlorines." *American Journal of Epidemiology* 152 (2000): 363–70.

Rose, Geoffrey. "Sick Individuals and Sick Populations." *International Journal of Epidemiology* 14 (1985): 32–38.

Rosenberg, Steve. "Limiting Burning of PVC Is Urged." *Boston Globe*, December 9, 2004. Available at http://www.besafenet.com/pvcnewspage1.htm. Retrieved June 8, 2005.

Rosner, David and Gerald Markowitz. *Deadly Dust: Silicosis and the Politics of Occupational Disease in Twentieth-Century America*. Princeton, N.J.: Princeton University Press, 1991.

Rosner, David and Gerald E. Markowitz. "From Dust to Dust: The Birth and Rebirth of National Concern about Silicosis." In *Illness and the Environment: A Reader in Contested Medicine*, ed. Stephen Kroll-Smith, Phil Brown, and Valerie Gunter, 162–74. New York: New York University Press, 2000.

Rudel, Ruthann A., Julia Brody, John D. Spengler, Jose Vellarino, Paul W. Geno, and Alice Yau. "Identification of Selected Hormonally Active Agents and Animal Mammary Carcinogens in Commercial and Residential Air and Dust Samples." *Journal of the Air and Water Management Association* 51 (2001): 499–513.

Rudel, Ruthann A., David E. Camann, John D. Spengler, Leo R. Korn, and Julia G. Brody. "Phthalates, Alkylphenols, Pesticides, Polybrominated Diphenyl Ethers, and Other Endocrine-Disrupting Compounds in Indoor Air and Dust." *Environmental Science and Technology* 37 (2003): 4543–553.

Rudestam, Kirsten. "The Importance of Place: Asthmatic Children's Perceptions of Inside and Outside Environments." Senior honors thesis, Center for Environmental Studies, Brown University, 2001.

Rudestam, Kirsten, Phil Brown, Christina Zarcadoolas, and Catherine Mansell. "Children's Asthma Experience and the Importance of Place." *Health* 8 (4) (2004): 423–44.

Ruzek, Sheryl Burt, Virginia L. Olesen, and Adele E. Clarke, eds. *Women's Health: Complexities and Differences*. Columbus: Ohio State University Press, 1997.

Safe, Stephen H. "Xenoestrogens and Breast Cancer." *New England Journal of Medicine* 337 (1997): 1303–304.

Salamon, R., C. Verret, M. A. Jutand, M. Begassat, F. Laoudj, F. Conso, and P. Brochard. "Health Consequences of the First Persian Gulf War on French Troops." *International Journal of Epidemiology* 35 (2006): 479–87.

Samet, Jonathan M. and Thomas A. Burke. "Turning Science Into Junk: The Tobacco Industry and Passive Smoking." *American Journal of Public Health* 91 (2001): 1742–744.

Samet, Jonathan M., Francesa Dominici, Frank C. Curriero, Ivan Coursac, and Scott L. Zeger. "Fine Particulate Air Pollution and Mortality in 20 U.S. Cities, 1987–1994." *New England Journal of Medicine* 343 (2000): 1724–729.

Samet, Jonathan, Scott Zeger, and Kiros Berhane. *The Association of Mortality and Particulate Air Pollution*. Boston: Health Effects Institute, 1995.

Schecter, A., P. Toniolo, L. C. Dai, L. T. B. Thuy, and M. S. Wolff. "Blood Levels of DDT and Breast Cancer Risk among Women Living in the North of Vietnam." *Archives of Environmental Contamination and Toxicology* 33 (1997): 453–56.

Schmitt, Eric. "Panel Criticizes Pentagon Inquiry on Gulf Illnesses." *New York Times*, January 8, 1997.

Schneider, Conrad. *Death, Disease, and Dirty Power: Mortality and Health Damage due to Air Pollution from Power Plants*. Boston: Clean Air Task Force, 2000.

Schuck, Peter. 1987. *Agent Orange on Trial: Mass Toxic Disasters in the Courts*. Cambridge, Mass.: Harvard University Press, 1987.

Schwartz, Joel. "Particulate Air Pollution and Chronic Respiratory Disease." *Environmental Research* 62 (1993): 7.

———. "Fine Particulate Air Pollution: Smoke and Mirrors of the '90s or Hazard of the New Millennium." Paper presented at the Annual Meeting of the American Public Health Association, Boston, November 14, 2000.
Schwartz, J., D. Slater, T. V. Larson, W. E. Pierson, and J. Q. Koenig. "Particulate Air Pollution and Hospital Emergency Visits for Asthma in Seattle." *American Review of Respiratory Disease* 147 (1993): 826-31.
Sclove, Richard. "Better Approaches to Science Policy." *Science* 279 (1998): 1283.
Scott, Wilbur. "Competing Paradigms in the Assessment of Latent Disorders: The Case of Agent Orange." *Social Problems* 35 (1988): 145-61.
———. "PTSD and Agent Orange: Implications for a Sociology of Veterans' Issues." *Armed Forces and Society* 18 (1992): 592-612.
Senier, Laura, Brian Mayer, and Phil Brown. "Report to Massachusetts Committee on Occupational Safety and Health and Boston Urban Asthma Coalition." Unpublished manuscript, 2005.
Sharpe, Michael and Alan Carson. "'Unexplained' Somatic Symptoms, Functional Syndromes, and Somatization: Do We Need a Paradigm Shift?" *Annals of Internal Medicine* 134 (9) (2001): 926-30.
Shenon, Philip. "Advisers Condemn Pentagon Review of Gulf Ailments." *New York Times*, November 8, 1996.
———. "Defense Secretary Vows Thorough Inquiry on Gulf War Illnesses." *New York Times*, March 6, 1997.
———. "Gulf War Panel Reviews Researcher's Ouster." *New York Times*, December 23, 1996.
———. "Half of Gulf-Illness Panel Now Calls Gas a Possible Factor." *New York Times*, August 19, 1997.
———. "House Committee Assails Pentagon on Gulf War Ills." *New York Times*, October 26, 1997.
———. "Oversight Suggested for Study of Gulf War Ills." *New York Times*, November 14, 1996.
———. "Panel Disputes Studies on Gulf War Illnesses." *New York Times*, November 21, 1996.
Shepard, Peggy M., Mary E. Northridge, Swati Prakash, and Gabriel Stover. "Preface: Advancing Environmental Justice Through Community-Based Participatory Action Research." *Environmental Health Perspectives* 110 (2002): 139-44.
Shogren, Elizabeth. "EPA Drops Its Case Against Dozens of Alleged Polluters." *Los Angeles Times*, November 6, 2003. Available at http://www.commondreams.org/headlines03/1106-01.htm.
Shostak, Sara. "Locating Gene-Environment Interaction: At the Intersections of Genetics and Public Health." *Social Science and Medicine* 56 (2003): 2327-342.
Shriver, Thomas E. "Environmental Hazards and Veterans' Framing of Gulf War Illnesses." *Sociological Inquiry* 71 (4) (2001): 403-20.
Shriver, Thomas E. and Sherry Cable. "Fault Lines and Frictions: Intramovement Conflicts in the Gulf War Illness Movement." Paper presented at the Annual Meeting of the Southern Sociological Society, New Orleans, March 26-29, 2003.
Shriver, Thomas E., Amy L. Chasteen, and Brent D. Adams. "Cultural and Political Constraints in the Gulf War Illness Social Movement." *Sociological Focus* 35 (2) (2002): 123-43.
Shriver, Thomas E., Amy Chasteen Miller, and Sherry Cable. "Women's Work: Women's Involvement in the Gulf War Illness Movement." *Sociological Quarterly* 44 (4) (2003): 639-58.
Shriver, Thomas E., Gary R. Webb, and Brent Adams. "Environmental Exposures, Contested Illnesses, and Collective Action: The Controversy Over Gulf War Illness." *Humboldt Journal of Social Relations* 27 (1) (2002): 73-105.

Shusterman, D., J. Lipscomb, R. Neutra, and K. Satin. 1991. "Symptom Prevalence and Odor-Worry Interaction Near Hazardous Waste Sites." *Environmental Health Perspectives* 94 (1991): 25–30.

Silent Spring Institute. *The Cape Cod Breast Cancer and Environment Study: Results of the First Three Years of Study.* Newton, Mass.: Silent Spring Institute, 1998.

———. *Grassroots Breast Cancer Advocacy and the Environment: A Report on Interviews with Grassroots Leaders.* Newton, Mass.: Silent Spring Institute, 2004.

Silvers, Anita, David Wasserman, and Mary Mahowald. *Disability, Difference, Discrimination: Perspectives on Justice in Bioethics and Public Policy.* Lanham, Md.: Rowman and Littlefield, 1998.

Snedeker, Suzanne M. "Pesticides and Breast Cancer Risk: A Review of DDT, DDE, and Dieldrin." *Environmental Health Perspectives* 109 (suppl. 1) (2001): 35–47.

Snow, David, E. Burke Rocheford, Steven Worden, and Robert Benford. "Frame Alignment Processes, Micromobilization, and Movement Participation." *American Sociological Review* 51 (1986): 464–81.

Snow, John. *On the Mode of Communication of Cholera.* London: John Churchill, 1855.

Spector, Malcolm and John Kitsuse. *Constructing Social Problems.* Menlo Park, Calif.: Cummings, 1977.

Spengler, John. "Persistent Organic Pollutants." Presentation given at the Silent Spring Institute, Newton, Mass., November 30, 2004.

Star, Susan Leigh and James R. Greisemer. "Institutional Ecology, 'Translations,' and Boundary Objects: Amateurs and Professionals in Berkeley's Museum of Vertebrate Zoology, 1907–39." *Social Studies of Science* 19 (1989): 387–420.

Starr, Paul. *The Social Transformation of American Medicine.* New York: Basic Books, 1988.

Steinbrook, Robert. "Science, Politics, and Federal Advisory Committees." *New England Journal of Medicine* 350 (2004): 1454–460.

Steingraber, Sandra. *Having Faith: An Ecologist's Journey to Motherhood.* New York: Perseus Books, 2001.

———. *Living Downstream: An Ecologist Looks at Cancer and the Environment.* Reading, Mass.: Perseus Books, 1997.

———. "Report from Europe: Precaution Ascending." *Rachel's Environment and Health News* 786 (March 4, 2004). Available at http://www.rachel.org.

Stellman, S. D. and Q. S. Wang. "Cancer Mortality in Chinese Immigrants to New York City: Comparison with Chinese in Tianjin and with White Americans." *Cancer* 73 (1994): 1270–275.

Stolberg, Sheryl Gay. "U.S. Reports Disease Link to Gulf War." *New York Times,* December 11, 2001.

Sullivan, Kimberly, Maxine Krengel, Susan P. Proctor, Sherral Devine, Timothy Heeren, and Roberta F. White. "Cognitive Functioning in Treatment-Seeking Gulf War Veterans: Pyridostigmine Bromide Use and PTSD." *Journal of Psychopathology and Behavioral Assessment* 25 (2) (2003): 95–103.

Swan, Shanna H., Katharina M. Main, Fan Liu, Sara L. Stewart, Robin L. Kruse, Antonia M. Calafat, Catherine S. Mao, J. Bruce Redmon, Christine L. Ternand, Shannon Sullivan, J. Lynn Teague, and the Study for Future Families Research Team. "Decrease in Anogenital Distance Among Male Infants with Prenatal Phthalate Exposure." *Environmental Health Perspectives* 113 (2005): 1056–61.

Symons, Jeremy. "How Bush and Co. Obscure the Science." *Washington Post,* July 13, 2003.

Szasz, Andrew. *Ecopopulism: Toxic Waste and the Movement for Environmental Justice.* Minneapolis: University of Minnesota Press, 1994.

Taylor, Dorceta E. "The Rise of the Environmental Justice Paradigm: Injustice Framing and the Social Construction of Environmental Discourse." *American Behavioral Scientist* 43 (2000): 508–80.

Tesh, Sylvia Noble. *Hidden Arguments: Political Ideology and Disease Prevention Policy.* New Brunswick, N.J.: Rutgers University Press, 1988.

Thomas, John K., Bibin Qin, Doris A. Howell, and Barbara A. Richardson. "Environmental Hazards and Rates of Female Breast Cancer Mortality in Texas." *Sociological Spectrum* 21 (2001): 359–75.

Thompson, H. J. "Effect of Amount and Type of Exercise on Experimentally Induced Breast Cancer." *Advances in Experimental Medicine and Biology* 322 (1992): 61–71.

Thornton, Joseph. "Biomonitoring of Industrial Pollutants: Health and Policy Implications of the Chemical Body Burden." *Public Health Reports* 117 (2002):315–23.

———. *Pandora's Poison: Chlorine, Health, and a New Environmental Strategy.* Cambridge, Mass.: MIT Press, 2000.

Tickner, Joel. "Emerging Contaminants and Environmental Health—Outlining a Precautionary and Preventive Science and Policy Framework." Lecture at Brown University Center for Environmental Studies, Providence, R.I., September 30, 2004.

———. *European Chemicals Management Initiatives.* Lowell, Mass.: Lowell Center for Sustainable Production, 2000.

———. "A Map toward Precautionary Decision Making." In *Protecting Public Health and the Environment: Implementing the Precautionary Principle,* ed. Carolyn Raffensperger and Joel Tickner, 162–86. Washington, D.C.: Island Press, 1999.

Tickner, Joel and Ken Geiser. "The Precautionary Principle Stimulus for Solutions- and Alternatives-Based Environmental Policy." *Environmental Impact Assessment Review* 34 (2004): 801–24.

Tickner, Joel and Nancy Myers. "Current Status and Implementation of the Precautionary Principle." *Science and Environmental Health Network* (2000). Available at http://www.sehn.org/ppcurrentstatus.html. Retrieved June 6, 2005.

Toxics Action Center and Maine Environmental Policy Institute. *Overkill: Why Pesticide Spraying for West Nile Virus May Cause More Harm Than Good.* Pamphlet. Boston: Toxics Action Center, July 2001.

Toxic Use Reduction Institute (TURI). *Outreach Successes and Challenges.* Lowell, Mass.: TURI, 2004. Available at http://www.turi.org/content/content/view/full/2125/. Retrieved June 6, 2005.

Trust for America's Health. *Health Tracking Network.* Washington, D.C.: Trust for America's Health, 2005. Available at http://healthyamericans.org/topics/index.php?TopicID = 28. Retrieved June 21, 2005.

———. *Nationwide Health Tracking: Investigating Life-Saving Discoveries.* Washington, D.C.: Trust for America's Health, 2005. Available at http://healthyamericans.org/reports/files/HealthTrackingBackgrounder.pdf. Retrieved June 21, 2005.

Tuite, James J., III. "Testimony to the Presidential Advisory Committee on Gulf War Veterans' Illnesses." Salt Lake City, Utah, March 18, 1997. Available at http://www.chronicillnet.org/PGWS/Tuite/SLC.html. Retrieved October 1, 2000.

———. "When Science and Politics Collide." Unpublished manuscript, 1995. Available at http://www.chronicillnet.org/PGWS/tuite/collide.html. Retrieved October 1, 2000.

Union of Concerned Scientists. *Scientific Integrity in Policymaking.* Cambridge, Mass.: Union of Concerned Scientists, 2004. Available at http://www.ucsusa.org/global_environment/rsi/page.cfm?pageID = 1363. Retrieved June 7, 2005.

Unwin, C. E., N. Blatchley, W. Coker, S. Ferry, M. Hotopf, L. Hull, K. Ismail, I. Palmer, A. David, and S. Wessely. "The Health of United Kingdom Servicemen Who Served in the Persian Gulf War." *The Lancet* 353 (1999): 169–78.

U.S. Army Environmental Hygiene Agency. *Final Report: Kuwait Oil Fire Health Risk Assessment.* Report no. 39-26-L192-91, May 5 to December 3, 1991. Washington, D.C.: U.S. Department of Defense, 1994.

U.S. Department of Energy. GrayLIT Network. 2003. Available at http://www.osti.gov/graylit.

U.S. Department of Health and Human Services. *Healthy People 2010: Understanding and Improving Health.* Washington, D.C.: U.S. Department of Health and Human Services, 2001.

U.S. Department of Veterans Affairs. *VA Creates Gulf War Advisory Committee.* Washington, D.C.: U.S. Department of Veterans Affairs, January 23, 2002. Available at http://www1.va.gov/rac-gwvi/docs/Pressrelease_ VACreatesGulfWarAdvisoryCommittee_Jan2002.doc. Retrieved July 21, 2005.

U.S. General Accounting Office (GAO). *Department of Veterans Affairs: Federal Gulf War Illnesses Research Needs Reassessment.* GAO-04-767. Washington, D.C.: U.S. GAO, June 1, 2004.

———. *Gulf War Illnesses: DOD's Conclusions about U.S. Troops' Exposure Cannot Be Adequately Supported.* GAO-04-821T. Washington, D.C.: U.S. GAO, June 1, 2004.

———. *Gulf War Illnesses: Federal Research Efforts Have Waned, and Research Findings Have Not Been Reassessed.* GAO-04-815T. Washington, D.C.: U.S. GAO, June 1, 2004.

U.S. General Accounting Office (GAO), Department of Veterans Affairs. *Report to the Chairman, Subcommittee on National Security, Emerging Threats, and International Relations, Committee on Government Reform, House of Representatives: Federal Gulf War Illnesses Research Strategy Needs Reassessment.* Washington, D.C.: U.S. GAO, 2004.

Vicini, James. "Judge Bars Mandatory Anthrax Shots for Troops." *Boston Globe*, October 28, 2004.

Vyner, Henry M. *Invisible Trauma: The Psychosocial Effects of the Invisible Environmental Contaminants.* Lexington, Mass.: Lexington Books, 1988.

Waldman, Peter. "EPA Bans Staff from Discussing Issue of Perchlorate Pollution." *Wall Street Journal*, April 28, 2003.

Warner, John. "Green Chemistry." Lecture at Brown University, Providence, R.I., September 30, 2005.

Warner, Marcella, Brenda Eskenazi, Paolo Mocarelli, Pier Mario Gerthoux, Steven Samuels, Larry Needham, Donald Patterson, and Paolo Brambilla. "Serum Dioxin Concentrations and Breast Cancer Risk in the Seveso Women's Health Study." *Environmental Health Perspectives* 110 (2002): 625–28.

Weinberg, Alvin. "Science and Transcience." *Minerva* 10 (12) (1972): 209–22.

Weiner, Mark. "CDC: Back Off on West Nile Spraying." *Syracuse Herald Journal*, April 5, 2001.

Weinhold, Bob. "Fuel for the Long Haul: Diesel in America." *Environmental Health Perspectives* 110 (2002): A458–A464.

Weisman, Carol S. *Women's Health Care: Activist Traditions and Institutional Change.* Baltimore: Johns Hopkins University Press, 1998.

Weiss, Robert. "HHS Seeks Science Advice to Match Bush View." *Washington Post*, September 17, 2002.

Weiss, S. T. 1998. "Environmental Risk Factors in Childhood Asthma." *Clinical and Experimental Allergy* 28 (suppl. 5) (1998): 29–34.

Wessely, Simon. "A Controlled Epidemiological and Clinical Study into the Effect of Gulf War Service on Servicemen and Women of the United Kingdom Armed Forces." 2000. Available at http://www.csa.com.

———. "Ten Years On: What Do We Know about the Gulf War Syndrome?" *Clinical Medicine* 1 (2001): 1–10.

Wessely, Simon, Chaichana Nimnuan, and Michael Sharpe. "Functional Somatic Syndromes: One or Many?" *The Lancet* 354 (1999): 936–39.

West, Dee, Sally Glaser, and Angela Prehn. *Status of Breast Cancer Research in the San Francisco Bay Area*. Union City: Northern California Cancer Center, 1998.

West Harlem Environmental Action, Inc. (WE ACT). *WE ACT History*. New York: WE ACT, 2003. Available at http://www.weact.org/history.html. Retrieved June 7, 2005.

Wheelwright, Jeff. *The Irritable Heart: The Medical Mystery of the Gulf War*. New York: Norton, 2001.

White, Roberta. "Service in the Gulf War and Significant Health Problems: Focus on the Central Nervous System." *Journal of Psychopathology and Behavioral Assessment* 25 (2) (2003): 77–83.

Whitman v. American Trucking Associations. No. 99-1257. 175 F.3d 1027 and 195 F.3d 4 (D.C. Circuit 1999).

Williams, Gareth. "The Genesis of Chronic Illness: Narrative Re-construction." *Sociology of Health and Illness* 6 (1984): 176–200.

Williams, Simon J. "Chronic Respiratory Illness and Disability: A Critical Review of the Psychosocial Literature." *Social Science and Medicine* 28 (1989): 791–903.

Wilson, Richard. "Introduction." In *Particles in Our Air: Concentrations and Health Effects*, ed. Richard Wilson and John Spengler, 1–14. Cambridge, Mass.: Harvard University Press, 1996.

Wing, Steve. "Limit of Epidemiology." *Medicine and Global Survival* 1 (1994): 74–86.

———. "Social Responsibility and Research Ethics in Community-Driven Studies of Industrialized Hog Production." *Environmental Health Perspectives* 110 (2002): 437–44.

Winkenwerder, William Jr., M.D., MBA, Assistant Secretary of Defense for Health Affairs. "Testimony on Pre- and Post-deployment Health Assessments," 2003. In *Federal Document Clearing House Congressional Testimony*. Bethesda, Md.: Congressional Information Service, July 9, 2004. Available from *LexisNexis™ Congressional* (online service).

Winn, Deborah. "The Long Island Breast Cancer Study Project." *Nature* 5 (2005): 986–94.

Wolfe, Jessica, Darin J. Erickson, Erica J. Sharkansky, Daniel W. King, and Lynda A. King. "Course and Predictors of Posttraumatic Stress Disorder Among Gulf War Veterans: A Prospective Analysis." *Journal of Consulting and Clinical Psychology* 67 (1999): 520–28.

Wolfe, Jessica, Susan P. Proctor, Jennifer Duncan Davis, Marlana Sullivan Borgos, and Matthew J. Friedman. "Health Symptoms Reported by Persian Gulf War Veterans Two Years after Return." *American Journal of Industrial Medicine* 33 (1998): 104–13.

Wolff, Mary S. and Paolo G. Toniolo. "Environmental Organochlorine Exposure as a Potential Etiologic Factor in Breast Cancer." *Environmental Health Perspectives* 103 (suppl. 7) (1995): 141–45.

Wolff, Mary S., Paolo G. Toniolo, Eric W. Lee, Marilyn Rivera, and Neil Dublin. "Blood Levels of Organochlorine Residues and Risk of Breast Cancer." *Journal of the National Cancer Institute* 7 (1993): 579–83.

Wolfson, Mark. *The Fight Against Big Tobacco: The Movement, the State, and the Public's Health*. Chicago: Aldine, 2001.

Women's Community Cancer Project. *Fact Sheet on the Precautionary Principle*. Cambridge, Mass.: Women's Community Cancer Project, 2001.

Writing Group for the Women's Health Initiative Investigators. "Risks and Benefits of Estrogen Plus Progestorone in Healthy Post-menopausal Women." *Journal of the American Medical Association* 288 (2002): 321–33.

Zavestoski, Stephen, Phil Brown, and Sabrina McCormick. "Gender, Embodiment, and Disease: Environmental Breast Cancer Activists' Challenges to Science, the Biomedical Model, and Policy." *Science as Culture* 13 (2004): 563–86.

Zeger, Scott L., Francesca Dominici, and Jonathan Samet. "Harvesting-Resistant Estimates of Air Pollution Effects on Mortality." *Epidemiology* 10 (1999): 171–75.

Ziem, Grace E. and Barry I. Castleman. "Threshold Limit Values: Historical Perspectives and Current Practice." *Journal of Occupational and Environmental Medicine* 31 (1989): 910–18.

Zierler, S. and N. Krieger. "Reframing Women's Risk: Social Inequalities and HIV Infection." *Annual Review of Public Health* 18 (1997): 401–36.

Zones, Jane S. "Profits from Pain: The Political Economy of Breast Cancer." In *Breast Cancer: Society Shapes an Epidemic*, ed. Anne S. Kasper and Susan J. Ferguson, 119–52. New York: St. Martin's, 2000.

INDEX

ACE. *See* Alternatives for Community and Environment
Actor-network theory, 55
Advocacy science, xxii, 26, 55
African Americans, 7, 260, 267; and breast cancer, 44, 91, 196; and lupus, 120; mortality rate from asthma, 101–2. *See also* Asthma; Race
Agency for Toxic Substances and Disease Registry, 233
Agent Orange, 11, 154, 257
Agyeman, Julian, 12
AIDS activism, 29, 89–90, 231
AIDS Coalition to Unleash Power, 90
Air pollution: and asthma controversies, 103, 105–11, 135–36; "confounding curves" issue, 110; court cases, 112–15, 249; diesel emissions, 109–10, 121, 257–58; and economics, 212; monitoring at peak traffic times and children's heights, 264, 270; mortality rates, 112; and public paradigms, 23, 117; regulatory disputes, 104, 106–7, 111–16; sources of, 118. *See also* Asthma science base; Particulate matter
AirBeat project, 128–29, 247
ALA. *See* American Lung Association
Alaska, 251–52, 276
Allergies, 3, 4 (table)
Alliance for a Healthy Tomorrow, xxii, 17, 81, 92, 204, 216, 239–40. *See also* Precautionary Principle Project

ALS (amyotrophic lateral sclerosis), 145, 159, 190, 191
Alternatives for Community and Environment (ACE), xxi, 15; AirBeat project, 128–29, 247; and asthma activism, 104, 117–38; author's research methodology, 40; collaborations with scientists, 128–34, 137–38; community empowerment through education, 122–24, 126; issues addressed by, 118–19, 247; local focus of, 132; national impact of, 133, 134; origins of, 118–19; and politicized illness experience, 123; Toxic Use Reduction Network grants, 225; and transportation issues, 120–21, 258
Altman, Rebecca Gasior, xxiii
Amaral, Joseph, xxiv
Amdur, Mary, 105, 260
American Academy of Allergy, Asthma, and Immunology, 197
American Cancer Society (ACS), 197; and asthma, 103, 107–8, 112; and dominant epidemiological paradigm, 20, 49
American Journal of Public Health, 204
American Legion, 142
American Lung Association (ALA), 103, 111, 197
American Trucking Association, Inc., et al., v. United States Environmental Protection Agency, 112
America's Children and Environment project, 243

AMVETS, 142
An Enemy of the People (Ibsen), 259
"Analytical blurring," 33, 55, 76
Anderson, Anne, xvii–xviii
Anderson, Jimmy, xvii–xviii
Anthrax vaccine, 145, 178, 184, 236
Antidotes, as potential source of Gulf War–related illnesses, 140, 145, 148, 190
Antimalaria medication, 178
Armed Forces Personnel Medical Readiness and Tracking Act of 2004, 191
Arsenic, 210
Asbestos, 9, 229
Asian women, and breast cancer, 67
Association for Birth Defect Children, 244
Asthma, 100–138, 233; author's research methodology, 40; doctors' focus on indoor exposures, 103, 106, 110, 135–36, 233; increased prevalence, 101–2; issue prominence, 10–11, 119; and managed care organizations, 234; mortality rates, 101–2, 106, 107, 109 (table); and NIEHS, 35; prevalence in low-income urban areas, 12, 102, 119; psychological effects, 124–25; public opinion on environmental causation, 3, 4 (table); public perception of risks, 193 (table), 195; and regulatory policy, 185–86; social discovery of, 116–17, 136, 137; statistics, 101–2; stigma attached to, 124. *See also* Air pollution; Asthma activism; Asthma policy; Asthma science base; Comparisons of asthma, breast cancer, and Gulf War–related illnesses
Asthma activism: in Boston (*see* Alternatives for Community and Environment; Boston Urban Asthma Coalition); and citizen-science alliances, 36–37, 127–33; and collective identity, 116, 123–28; and community-based participatory research, 13, 110–11; connection to other urban issues, 119–20; education campaigns, 118, 122–23; as embodied health movement, 27; and energy policy, 257–58; and environmental justice groups, 12–13, 110–11, 116–36; future prospects, 133–36; in New York (*see* West Harlem Environmental Action); power of social movement, 198; public scrutiny of research and regulation, 11; and social structural inequalities, 117–18; summary of conclusions, 136–38; and transportation issues, 120–22
Asthma policy: citizen participation policy, 184 (table), 186; detection/treatment/compensation policy, 184 (table), 185; explanations for policy outcomes, 192–201; regulatory policy, 111–16, 184 (table); research funding, 104, 183, 184 (table), 185
Asthma science base, 105–16; community-based participatory research, 13, 110–11; confidentiality issues, 113–14; cost-benefit analysis, 112–13; court cases, 112–15; diesel emissions, 109–10, 121; dominant epidemiological paradigm, 105–6, 135–36; environmentally attributable fraction of exacerbations, 108–9; etiology vs. exacerbation, 106; future prospects, 133–36; indoor environment vs. outdoor air pollution, 103, 106, 135–36; investigations of causation, 105–11; natural experiments, 108; particulate dose-response curve, 107, 112; regulatory disputes, 111–16; research methods and data issues, 105–10; sources of support for environmental causation hypotheses, 196–97; standards of proof, 129–30; strength of science base, 105–11
Astra-Zeneca, 44, 86
Atlanta Summer Olympics (1996), 108
Avon. *See* Breast Cancer Walk

Balint, Michael, 171
Balshem, Martha, xx
Bay Area Working Group on the Precautionary Principle, 91
Bay View/Hunter's Point Community Advocates, 82, 91
Be Safe campaign, 221–22
Beatrice Foods, xvi, xviii–xix
BEHC. *See* Boston Environmental Hazard Center
Benzene, 9
"Big Dig" highway project, 121
Bingham, Eula, 147, 155

Binns, James H., Jr., 144
Biomedical model, 64, 94–95
Biomonitoring, 268–69. *See also* Body burden of environmental contaminants
Biomonitoring Resource Center, 278
Birth Defect Research for Children, 244
Birth defects, xviii–xix, 3, 4 (table), 148, 223
Black lung, 229
Black Panther Party, 7
Body burden of environmental contaminants, 56, 72, 242, 265–70
Body Burden: Pollution in People report, 72–73
Boston, MA, 225. *See also* Alternatives for Community and Environment
Boston Environmental Hazard Center (BEHC), xxi, 16, 40, 140, 177
Boston Public Health Commission, 104
Boston University School of Public Health, 16, 263, 275. *See also* Boston Environmental Hazard Center
Boston Urban Asthma Coalition, 103–4
Boundary movements, 31–33, 45, 55, 75, 76, 98–99
Boundary objects, 32, 76
Brain tumors, 4 (table)
Brazil, 68
BRCA-1 and BRCA-2 genes, 47, 65, 76
Breast and Cervical Cancer Mortality Prevention Act of 1990, 45
Breast and Cervical Cancer Prevention and Treatment Act of 2000, 188
Breast cancer, 43–99; author's research methodology, 40; doctors' dismissal of environmental causation, 233; issue prominence, 10–11; public opinion on environmental causation, 3; public paradigm, 44 (*see also* Environmental breast cancer movement); public perception of risks, 193 (table), 195; Shop for the Cure campaign, 44, 45; social and scientific construction of epidemic, 43–47; statistics, 44. *See also* Breast cancer activism; Breast cancer policy; Breast cancer science base; Comparisons of asthma, breast cancer, and Gulf War–related illnesses
Breast Cancer Action, 82, 90–91, 96, 219

Breast cancer activism, 44–47, 88–89, 198; acting on science, 55–57; boundary movement activity, 45, 55, 75, 76; citizen-science alliances, 36–37, 52, 55, 72–75, 92–95; community involvement in research and data collection, 71–76; contrast to Gulf War veterans' activism, 162; and corporate actors, 44–45; as embodied health movement, 27; and environmental activism, 84–85, 88; and environmental breast cancer movement, 13, 88–89; history of, 45–46; issues addressed, 77–85; links with other social movements, 88–92; on Long Island, 77–79; in Massachusetts, 80–81 (*see also* Silent Spring Institute); participation in state and federal funding review panels, 14, 83; personal, scientific, and political frameworks, 86–88; and politicized illness experience, 30–31; power of social movement, 198–99; public scrutiny of research and regulation, 11, 83; and research agenda, 75, 79–80, 82; and research funding, 62, 78, 80, 82–84, 199; in San Francisco Bay area, 81–86; successes of, 45. *See also* Breast cancer science base; Environmental breast cancer movement; Silent Spring Institute
Breast Cancer Awareness Month, 44, 45, 76, 86
Breast Cancer Centers of Excellence, 189
Breast Cancer Fund, 76, 82–84, 96, 219, 223–24, 238
Breast cancer policy: acting on science, 55–57; citizen participation policy, 184 (table), 189; detection/treatment/compensation policy, 184 (table), 187–88; explanations for policy outcomes, 192–201; regulatory policy, 184 (table), 188; research funding policy, 184 (table), 186–87
Breast cancer science base, 47–76; community involvement in research and data collection, 48, 71–76; control group problem, 69–70; dominant epidemiological paradigm, 48–51; endocrine disruptor hypothesis (*see* Endocrine disruptor hypothesis); funding issues, 61–62; genetic

Breast cancer science base (*continued*)
factors, 47; genetics and lifestyle bandwagons, 51, 64–65; and "gray literature," 54–55, 62; immigration-related data, 67; individual vs. environmental risk factors, 47–51, 63–71; interpreting science, 53–55; Long Island Breast Cancer Study Project, 53, 69, 75, 78, 79; mainstream skepticism on environmental causation, 68–69; mixed results of research on environmental factors, 14, 67–71; occupational studies, 70; paradigm struggle, 48, 49 (fig), 73–74; policymaking and (acting on science), 55–57; prevention vs. treatment (upstream vs. downstream approaches), 48, 58–63; research agenda, 51–53; research methods and measurement issues, 60–63, 73–74; sources of support for environmental causation hypotheses, 197; standards of proof, 53–54, 73, 98; traditional vs. innovative approaches to environmental factors, 66, 74, 75; weight of evidence, 53, 54, 62. *See also* Silent Spring Institute
Breast Cancer Walk, 76
Breast milk, 267–70
Brody, Julia, 73
Brooklyn, NY. *See* El Puente
Brown, George, 114
Brown, Jesse, 191
Brown University, 91, 249
Browner, Carol, 112–13
Buffalo Creek flood, xix–xx
Building Healthy Communities from the Ground Up: Environmental Justice in California, 221
Burawoy, Michael, 41
Bush, George W., 135; antiregulatory stance, 210, 249–53; censorship of science, 250–51, 253, 255; denial of global warming, 250, 253; and energy companies, 114, 115; hostility toward environmentalism, 252–53; "loyalty tests" for advisors, 250; mercury trading system, 254; No Child Left Behind, 252; opposition to stem cell research, 245, 255; purges of scientific advisory boards, 250–51; rejection of science-based decision making, 210, 249–55
Buyer, Stephen, 170

California. *See* Marin County Breast Cancer Watch; San Diego, CA; San Francisco Bay area
California Environmental Contaminant Biomonitoring Program, 269
Campaign for Safe Cosmetics, 83, 84, 222–23, 278
Canada, 148
Cancer: and dissident scientists, 260; and dominant epidemiological paradigm, 20; and genetic determinism, 20, 65; and individual responsibility, 20; and insufficient study of other diseases, 262; public opinion on environmental causation of cancer in children, 3, 4 (table). *See also* Breast cancer; Woburn childhood leukemia case
"Cancer establishment," 20, 44, 46, 286n38. *See also* American Cancer Society; National Cancer Institute
Cape Cod Breast Cancer and Environmental Study Atlas, 74
Cape Cod, MA, 74, 80
Carcieri, Donald, xxiv
Carolina Breast Cancer Study, 75
Carson, Alan, 176
Carson, Rachel, 47, 204, 220, 259
CDC. *See* Centers for Disease Control
Center for Health, Environment, and Justice, 6, 205–6, 221–22, 224, 278
Centers for Children's Environmental Health and Disease Prevention Research, 186, 247–48
Centers for Disease Control (CDC): asthma projects, 186; body burden reports, 56, 72, 242; and Bush administration, 254; and children's environmental health, 218; and community-based participatory research, 247–48; and dominant epidemiological paradigm, 48; focus on terrorism preparedness, 244; health tracking studies, 243; and lead hazards, 254; and Long Island Breast Cancer Study Project, 78; and West Nile virus, 217; and Woburn childhood leukemia case, xviii
Children: lack of food choices in urban environments, 21, 120; and precautionary principle, 218–19; psychological

effects of asthma, 124–25; public opinion on environmental causation of cancer in children, 3, 4 (table); and Roxbury Environmental Empowerment Project (REEP), 123; vulnerability to toxins, 218; Woburn childhood leukemia case, xvi–xix, 6, 37, 264. *See also* Asthma; Schools; Teenagers, as activists
Children's and Families' Protection Act of 1996, 218
Childress, Adele, 35
Chloracne, 9
Cholera epidemic, 258–59
Chronic fatigue syndrome, 145, 172, 175, 198
Chronic multisymptom illnesses. *See* Medically unexplained physical symptoms
Citizen participation policy, 182; asthma, 184 (table), 186; breast cancer, 184 (table), 189; Gulf War–related illnesses, 184 (table), 191
Citizen-science alliances, xiv, 13–14, 75; and asthma, 36–37, 127–34, 137; benefits of, 94–95, 137–38, 264–65; as boundary movement activity, 33; and breast cancer, 52, 55, 72–75, 92–95; changing mutual perceptions, 93–94; and critical epidemiology, 37–39; defined/described, 33–37; differing lay and professional perspectives, 79, 129–30, 134–35, 160; and embodied health social movements, 29; formal and informal alliances, 33–34; and Gulf War–related illnesses, 159–63, 168–69, 178–79; obstacles to, 94, 130; and popular epidemiology, 34; and precautionary principle, 224–26; and public paradigms, 36; and value structure for science, 94–95; weakness of alliances for GWRIs, 161–62, 178–79. *See also* Community-based participatory research; Silent Spring Institute
A Civil Action (film), xix, 6
A Civil Action (Harr), xix
Clarke, Lee, xx
Clean Air Act of 1970, 183, 184 (table), 185, 198, 210, 251
Clean Air for Barrio Children's Health program, 118
Clean Water Act, 252
Clinical medicine. *See* Medical practice

Clinton, Bill, 188, 197
Coal mining, 111
Cohort studies, and asthma, 105–6
Colborn, Theo, 53, 241, 263
Collaborative on Health and Environment, 261–62, 278
Collective identity, 26, 27, 30; and asthma, 116, 123–28; and Gulf War–related illnesses, 141–42
Collective illness experience, 24, 25 (fig), 123–28. *See also* Illness experience; Politicized illness experience
Collins, Terry, 53
Colombia, 68
Columbia Center for Children's Environmental Health, 123, 133, 186, 248
Columbia University School of Public Health, 133
Committee on Measuring the Health of Gulf War Veterans, 144
Commonweal, 261, 265, 269, 278
Communities for a Better Environment, 91, 264, 266, 273, 279
Community Action Against Asthma, 110
Community Asthma Prevention Program (Philadelphia), 118
Community Water Rights Project, 221
Community-Based Initiative to Protect Infants in Northern Manhattan, 221
Community-based participatory research, 13–14, 33–36; and asthma, 110–11, 134, 186; author's experiences, 273–74; and breast cancer, 48, 71–76; differing lay and professional perspectives, 135; and informed consent, 267; and military, 257; and policymaking, 246–49
Comparisons of asthma, breast cancer, and Gulf War–related illnesses, 14, 141, 162–63, 180–201; factors affecting policy outcomes, 192–201, 193 (table); policy outcomes, 182–91, 184 (table); resistance to looking for environmental factors, 180–81. *See also* Policy
Compensation/treatment policy. *See* Detection/compensation/treatment policy
Concerned Citizens of Tillery, 91, 260
Congress: and breast cancer research funding, 187; and dominant epidemiological paradigm, 48; and EPA, 115; grant

Congress (*continued*)
    mechanisms, 62; and Gulf War–related illnesses, 143, 144, 155, 170, 190; health tracking studies, 242–43
Constituency-based health movements, 26–27
Consumer Advocates in Research and Related Activities program, 189
Contested illness perspective, xiv–xvi, 6–8; and citizen-science alliances, 264–65; and clinical medicine, 232–37; and community-based participatory research, 246–49; "contested causation," 172; and corporate practices, 237–39; and critical epidemiology, 263–64; endocrine disruptor research and surveillance, 240–41; grassroots body burden testing, 265–70; growing acceptance of, 229–32; historical examples, 229; improved monitoring and surveillance, 242–45; improvements in military medicine and VA care, 256–57; and informed consent, 236–37; lead poisoning, 7; medically unexplained physical symptoms (MUPS), 172–77; political control and censorship of science, 249–55; precautionary principle implementation, 239–40; and research funding, 245–46; research implications, 258–71; tobacco industry, 7; types of controversies, 14. *See also* Asthma; Breast cancer; Gulf War–related illnesses
Contested Illnesses Research Group, 276
Contextual stress perspective, 167–68, 171
Control groups: and breast cancer, 61, 62, 69–70; and military, 149
Corporate actors: and asthma activism, 104, 111; and breast cancer activism, 44–45, 82, 85, 97; and Bush, 114, 210; control over research, 38; controversy framed as assault on progress, 15; cost-benefit analyses, 210, 212; economic and political clout of, 59; economics of precautionary principle, 210–12, 237; emphasis on individual responsibility, 9; implications of contested illness perspective, 237–39; influence over regulatory process, 59, 63, 104, 111, 114; manufacturers' test data as "gray literature," 54; and market pressures, 238; New Source Review (NSR), 114–15; opposition to Toxics Use Reduction Act, 215; "pink" marketing campaigns, 82, 85; and radiation exposure, 9–10; research on alternative materials, 238; resistance to regulation, 106–7, 111–16, 180; and subsidized philanthropy, 85; voluntary changes, 237
Corvello, Gail, xxiv
Cosmetic companies, 62, 83, 238. *See also* Campaign for Safe Cosmetics
Couch, Stephen R., xx
Court cases. *See* Litigation
Critical epidemiology, xiv, 37–39, 181, 263–64
Cultural shaping of illness, 174, 176
"Cultures of action," 76

D'Amato, Alphonse, 78
Daniels, Arlene Kaplan, 236
Databases for tracking population health. *See* Health tracking
Daubert v. Merrell Dow, 253
Davis, Devra, 5–6, 259–60, 264
DDE, 61, 67
DDT, 61, 67, 68
Deep South Center for Environmental Justice, 278
DEET, 148
Defense Medical Surveillance System, 244
Delaney Clause, 208
Democratization of science and medicine, 13–14, 224–26, 231, 266–67. *See also* Citizen-science alliances
Denmark, 210–11
DEP. *See* Dominant epidemiological paradigm
Department of Defense (DOD): and breast cancer, 51, 53, 83, 187; denial of toxic exposures in Gulf War, 144, 152–53; disjunction with VA, 245, 256; and dominant epidemiological paradigm, 48; *Effectiveness of Medical Research Initiatives Regarding Gulf War Illnesses* report, 152; Gulf War–related illnesses and record-keeping, 140; missing information, misinformation, and secrecy, 153,

155; predeployment health tracking, 191, 244; research funding for GWRIs, 189; stress-based paradigm for Gulf War–related illnesses, 146

Department of Energy (DOE), 55

DES. *See* Diethylstilbestrol

Detection/compensation/treatment policy, 182; asthma, 184 (table), 185; breast cancer, 184 (table), 187–88; Gulf War–related illnesses, 184 (table), 190–91

Devillars, John, 129

Dibutyl phthalate, 80, 238

Diesel emissions, 109–10, 121, 257–58

Diethyl hexyl phthalate, 238

Diethyl phthalate, 80

Diethylstilbestrol (DES), 9, 241

Dioxins, 2, 56, 68

Disabled American Veterans, 142

Dockery, Douglas, 105, 107, 108, 113

Doctors. *See* Medical practice

Dominant epidemiological paradigm (DEP), xiv, xvi, 106; actors involved in contestation, 21; and asthma, 103, 105–6, 135–36; and breast cancer, 44, 47–53, 58–76; challenges to, 21, 28–29, 50, 73, 98 (*see also* Public paradigm); consequences of, 19; defined, 17–23; focus on individual lifestyle as key to disease prevention, 20–21, 47–51, 65–67, 106; focus on indoor environment in asthma causation, 103, 106, 135–36, 233; and Gulf War–related illnesses, 140–41, 150, 166–67, 169–71; and "hidden arguments," 63–64; organizations, agencies, and research traditions, 20; postdiscovery period, 21, 22 (fig); and strength of social movements, 200

Donora, PA, 106

Draw-A-Breath, 124, 234–35

Dumanoski, Dianne, 53, 241, 263

EBCM. *See* Environmental breast cancer movement

Ecology Center (Ann Arbor, Michigan), 221

Economic impact of precautionary principle, 210–12, 237

Eddington, Patrick, 158

Edelstein, Michael, xx

Effectiveness of Medical Research Initiatives Regarding Gulf War Illnesses report, 152

El Puente, 110–11

Electric Power Research Institute, 112

Embodied health social movements, 26–31

ENACT. *See* Environmental Neighborhood Awareness Committee of Tiverton

Endocrine Disrupter Screening and Testing Advisory Committee, 60, 188, 240

Endocrine disruptor hypothesis, 13, 53, 263–64; and cosmetics, 83; and critical epidemiology, 38; and prevention vs. treatment, 59–60, 62; as public paradigm, 23, 24; research and surveillance policies, 240–41; and wildlife studies, 65

Environmental activism, 84–85, 88

Environmental breast cancer movement (EBCM), 13, 76–77, 84–85; and AIDS movement, 89–90; as boundary movement, 98–99; and citizen-science alliances, 92–95; and conflicting public paradigms, 24; criticism of general movement successes, 46; criticism of mammography, 32, 46; emergence of, 46; and environmental health movement, 92–95; and environmental justice movement, 90–92, 279; goals of, 44; and mainstream breast cancer movement, 85, 86, 88–89, 186; personal, scientific, and political frameworks, 86–88; and precautionary principle, 204, 205, 219–20; as public paradigm, 44; and research funding, 187; science base for environmental causes, 194–95; summary of conclusions, 96–99; and women's health movement, 89

Environmental causation hypothesis, 2; and clinical medicine, 232–35; patient-centered medicine, 235–36; public opinion on, 3–6; public reaction to environmental health effects, 8–10; research funding, 245–46; research silences, 262; resistance to looking for environmental factors, 46, 52, 59, 68–69, 180, 230–31, 233. *See also* Asthma science base; Breast cancer science base; Contested illnesses; Endocrine disruptor hypothesis; Gulf War–related illnesses science base

Environmental health activism. *See* Asthma activism; Environmental breast cancer movement; Gulf War–related illnesses activism; Precautionary principle; Public paradigm

Environmental health movement: and environmental breast cancer movement, 92–95; future directions, 271–80; and precautionary principle, 203, 218–19

Environmental inequality model, 12

Environmental Justice and Community-Based Participatory Research programs, 34, 263

Environmental Justice Coalition for Water, 221

Environmental justice movement, xv, 231; and asthma activism, 12–13, 103–4, 110–11, 116–36; and breast cancer activism, 82, 90–92; as constituency-based health movement, 27; differing lay and professional perspectives, 134–35; and environmental breast cancer movement, 90–92, 279; and precautionary principle, 204–5, 207, 220–22; as public paradigm, 23, 117. *See also* Alternatives for Community and Environment; Bay View/Hunter's Point Community Advocates; West Harlem Environmental Action

Environmental Justice: Partnerships for Communication, 186, 269

Environmental Neighborhood Awareness Committee of Tiverton (ENACT), xxiv, 274

Environmental Protection Agency (EPA): America's Children and Environment project, 243; and Bush, 115, 135; and censorship of science, 250–51, 253, 254; and children's environmental health, 218; and community-based participatory research, 247–48; court cases, 112–15, 249; and dioxins, 2; and dominant epidemiological paradigm, 48; Endocrine Disrupter Screening and Testing Advisory Committee, 60, 188, 240; and "gray literature," 54; and Long Island Breast Cancer Study Project, 78; and New Source Review, 114–15; and precautionary principle, 209, 218; regulation of air pollution, 111–16, 135, 185; and research funding, 183, 185, 245; Toxic Release Inventory, 68; and Woburn childhood leukemia case, xvii–xviii

Environmental Working Group (EWG), 72, 263, 265–68

Environmentally Compromised Home Opportunity program, xxiv

EPA. *See* Environmental Protection Agency

Epidemiology. *See* Critical epidemiology; Dominant epidemiological paradigm; Popular epidemiology; Population-based approach to disease; Public paradigm; Social-structural approach to epidemiology

Epstein, Steven, 29, 33

Erikson, Kai, xix–xx

Erin Brokovich (film), 6

Estée Lauder, 238

European Union: ban on PBDEs, 211; and cosmetics, 223; policy on toxicity information, 60; and precautionary principle, 205, 207, 210–11, 223; regulation of phthalates, 62, 238

Evans, Lane, 155

Everything in Its Path (Erikson), xix–xx

EWG. *See* Environmental Working Group

ExxonMobil, 135

Fair Labor Association, 238

FDA. *See* Food and Drug Administration

Federal Data Quality Act, 250

Felag, Walter, xxiv

Ferguson, Susan, 85

Feussner, Jack, 164

Fibromyalgia, 145, 172, 175

Fields of movement approach, 33

First National Conference on Precaution, 205–6

Flame retardants, 211, 267–68

Floyd, Hugh H., 175

Food, Drug, and Cosmetics Act of 1938, 208

Food Additives Amendment of 1958, 208

Food and Drug Administration (FDA), 83

Food Quality Protection Act of 1996, 188, 218, 240

For a Cleaner Environment, xviii–xix

France, 148–49
Freedom of Information Act, 113–14
Freidson, Elliot, 8
Functional somatic syndromes. *See* Medically unexplained physical symptoms
Funding for research, 245–46; asthma, 104, 183, 184 (table), 185; breast cancer, 53, 61–62, 184 (table), 186–87; Gulf War–related illnesses, 140, 150, 151 (fig), 168, 177, 184 (table), 189–90

Gay and lesbian health movement, 27
Geiser, Ken, 214, 216
Genetic factors in cancer, 20, 47, 65
Genetic testing, 76, 85
Gerrity, Tim, 158
Gibbs, Lois, 6, 221
GIS mapping, 270; and asthma research, 134; and breast cancer research, 74, 77, 79, 80
Global warming, 250–51, 253
Globalization, 238–39
Gottlieb, Robert, 120
Government actors, 106; alliances between state governments and environmental organizations, 114–15; and Alternatives for Community and Environment (ACE), 129; and asthma, 104; and boundary movements, 32; and breast cancer, 57, 97; disputes between government bodies over Gulf War–related illnesses, 155; distrust of government by Gulf War veterans, 154–56; and Gulf War–related illnesses, 143–44, 151–53, 155; and industry opposition to regulation, 59, 135; local governments, 223–24; loss of public trust following slow responses, 8–10; political control and censorship of science, 249–55; and postdiscovery period, 22 (fig); and radiation exposure, 9–10; regulatory disputes, 104, 106–7; support for programs with public involvement, 35. *See also* Bush, George W.; Congress; Dominant epidemiological paradigm; Funding for research; Policy; Regulatory policy; *specific agencies*
Grassroots body burden testing, 242, 265–70

Gray, Gregory, 176
"Gray literature," 54–55, 62
GrayLIT Network, 55
Great Lakes Chemical Corp., 211, 268
Green Cleaners in Boston Schools, 225
Greenaction, 221
Greenpeace, 222, 268
Griffith, J., 68
Groupe L'Oreal, 238
Gulf War–related illnesses (GWRIs), 14, 139–79; ALS, 145, 159, 190, 191; author's research methodology, 40; and collective identity, 141–42; controversies, 150–66 (*see also* Gulf War–related illnesses science base); difficulties in getting recognition, 141–42, 154, 166–72, 233; difficulties in getting treatment, 141–44; and distrust of government, 154–56; and dominant epidemiological paradigm, 140–41, 150, 166–67, 169–71; issue prominence, 11, 170–71; lack of communication between VA and DOD, 245, 256; medically unexplained physical symptoms (MUPS), 172–77; and military culture, 162–63; public perception of, 193 (table), 196; record-keeping issues, 140; social discovery of, 141–44; statistics, 139–40; summary of conclusions, 177–79; symptoms, 139–41, 147, 148. *See also* Comparisons of asthma, breast cancer, and Gulf War–related illnesses; Gulf War–related illnesses activism; Gulf War–related illnesses policy; Gulf War–related illnesses science base
Gulf War–related illnesses activism, 13, 194–95; and citizen-science alliances, 36–37, 159–63; contrast to asthma and breast cancer activism, 162–63; as embodied health movement, 27; fragmentation among veterans, 163–66; and government attribution of responsibility, 167–68; loss of trust, 160–61; and military culture, 162–63, 179; obstacles to, 163–66; opposition to contextual stress perspective, 167–68; public scrutiny of research and regulation, 11; weakness of social movement, 198–99

Gulf War–related illnesses policy: citizen participation, 184 (table), 191; compensation/treatment, 190–91; detection/treatment/compensation policy, 184 (table); explanations for policy outcomes, 192–201; improvements in military medicine and VA care, 256–57; and informed consent, 236–37; regulatory policy, 184 (table), 191; research funding, 140, 150, 151 (fig), 168, 177, 184 (table), 189–90

Gulf War–related illnesses science base, 145–50, 195; ambiguous nature of illnesses, 144–45; contextual stress perspective, 167–68, 171; dose-response effects, 148; environmental causes vs. stress, 140–41, 146–48, 166–72; lack of definitive connections to specific exposures, 145, 148; lack of evidence for unique syndrome, 150, 174; lay involvement in science, 159–63, 168–69, 178–79; missing information, misinformation, and secrecy, 151–53; multiple potential sources, 140, 145–46, 148; record-keeping issues, 148; research funding, 140, 150, 151 (fig), 168, 177, 184 (table), 189–90; research methods and measurement issues, 145, 148, 151–53, 169; sources of support for environmental causation hypotheses, 197–98; standards of proof, 168–69; validity of self-reported symptoms, 147, 148, 169; veterans' distrust of scientific process, 156–59

Haley, Robert, 144, 147, 148, 189
Hansen, Johnni, 70
Haraway, Donna J., 52
Harr, Jonathan, xix
Hartford Jewish Ledger, 85–86
Hartsock, Nancy, 52
Harvard Six Cities study. *See* Six Cities asthma study
Haverhill Environmental League, 263
Health Care Without Harm, 234, 278
Health Effects Institute, 112
Health inequalities approach to epidemiology, 37, 263. *See also* Social-structural approach to epidemiology

Health social movements: boundary movements, 31–33; constituency-based health movements, 26–27; defined, 26; and democratization of science and medicine, 231; embodied health social movements, 27–30; factors affecting strength of, 192, 193 (table); health access movements, 26; and politicized illness experience, 30–31; relative importance in policy outcomes, 198–201

Health tracking: lack of tracking in the U.S., 5, 261; and military deployments, 152, 178, 179, 191, 244; Millennium Cohort Study, 152, 244; in Scandinavian countries, 5, 56; U.S. initiatives, 3, 152, 242–43
Health voluntaries, 20, 49, 103, 197
Health-Link, 263
Health-Track, 3, 5, 242
Healthy Boston Schools Janitorial Project, 225
Healthy Food, Healthy Schools, and Healthy Communities project, 120
Healthy Hair Show project, 225
Healthy Home Healthy Child campaign, 123
Healthy People 2000 report, 35, 102
Healthy People 2010 report, 35, 102
Hodgson, Michael, 154
Home equity loan programs, xxiv, 274
Hormone replacement therapy, 241
Housing and Urban Development (HUD), 104, 218
Hubbard, Ruth, 52
HUD. *See* Housing and Urban Development
Huff, James, 69, 255
Hunter, David, 67, 69
Huntington Breast Cancer Action Coalition, 78–79
Hurricane Katrina, 265, 275
Hyams, Kenneth, 170

Iatrogenic diseases, 8, 140
Ibsen, Henrik, 259
Identity, and illness experience, 30. *See also* Collective identity; Illness experience
Identity-based social movements, 28
"If You Live Uptown, Breathe at Your Own Risk" campaign, 121–22

Illness experience: and asthma, 123–28; and breast cancer, 30–31, 87–88; collective illness experience, 24, 25 (fig), 123–28; and emergence of public paradigms, 24, 25 (fig); and Gulf War–related illnesses, 141–44; and medically unexplained physical symptoms, 176; politicized illness experience, 24, 25 (fig), 30–31, 87–88, 123–28, 176. *See also* Embodied health social movements
Immigration, and breast cancer, 67
Imperial Chemical Industries, 86
Individual responsibility for health factors: and asthma, 106; and breast cancer, 47–51, 65–67, 95, 97; and cancer establishment, 20–21; corporate emphasis on, 9, 97; and critical epidemiology, 37–38; and dominant epidemiological paradigm, 20–21; and research, 64. *See also* "Risk factor" approach to epidemiology
Indoor environment, and asthma, 80, 103, 106, 110, 135–36, 233
Infertility, 4 (table)
Informed consent, 236–37, 266–67
Inoculations, as potential source of Gulf War–related illnesses, 140, 145, 177–78
Institute of Medicine, 144, 145, 148–49, 152, 154, 155
Interpretive flexibility in science controversies, 53–54
Iraq War, 11, 178
Italy, 68

John Snow, Inc., 34, 263
Johns Manville Company, 9
Johnson & Johnson, 222
Joyner-Kersee, Jackie, 124
Just sustainability paradigm, 12

Kasper, Anne, 85
Kennedy, Donald, 250
Khamisiyah depot bombing, 152–53, 162, 169
King, Samantha, 85
Klawiter, Maren, 30–31
Komen Foundation, 197
Krieger, Nancy, 68
Krimsky, Sheldon, xxii, 23, 38, 260
Kroll-Smith, Steve, xx, 175

Lagakos, Steven, xviii
Landrigan, Philip, 108
Lariam, 178
Latinos, 7, 234, 267. *See also* West Harlem Environmental Action
Latour, Bruno, 55
Lave, Lester, 105
Lead poisoning, 59; and Bush administration, 250, 254; as contested illness, 7; and dissident scientists, 260; Tribal Efforts Against Lead, 246–47
Learning disabilities, 4 (table), 9
Lerner, Michael, 278
Leukemia in children, 4 (table). *See also* Woburn childhood leukemia case
Levine, Adeline Gordon, xvii, xx
Levins, Richard, 213, 227
Levy, Jon, 127
Lifestyle factors in disease prevention. *See* Individual responsibility for health factors
Linking Breast Cancer Activism and Environmental Justice, 279
Litigation: and air pollution, 112–15, 249; Daubert challenges, 253; Woburn childhood leukemia case, xviii
Living Downstream (Steingraber), 202, 264
Loka Institute, 34
"London fog" of 1952, 106
Long Island Breast Cancer Network, 77
Long Island Breast Cancer Study Project, 53, 69, 75, 78, 79
Long Island, NY, 77–79, 92–93
Lord, Audre, 220
Lou Gehrig's disease. *See* ALS
Louisiana Bucket Brigade, 265
Louisville Charter for Safer Chemicals, 209
Love Canal, 37, 206, 264
Love Canal: Power, Politics, and People (Levine), xvii, xx
Lowell Center for Sustainable Production, 203, 214, 216–17, 238
Lung cancer, 103, 145

Mailman School of Public Health, 134
Malathion, 217
Mammography, criticism of, 32, 46, 286n38

Mammography machines, 32, 76, 85–86
Managed care organizations, 234
Marcus, George E., 41
Marin Cancer Project, 82
Marin County Breast Cancer Watch, 82, 84, 92
Markowitz, Gerald, 59
Martin, Andrea, 73
Mashapaug Pond, 274
Massachusetts: and precautionary principle, 213–16, 225. *See also* Boston, MA; Cape Cod, MA; Children's and Families' Protection Act of 1996; Silent Spring Institute; Toxic Use Reduction Act; Toxic Use Reduction Institute
Massachusetts Bay Transit Authority, 186
Massachusetts Breast Cancer Coalition, 80–81, 92, 219
Massachusetts Coalition for Occupational Safety and Health, 225
Massachusetts Public Interest Group, xxi
Materials Matter: Toward a Sustainable Materials Policy (Geiser), 216
Mayer, Brian, xxii
McCormick, Sabrina, xxii
Medical philanthropy, 20
Medical practice, 8, 19, 232–37. *See also* Dominant epidemiological paradigm; Iatrogenic diseases
Medical sociology, 30
Medical supply companies, 85–86
Medically unexplained physical symptoms (MUPS), 172–77, 230, 233
Mercury contamination, 73, 234, 254, 268
Methyl paraben, 80
Mexico, 68
Microsoft, 222
Mikkelsen, Edwin, xvi–xvii
Military: and informed consent, 236–37; lack of predeployment health data, 244, 256; military culture, 162–63, 179; Millennium Cohort Study, 152, 244. *See also* Department of Defense; Gulf War–related illnesses; Veterans Administration
Millennium Cohort Study, 152, 244
Millikan, R., 68
Mills, C. Wright, 23, 231
Monsanto Chemical, 259

Morello-Frosch, Rachel, xxiii, xxiv, 91, 220
Mortality rate, 200; and air pollution, 112; and asthma, 101–2; and breast cancer, 192, 195, 196; and Gulf War–related illnesses, 196; and risk perception, 195
Mount Sinai School of Medicine, 265
Moyers, Bill, 6, 59, 73, 266
Multiple chemical sensitivity, 145, 172, 175, 198
Multisited ethnography, 41
Multnomah County Board of Commissioners, 224
MUPS. *See* Medically unexplained physical symptoms
Myers, John Peterson, 53, 241, 263
Myhre, Jennifer R., 85

NAAQS. *See* National Ambient Air Quality Standards
Najem, G. R., 68
Nanoparticles, 114–15
Narrative reconstruction, and asthma, 116–17
National Academy of Science, 54
National Agenda to Protect Children's Health, 218
National Ambient Air Quality Standards (NAAQS), 111, 183, 185
National Breast Cancer Coalition (NBCC), 37, 62, 72, 81, 88
National Cancer Institute (NCI): and breast cancer, 46, 48, 52, 59; and dominant epidemiological paradigm, 20; and Long Island Breast Cancer Study Project, 78; research funding, 62, 245
National Center for Environmental Health, 250
National Children's Study, 242–43
National Environmental Policy Act of 1969, 208
National Gulf War Resource Center, 142, 156, 168
National Institute of Environmental Health Sciences (NIEHS), 16, 263; and asthma, 118, 120, 186; and breast cancer, 51, 52; and Breast Cancer Fund, 83; Center for Excellence, 92; and children's environmental health, 218; and community-

based participatory research, 35, 247–48; Environmental Justice and Community-Based Participatory Research programs, 34, 263; Environmental Justice: Partnerships for Communication program, 186, 269; research funding, 245; Roxbury Lupus Project, 120; and Silent Spring Institute, 91; support for programs with public involvement, 35, 92; Translational Research Program, 35; and WE ACT, 133

National Institute of Occupational Safety and Health, 35, 248

National Institutes of Health (NIH): asthma as priority, 102, 185; and breast cancer, 61, 187, 189; and Gulf War–related illnesses, 146; research funding, 245. *See also* National Cancer Institute

National Report on Human Exposure to Environmental Chemicals report, 72

National Research Council (NRC), 54, 115

National Resources Defense Council, 114–15

National Toxicology Program, 54, 255

National Toxics Campaign, xxi

Nationwide Health Tracking Act, 243

Native Americans, 34, 246–47, 267

NBCC. *See* National Breast Cancer Coalition

NCI. *See* National Cancer Institute

Needleman, Herbert, 7, 260

New England Journal of Medicine, 69

New Source Review (NSR), 114–15

New York. *See* Long Island, NY; West Harlem Environmental Action

New York Metropolitan Transit Authority, 122

New York Times, 124, 135

NIEHS. *See* National Institute of Environmental Health Sciences

NIH. *See* National Institutes of Health

Nike, 239

Nitrous oxide, 109

No Child Left Behind, 252

North Carolina, 91, 260

North River Sewage Treatment Plant, 119

NorthWest Environment Watch, 269

Not Too Pretty: Phthalates, Beauty Products, and the FDA, 223

NSR. *See* New Source Review

Occupational Safety and Health Act of 1970, 208

Occupational safety and health movement, 9, 205, 231

Occupational studies of breast cancer, 70

Oil well fires, 145, 146, 148, 162

Oklahoma, 246–47

Olympic games (1996), 108

Operation Desert Shield/Storm Association, 142

Oregon Center for Environmental Health, 224

Organochlorines, 61, 67, 68, 79

Organophosphates, 148, 217, 242

Our Stolen Future (Colborn et al.), 53, 241, 263–64

Packard, Randall M., 235

PAHs, 67, 79

Paradigms, 17–18. *See also* Dominant epidemiological paradigm; Public paradigm

Parkinson's disease, 4 (table)

Particulate matter, 108–10; and asthma, 103, 106–11; "confounding curves" issue, 110; particulate dose-response curve, 107; PM_{10}, 111; $PM_{2.5}$, 103, 107–8, 109 (table), 110, 111; regulatory disputes, 111–16; ultrafine particulates, 115–16

Pastor, Manuel, Jr., 220

Patient-centered medicine, 235–36

PBDEs, 5, 211, 267–68, 269

PCBs: and breast cancer research, 61, 68, 79; CDC body burden report, 56, 242; slow government response to dangers of, 9

Pellow, David N., 12

Pelosi, Nancy, 83

Perchlorate pollution, 255

Perera, Frederica, 127

Permethrin, 148, 242

Persian Gulf Expert Scientific Committee, 155, 191

Persian Gulf War Illness Compensation Act of 2001, 170, 190

Persian Gulf War Veterans Act of 1998, 190

Persian Gulf War Veterans' Benefits Act of 1994, 170, 190

Pesticide Action Network of North America, 269

Pesticides, 36, 56, 216–18
Pew Charitable Trusts, 3
Pharmaceutical companies, 44, 63, 85–86
Philadelphia, PA, 118
Phthalates, 242; banned in Denmark, 210–11; CDC body burden report, 56; and cosmetics, 223, 238; regulation of, 62
Physicians for Social Responsibility, 233–34
Picou, Steven, xx
"Pink" marketing campaigns, 82, 85
Piperonyl butoxide, 217
Policy, 182–201; and asthma, 104, 183–86, 192–201; and breast cancer, 55–57, 186–89, 192–201; Bush administration's rejection of science-based decision making, 249–55; and community-based participatory research, 246–49; DEP and the scientization of decision making, 19; endocrine disruptor research and surveillance, 240–41; and Gulf War–related illnesses, 189–201; improved monitoring and surveillance, 242–45; improvements in military medicine and VA care, 256–57; political control and censorship of science, 249–55; and precautionary principle, 239–40; prevalence and public perception of risks, 192, 193 (table), 195–96; research funding, 245–46; science base for environmental causes, 192, 193 (table), 194–95, 200 (table); sources of support for environmental causation hypotheses, 192, 193 (table), 196–98; stages of health social movements, 182, 183 (fig); strength of health social movements, 193 (table), 198–201; and transportation issues, 257–58. *See also* Citizen participation policy; Detection/compensation/treatment policy; Funding for research; Policy; Precautionary principle; Regulatory policy
Political-economic production of disease, 64
Politicized illness experience, 24–25; and asthma, 123–28; and breast cancer, 30–31, 87–88; and medically unexplained physical symptoms, 176
Pollution. *See* Air pollution; Indoor environment, and asthma; Particulate matter; *specific materials*

Pope, C. Arden, 107–8
Popular epidemiology, 17, 33–34, 181, 264
Population-based approach to disease, 64–66
Pork industry, 260
Portland City Council, 224
Postmodern illnesses, 174–76, 230
Posttraumatic stress disorder, 146–48, 167, 178
Poverty, and asthma, 117m12m102, 118, 119
Power production, 111, 114–15, 254
Precautionary principle, 202–27; Be Safe campaign, 221–22; burden of proof, 207–8; in California, 219, 221; children's environmental health activities, 218–19; and cosmetics, 222–23; creation and diffusion of, 204–13; democratization of information and science, 224–26; and economics, 210–12, 237; and environmental breast cancer movement, 44, 57, 71, 91, 97, 204, 205, 219–20; and environmental justice groups, 204–5, 207, 220–22; in Europe, 206, 207, 210–12, 223; implementation of, 239–40; "late lessons from early warnings," 8, 205, 241; legislation on toxins, 206, 208–9, 213, 215, 218; in Massachusetts, 203, 213–16, 219–20, 225; in Michigan, 221; need for research on alternative materials, 213; in New York, 221; and occupational health and safety, 205, 208; in Oregon, 224; as public paradigm, 23, 202–13, 216; Quijano's list of types of actions associated with, 207–8, 209; in San Francisco, 219; in Scandinavian countries, 56; summary of conclusions, 226–27; toxic waste activism, 205–6; toxins reduction approach, 205; traditional vs. cumulative risk assessment, 209; and TURI, 203; and West Nile virus, 216–18; Wingspread Statement, 204
Precautionary Principle Project, xxii, 17, 81, 92, 204, 214–15. *See also* Alliance for a Healthy Tomorrow
Presidential Advisory Committee (Gulf War–related illnesses), 146, 147, 153, 155, 167
Primary prevention, defined, 58

Principi, Anthony J., 144, 178
Proctor, Susan P., 147
Proctor & Gamble, 238
Project BioShield Act of 2004, 178
Project LEAD, 37, 72
Prostate cancer, 4 (table)
Providence School Asthma Partnership, 124, 234–35
Public Health Reports, 204
Public hypothesis, 23
Public opinion on environmental causation of disease, 3–6, 4 (table)
Public paradigm, xiv, 17, 181; and asthma, 117; and breast cancer, 44, 98; conflicting paradigms, 24; defined, 23–26; examples, 23, 36. *See also* Asthma activism; Environmental breast cancer movement; Precautionary principle
Public perception of risks, 193 (table), 195–96
Public reaction to environmental health effects, 8–10
Pulido, Laura, 12
PVC contamination, 59, 222, 234
Pyridostigmine bromide. *See* Antidotes, as potential source of Gulf War–related illnesses

Quijano, Romeo, 207–8, 209

Race: and asthma, 12, 101–2, 119; and breast cancer, 68, 82, 90; and constituency-based health movements, 26–27; and public paradigms, 23; "transit racism," 121, 127; unequal healthcare access, 32, 90
Race for the Cure, 85
Radiation exposure, 9–10, 236
Raloxifene, 219
RAND Corporation, 144, 148
Rapp, Rayna, 41
Ray, Rayka, 33
REEP. *See* Roxbury Environmental Empowerment Project
Registration, Evaluation, and Authorisation of Chemicals (EU policy), 211
Regulatory Modeling System for Aerosols and Acid Deposition, 109
Regulatory policy: and asthma, 134, 184 (table), 185–86; and breast cancer, 184 (table), 188; and Bush administration, 249–53, 255; defined, 182; and globalization, 238–39; and Gulf War–related illnesses, 184 (table), 191; industries' influence over, 59, 63, 104, 114; opposition of government and corporate actors, 106–7, 111–16, 135. *See also* European Union; *specific materials*
Reich, Michael, xx
Republican politicians, 78
Research Advisory Committee on Gulf War Veterans' Illnesses, 144, 178
Revlon, 238
Rhode Island. *See* Draw-A-Breath; Environmental Neighborhood Awareness Committee of Tiverton
Rhode Island Department of Environmental Management, 275
Risk assessment, traditional vs. cumulative, 209. *See also* Precautionary principle
"Risk factor" approach to epidemiology, 37–38, 63–71. *See also* Individual responsibility for health factors
Rockefeller, John D., IV, 167
Rose, Geoffrey, 64
Rosner, David, 59
Roswell, Robert, 170
Roxbury Environmental Empowerment Project (REEP), 122–23
Roxbury Lupus Project, 120

Sadd, James, 220
Safe Drinking Water Act, 188, 240
San Diego, CA, 118
San Francisco Bay area: breast cancer activism, 81–86, 219; and citizen-science alliances, 92; and precautionary principle, 219, 223–24, 239
Sanders, Bernard, 167, 197
Sargent, Be, 220
Sarin gas, 148, 152, 189, 257. *See also* Antidotes, as potential source of Gulf War–related illnesses
Scandinavia, 5, 56
Schaeffer, Eric, 251
Schlichtmann, Jan, xvi

Schools: asthma and absenteeism, 102; on contaminated land, 274; and pesticides, 218; school lunch programs, 21, 50; and toxic cleaning products, 225
Schwartz, David, 248
Schwartz, Joel, 108, 112
Science: advocacy science, xxii, 26, 55; and boundary movements, 31; Bush administration's rejection of science-based decision making, 210, 249–50; consequences of withholding information, 259; critical epidemiology, 37–39, 263–64; democratization of, 13–14, 224–26, 231, 266–67; and informed consent, 266–67; loss of public trust, 10; political control and censorship, 249–55; and postdiscovery period, 22 (fig); "research right to know," 266; research silences, 261–62; resistance to looking for environmental factors, 230–31; value structure, 94–95. *See also* Asthma science base; Breast cancer science base; Citizen-science alliances; Dominant epidemiological paradigm; Gulf War–related illnesses science base; Scientists
Science and Environment Health Network, 224
Science to Achieve Results (STAR) grants, 245
Scientific Integrity in Policymaking, 255
Scientists: advocacy science, xxii, 26, 55; Bush administration "loyalty tests," 250; conflicts of interest and censorship, 25; and critical epidemiology, 37–39, 263–64; critical orientation, 25–26; dissident scientists, 258–61; and policymaking, 57. *See also* Citizen-science alliances
Scotchgard stain repellant, 73
Seattle, WA, 108
Secondary prevention, defined, 58
Senier, Laura, xxii
Seskin, Eugene, 105
Seveso, Italy, dioxin contamination, 68
Sharpe, Michael, 176
Shays, Christopher, 155, 197
Shelby, Richard, 113
Shelby Amendment on Data Access, 113–14, 250

Shop for the Cure campaign, 44, 45
Shriver, Thomas, 174
Sierra Club, 114
Sightline, 269
Silent Spring (Carson), 47, 204, 259
Silent Spring Institute, xxi, 16, 70–71, 279; author's research methodology, 40; Cape Cod study, 80; and citizen-science alliances, 34, 73, 92, 189, 249; and endocrine disruptor hypothesis, 60; and environmental justice movement, 91; founding of, 80; funding of Cape Cod study, 53; identification of indoor contaminants, 80; public communication, 74, 266; researchers' views on weight of evidence, 55
Sistahs United, 91
Sisters in Action for Reproductive Empowerment, 223
"Situated knowledge," 52
Six Cities asthma study, 107, 112, 113
Skanska, 212
Snedeker, Suzanne M., 68
Snow, John, 258–59
Social Accountability International, 238
Social discovery of disease clusters, 24, 25 (fig), 141–44
Social production of disease, 64
Social research capital, 247
Social-structural approach to epidemiology, 23, 26–27, 37–38, 64, 263
Society of Environmental Journalists, 268
Southampton Breast Cancer Coalition, 78
Southeast Community Research Center, 278
Southern California Children's Health Study, 108
Southwest Network for Environmental and Economic Justice, 278
Soviet Union, 253
Spencer, Peter, 158
Spengler, John, 105, 127, 268–69
Srinivasan, Shobha, 35
Stamp Out Breast Cancer Program (breast cancer postage stamp), 187, 245
Standards of proof: and asthma, 129–30, 134; and breast cancer, 53–54, 73, 98; and Gulf War–related illnesses, 165

"Standpoint" theory, 52
State of the Evidence: What is the Connection Between the Environment and Breast Cancer? report, 96
Status quo. *See* Dominant epidemiological paradigm
Steingraber, Sandra, 202, 261, 264
Strategies to Protect the Health of Deployed U.S. Forces report, 152, 245
Stress, and Gulf War–related illnesses: contextual stress perspective, 167–68, 171; and dominant epidemiological paradigm, 140–41, 166–67; government acceptance of stress-based paradigm, 146; stigma attached to stress as a cause of health problems, 167; struggle to legitimate GWRIs, 166–72
Subaltern struggles, 12
Subsidized philanthropy, 85
Sulfur dioxide, 109
Superfund sites, 233
Susan G. Komen Breast Cancer Foundation, 75
Sweden, 5, 211
Synthetic hormones, 65. *See also* Endocrine disruptor hypothesis

Tamoxifen, 66, 86, 219
Teenagers, as activists, 120–21, 223, 264
Tellus Institute, 263
Terrorism, 243, 244
Tertiary prevention, 58, 61
Tesh, Sylvia Noble, 63
The Netherlands, 34
Third National Report on Human Exposures to Environmental Chemicals report, 56
Thompson, Tommy, 254
3M Company, 73
Tickner, Joel, 214
Tillery, NC, 91, 260
Tiverton, RI. *See* Environmental Neighborhood Awareness Committee of Tiverton
Tobacco industry, 7
Tomatis, Lorenzo, 69
Toward Tomorrow Project, 278–79
Toxic Release Inventory, 68
Toxic Substances Control Act, 266

Toxic Tour, 91, 248
Toxic Use Reduction Institute (TURI), xxii, 17, 203, 213–16, 225, 237, 239
Toxic Use Reduction Network, 225
Toxic waste activism, 206–7. *See also* Love Canal; Woburn childhood leukemia case
Toxics Action Center, xxi, xxiii, 263, 274, 278
Toxics Use Reduction Act, 206, 213, 215
Toys, phthalates in, 210–11
Trade Secrets (television special), 6, 59
Train, Russell, 253
"Transit racism," 121, 127
Transit Riders' Union, 121
Translational Research Program, 35
Transportation, 109–10, 120–22, 257–58
Tribal Efforts Against Lead, 246–47
Trucking industry, 111, 112–13
Truman, John, xviii
Trust for America's Health, 3, 242, 243
Tucker, Jonathan, 155
Tuite, James J., III, 146
TURI. *See* Toxic Use Reduction Institute

Ultrafine particulates, 115–16
Umbilical cord blood, 268
Unifirst, xviii–xix
Unilever, 238
Union of Concerned Scientists, 255
United Farm Workers, 36
United Kingdom, 148
University of Massachusetts–Lowell, 213
Uranium, 148, 189
Urban Habitat, 91
U.S. v. Duke Energy, 115
Utah steel mill closure, 108

VA Medical Center (Boston), 16. *See also* Boston Environmental Hazard Center
Vaccinations. *See* Inoculations, as potential source of Gulf War–related illnesses
Veteran service organizations, 142, 163–66, 197–98
Veterans Administration: and ALS, 159, 191; disjunction with DOD, 245, 256; disputes with other government bodies, 155; record-keeping issues, 140, 256; research agenda, 146–47, 150, 177; War-Related Illness and Injury Study Centers, 178, 256

Veterans' Compensation Rate Amendments of 2001, 170
Veterans Millennium Health Care and Benefits Act, 191
Veterans of Foreign Wars, 142
Veterans Programs Enhancement Act of 1998, 190, 191
Vietnam, and breast cancer research, 68
Vietnam veterans, 163
Vietnam War, 11, 154, 167
Vyner, Henry, xx

W. R. Grace Chemicals, xvi, xviii–xix
War-Related Illness and Injury Study Centers, 178, 256
Washington Toxics Coalition, 269
WE ACT. *See* West Harlem Environmental Action
Weight of evidence, 53, 54, 62
Wessely, Simon, 172, 174
West Harlem Environmental Action (WE ACT), xxi, 15–16; and asthma activism, 104, 117–38; collaborations with scientists, 129–34, 137–38; community empowerment through education, 123, 124, 126; and community-based participatory research, 248; and cosmetics, 223; and environmental breast cancer movement, 91; issues addressed by, 119; local focus of, 132; national impact of, 133, 134; national networking, 126–27; origins of, 119; and politicized illness experience, 123; and precautionary principle, 221; and transportation issues, 121–22
West Islip group, 78

West Nile virus, 216–18
When Smoke Ran Like Water (Davis), 5–6, 260, 264
Whitman, Christine Todd, 251
Whitten, Jamie L., 259
Wignall, Stephen, 170
Wilson, Richard, 105
Wing, Steve, 38, 260, 263
Wingspread Statement, 204
Winkenwerder, William, 244
Winn, Deborah, 79
Woburn childhood leukemia case, xvi–xix, 6, 37, 264
Wolff, Mary S., 67
Wolfson, Mark, 231
Women's Cancer Resource Center, 82–83
Women's Community Cancer Project, 81, 220
Women's health movement, 27, 88–89, 198, 231
Woonasquatucket River Watershed Council, 274
Worldwide Responsible Apparel Production, 238

Xenoestrogens, 10. *See also* Endocrine disruptor hypothesis

Young, Bruce, xvii–xviii
Young Lords Organization, 7
Yupiks, 276

Zavestoski, Steven, xxvi–xxvii
Zelen, Marvin, xviii
Zones, Jane S., 85
Zoots Dry Cleaners, 237